An Introduction to Behavioural Ecology

SECOND EDITION

J.R. Krebs FRS, Lecturer in
Zoology at the Edward Grey
Institute of Field Ornithology,
Department of Zoology, Oxford
and E.P. Abraham
Fellow of Pembroke College

N.B. Davies Lecturer in
Zoology at the
University of Cambridge
and Fellow of
Pembroke College

Drawings by Jan Parr

Blackwell Scientific Publications

OXFORD LONDON EDINBURGH

BOSTON PALO ALTO MELBOURNE

First published 1981
Reprinted 1982 (twice), 1983, 1985
Second edition 1987

Distributed in the USA by
Sinauer Associates, Inc., Publishers
Sunderland, MA 01375

Krebs, J.R.
  An Introduction to behavioural ecology. — 2nd ed.
  1. Animal behavior   2. Animal ecology
  3. Behavior evolution
  I. Title   II. Davies, N.B.   III. Parr, Jan
  591.51   QL751

ISBN 0−632−01498−9

Set in 10/12.5 Melior by Setrite Typesetters
Printed in Great Britain at the
Alden Press, Osney Mead, Oxford

Contents

Acknowledgements

1st Edition

This book is based on lectures given by us at Cambridge and Oxford Universities and we thank our students for providing a stimulating and critical audience.

We thank especially Tim Birkhead for reading the first draft of the whole manuscript and the following for comments on particular chapters: Anthony Arak, Patrick Bateson, Jane Brockmann, Tim Clutton-Brock, William Foster, Peter de Groot, Paul Harvey and Geoff Parker.

For sending manuscripts and allowing us to quote unpublished work we thank: Jeffrey Baylis, Lew Oring, Richard Wrangham, Robert Hinde, Dan Rubenstein, Peter de Groot, Uli Reyer, Bob Metcalf, Ron Ydenberg, Ric Charnov, Haven Wiley, Clive Catchpole and Malte Andersson.

Finally we thank Robert Campbell for his encouragement and enthusiasm during the preparation of this book.

2nd Edition

We thank Tim Clutton-Brock, William Foster, Paul Harvey, Nadav Nur and Alan Grafen for their valuable comments. Finally we thank Sara Trevitt and Robert Campbell of Blackwell Scientific Publications for their help in the preparation of the manuscript.

Introduction

This brief introduction describes the organization and contents of our book. The book is about the survival value of behaviour. We call this subject 'behavioural ecology' because the way in which behaviour contributes to survival and reproduction depends on ecology. If, for example, we want to answer the question 'How does living in a group contribute to an individual's survival?', we have to start thinking in terms of the animal's ecology; the kind of food it eats, its enemies, its nesting requirements and so on. These ecological pressures will determine whether grouping is favoured or penalized by selection. Behavioural ecology is not only concerned with the animal's struggle to survive by exploiting resources and avoiding predators, but also with how behaviour contributes to reproductive success. Much of the book is therefore about competition between individuals for the chance to reproduce and pass on their genes to future generations.

The book emphasizes the theoretical background to each subject discussed, but we prefer to illustrate the theory with examples after a very brief general introduction, rather than developing long, abstract, theoretical arguments. Although none of the ideas we have used are difficult to understand we have placed some of the more complicated arguments and details in boxes which can be ignored if the reader is in a hurry.

Chapter 1 is a general introduction to the book, in which we distinguish between different kinds of questions that one can ask about behaviour. In particular we emphasize the difference between questions about survival value or function and those concerned with causal mechanisms. We show that natural selection should favour individuals who are best able to propagate their genes to future generations.

In chapter 2 we discuss how to test hypotheses for the adaptive advantage of behaviour. One method is comparison among species. The rationale here is that differences between species in behaviour can be correlated with differences in their ecology. From these correlations, inferences can be drawn about the adaptive significance of behavioural traits. We illustrate this approach with reference to social organization in weaver birds, antelope and primates. The second method is to perform experiments, for example to change behaviour and measure the consequences this has for the individual's chances of survival and for its reproductive success.

In chapter 3 we focus on the individual. Animals are viewed as making 'decisions' between alternative courses of action and the decisions can be analysed in terms of their costs and benefits. A powerful tool in this approach is optimality theory, which allows us to test hypotheses about the importance of various costs and benefits by predicting their effects on the animal's decision rules. By considering the basic decisions underlying behaviour patterns, we show how the same models can be used to understand what at first sight seem very different problems, such as feeding and searching for mates. Chapter 4 looks at decisions over evolutionary time, and how these change during arms races between predators and prey, and brood parasites and hosts.

In the next three chapters we consider how individuals should behave when they have to compete with others for scarce resources such as food, territories or mates. We discuss how competitors should be distributed in relation to resource distribution and abundance (chp. 5) and the costs and benefits of living in groups (chp. 6). In chapter 7 we introduce the idea of game theory as a technique for analysing how individuals behave in contests for resources.

Chapters 8, 9 and 10 are concerned with sexual reproduction. A consideration of the basic differences between males and females leads to the idea that members of one sex (usually male) may compete for access to members of the other (chp. 8). This is the theory of sexual selection. The differences between male and female also suggest that the interests of the two sexes during reproduction often differ (the theory of sexual conflict). Chapter 9 discusses how these battles within and between the sexes are influenced by ecology. Here we rely heavily on the comparative approach, correlating differences between species in sexual strategies with differences in ecology. From differences between species we turn to differences between individuals (chp. 10). We introduce the idea that different individuals within a species sometimes adopt different sexual strategies. These differences may be related to age or size, or they may simply be equally profitable, alternative, ways of achieving the same end.

In chapter 11 we examine how altruistic behaviour can evolve. We then illustrate the theoretical arguments with reference to 'helpers'; individuals that help others to rear young instead of producing their own. Chapter 12 deals with birds, mammals and fish, and chapter 13 is devoted entirely to the social insects, where helping reaches its most sophisticated level of development. Many of the earlier chapters refer to communication as a behavioural mechanism of competition for resources and social interaction. In chapter 14 we tie together these threads in a general discussion of animal signals. We follow the pattern set in earlier chapters by considering both ecological constraints and intraspecific selection pressures. In the final chapter we reassess the view that the survival value of behaviour can be understood within a neo-Darwinian framework using methods such as optimality models and game theory.

Finally a word about the style of presentation. We generally use convenient and informal shorthand rather than traditional formal scientific style. A phrase such as 'Offspring are selected to demand more food than the parent wants to give' is short for 'During the course of evolution selection acting on genetic differences in the begging behaviour of offspring will have favoured an increase in the intensity of begging. This increase will have been favoured to the extent where the level of begging by any individual offspring exceeds the optimum level for the parent'.

Some readers may wonder whether our informal shorthand, together with catchy descriptive labels for various behaviour patterns such as 'manipulation' and 'sneaker', are a sign of sloppy thinking. There is no doubt that loose terminology can indicate imprecise thinking and half-formulated ideas. But it is equally easy to conceal woolly arguments behind an obfuscating screen of scientific jargon. We have used a simple direct style in order to make our arguments clear and not because behavioural ecology is a woolly subject. This point is nowhere better illustrated than by George Orwell in his brilliant essay 'Politics and the English Language' (1946). He translates the following well-known verse from *Ecclesiastes*, into modern English. 'I returned and saw under the sun, that the race is not always to the swift, nor the battle to the strong, neither yet bread to the wise, nor yet riches to men of understanding, nor yet favour to men of skill; but time and chance happeneth to them all.'

And now the translation.

'Objective consideration of contemporary phenomena compels the conclusion that success or failure in competitive activities exhibits no tendency to be commensurate with innate capacity, but that a considerable element of the unpredictable must invariably be taken into account.'

This translation is not only tired and ugly, lacking the fresh, vivid imagery of the biblical passage, but it replaces precise illustrations with woolly generalization. While we cannot hope to emulate the clarity and brilliance of the writer of *Ecclesiastes*, or indeed of George Orwell, we hope we have avoided the worst excesses of the Orwellian parody and presented our ideas in simple but precise language.

Chapter 1. Natural Selection, Ecology and Behaviour

Questions about behaviour

In this book we will explore the relationships between animal behaviour, ecology and evolution. We shall describe how animals behave under particular ecological conditions and then ask 'Why has this behaviour evolved?' For example, we shall attempt to understand why some animals are solitary while others go around in groups and why most individuals court before they copulate. Why do some birds have songs consisting of pure whistles while others produce buzzes and trills? We shall also ask some precise, quantitative questions such as why do sunbirds defend territories containing 1600 flowers and why does the male dungfly copulate for on average 41 min?

Niko Tinbergen, one of the founders of ethology, emphasized that there are several different ways of answering the question 'Why?' in biology. These have come to be known as Tinbergen's four questions (Tinbergen 1963). For example, if we asked why starlings, *Sturnus vulgaris*, sing in the spring, we could answer as follows.

1 In terms of *survival value or function*. Starlings sing to attract mates for breeding.

2 In terms of *causation*. Because increasing daylength triggers off changes in hormone levels in the body, or because of the way air flows through the syrinx and sets up membrane vibrations. These are answers about the internal and external factors which cause starlings to sing.

3 In terms of *development*. Starlings sing because they have learned the songs from their parents and neighbours.

4 In terms of *evolutionary history*. This answer would be about how song had evolved in starlings from their avian ancestors. The most primitive living birds make very simple sounds, so it is reasonable to assume that the complex songs of starlings and other song birds have evolved from simpler ancestral calls.

It is important to distinguish these various kinds of answer or otherwise time can be wasted in sterile debate. If someone said that swallows migrate south in the autumn because they are searching for richer food supplies while someone else said they migrated because of decreasing daylength, it would be pointless to argue about who was correct. Both answers may be right, the first is in terms of survival value or function and the second is in terms of causation. Factors influencing survival value are sometimes called 'ultimate' while causal factors are referred to as 'proximate'. It is these two answers that are the most frequently muddled up and so to make the distinction clear we will discuss an example in detail.

REPRODUCTIVE BEHAVIOUR IN LIONS

In the Serengeti National Park, Tanzania, lions (*Panthera leo*) live in prides consisting of between 3 and 12 adult females, from 1 to 6 adult males and several cubs. The group defends a territory in which it hunts for prey, especially gazelle and zebra. Within a pride all the females are related; they are sisters, mothers and daughters, cousins, and so on. All were born and reared in the pride and all stay there to breed. Females reproduce from the age of 4 to 18 years and so enjoy a long reproductive life.

For the males, life is very different. When they are 3 years old, young related males (sometimes brothers) leave their natal pride. After a couple of years as nomads they attempt to take over another pride from old and weak males. After a successful take-over they stay in the pride for 2 to 3 years before they, in turn, are driven out by new males. A male's reproductive life is therefore short.

The lion pride thus consists of a permanent group of closely related females and a smaller group of separately interrelated males present for a shorter time. We will consider three interesting observations about reproductive behaviour in a pride (Bertram 1975).

1 Lions may breed throughout the year but although different prides may breed at different times, within a pride all the females tend to come into oestrus at about the same time. The mechanism, or causal explanation, may be the influence of an individual's pheromones on the oestrus cycles of other females in the pride. A similar phenomenon occurs in schools, where girls living in the same dormitory may also synchronize their menstrual cycles, perhaps due to the effect of pheromones (McClintock 1971).

The function of oestrus synchrony in lionesses is that different litters in the pride are born at the same time and cubs born synchronously survive better. This is because there is communal suckling and with all the females lactating together a cub may suckle from another female if its mother is out hunting (Fig. 1.1). In addition, with synchronous births there is a greater chance that a young male will have a companion when

Fig. 1.1. Top: when a new male takes over a lion pride, he kills the young cubs fathered by the previous males. Bottom: a female suckles her sister's cub alongside her own.

it reaches the age at which it leaves the pride. With a companion a male is more likely to achieve a successful take-over of another pride (Bygott *et al.* 1979).

2 A lioness comes into heat every month or so when she is not pregnant. She is on heat for 2 to 4 days during which time she copulates once every 15 min thoughout the day and night. Despite this phenomenal rate of copulation the birth rate is low. Even for those cubs that are born, only 20 per cent will survive to adulthood. It can be calculated that there are 3000 copulations for each offspring that attains the adult stage.

The causal explanation for why lion matings are so unsuccessful is not the failure of the male to ejaculate but rather the high probability of ovulation failure by the female or a high rate of abortion. But why are females designed in this apparently inefficient way?

One hypothesis is that it may be advantageous to the female to be receptive even at times when conception is unlikely, because this means that each copulation is devalued. For a male there is only a 1:3000 chance that a given copulation will produce a surviving cub and so it is not worth fighting with other males in the pride over a single mating opportunity. Given that males may also kill cubs that are not their own (see below), it may pay a female to mate with all the males in the pride to increase paternity uncertainty. Ideally a female may give each male a sufficient chance of being the father of her cubs that it does not pay him to kill them!

3   When a new male, or group of males, takes over a pride they sometimes kill the cubs already present (Fig. 1.1). The causal explanation for this behaviour may be the unfamiliar odour of the cubs which induces the male to destroy them. A similar effect, known as the Bruce Effect, occurs in rodents where the presence of a strange male prevents the implantation of a fertilized egg or induces abortion.

The advantage of the infanticide for the male that takes over the pride is that killing the cubs fathered by a previous male brings the female into reproductive condition again much quicker and so hastens the day that he can father his own offspring. If the cubs were left intact then the female would not come into oestrus again for 25 months. By killing the cubs she becomes ready for mating after only 9 months. Remember that a male's reproductive life in the pride is short, so any individual that practices infanticide when he takes over a pride will father more of his own offspring and therefore the tendency to commit infanticide will spread by natural selection.

The take-over of a pride by a new coalition of adult males also contributes to the reproductive synchrony of the females; because all the dependent offspring are either killed or evicted during the take-over, the females will all tend to come into oestrus again at about the same time (Packer & Pusey 1983a). Interestingly, the heightened sexual activity of the females is most marked during the first few months after a take-over. The females play an active role in soliciting copulations from several males and this appears to elicit competition between different male coalitions for the control of the pride, with the result that larger coalitions eventually become resident. This is of adaptive advantage to the female because she needs protection from male harassment of her cubs for over 2 years in order to rear her cubs successfully (3.5 months gestation plus 1.5 to 2 years with dependent young) and only large male coalitions are likely to remain in the pride for more than 2 years. High sexual activity in females may therefore incite male-male competition and so result in the best protectors taking over the pride (Packer & Pusey 1983b).

The differences between the causal and functional explanations of these three aspects of reproductive behaviour in the lions are summarized in Table 1.1.

Natural selection

Throughout this book we will be focusing on functional questions about behaviour. Our aim is to try and understand how an animal's behaviour is adapted to the environment in which it lives. When we discuss adaptations we are referring to changes brought about during evolution by the process of natural selection. For Charles Darwin, adaptation was an obvious fact. It was obvious to him that eyes were well designed for vision, legs for running, wings for flying and so on. What he attempted to explain was how adaptation could have

Table 1.1. Summary of causal and functional explanations for three aspects of reproductive behaviour in lions. From Bertram 1975 and Packer & Pusey (1983a,b)

Observation	Causal explanations	Functional explanations
1. Females are synchronous in oestrus	Chemical cues? Take-overs by males	Better cub survival Young males survive better and have greater reproductive success when they leave pride if in a group
2. High rate of copulation	Female infertility Time of ovulation concealed	Each copulation of less value to a male Increased paternity uncertainty may protect cubs from males Elicits competition between male coalitions, so females get best protectors taking over pride
3. Young die when new males take over pride	Abortion (?chemical) Take-over males kill or evict young	Females come into oestrus quicker Male removes older cubs which would compete with his young

arisen without a creator. His theory of natural selection, published in the *Origin of Species* in 1859, can be summarized as follows.

1   Individuals within a species differ in their morphology, physiology and behaviour (*variation*).

2   Some of this variation is *heritable*; on average offspring tend to resemble their parents more than other individuals in the population.

3   Organisms have a huge capacity for increase in numbers; they produce far more offspring than give rise to breeding individuals. This capacity is not realized because the number of individuals within a population tends to remain more or less constant over time. Therefore there must be *competition* between individuals for scarce resources such as food, mates and places to live.

4   As a result of this competition, some variants will leave more offspring than others. These will inherit the characteristics of their parents and so evolutionary change will take place by *natural selection*.

5   As a consequence of natural selection organisms will come to be *adapted* to their environment. The individuals that are selected will be those best able to find food and mates, avoid predators and so on.

When Darwin formulated his idea he had no knowledge of the mechanism of heredity. The modern statement of the theory of natural selection is in terms of genes. Although selection acts on differences in survival and reproductive success between individual organisms, or phenotypes, what changes during evolution is the relative frequency of genes.

We can restate Darwin's theory in modern genetic terms as follows.

1   All organisms have genes which code for protein synthesis. These proteins regulate the development of the nervous system, muscles and structure of the individual and so determine its behaviour.

2   Within a population many genes are present in two or more alternative forms, or alleles, which code for slightly different forms of the same protein. These will cause differences in development and so there will be variation within a population.

3   There will be competition between the alleles of a gene for a particular site (*locus*) on the chromosomes.

4   Any allele that can make more surviving copies of itself than its alternative will eventually replace the alternative form in the population. Natural selection is the differential survival of alternative alleles.

The individual can be regarded as a temporary vehicle or

survival machine by which genes survive and are replicated (Dawkins 1976). Because selection of genes is mediated through phenotypes, the most successful genes will be those which promote most effectively an individual's survival and reproductive success (and that of relatives, see later). As a result we would therefore expect individuals to behave so as to promote gene survival.

Before we discuss how thinking about genes can help us to understand the evolution of behaviour, we should examine the evidence that gene differences can cause differences in behaviour.

Genes and behaviour

Behavioural ecology is concerned with the evolution of adaptive behaviour in relation to ecological circumstances. Natural selection can only work on genetic differences and so for behaviour to evolve (a) there must be, or must have been in the past behavioural alternatives in the population, (b) the differences must be, or must have been, heritable; in other words a proportion of the variation must be genetic in origin, and (c) some behavioural alternatives must confer greater reproductive success than others.

Three main methods have been used to study the ways in which genes influence behaviour.

(a) Use of genetic mutants

Seymour Benzer (1973) used mutagens (radiation or chemicals) to produce genetic mutations that change behaviour in the fruit fly, *Drosophila*. In one mutant, known colourfully as 'stuck', the male fails to disengage from the female after the normal 20 min period of copulation. Mutation of another gene produces 'coitus interruptus' males which disengage after only 10 mins and fail to produce any offspring. Benzer was able to trace the cause of these mutations and show that they resulted from abnormalities in the sensory receptors, nervous system or muscles of the flies. Mutants which exhibit deficiencies in learning ability have also been isolated (Dudai & Quinn 1980). Normal *Drosophila* learn to selectively avoid an odour which is associated with an electric shock. Mutant 'dunce' flies do not learn to avoid the shock though they show normal behaviour in other respects and can learn visual tasks. Dunce flies are produced by an abnormality of the gene coding for the enzyme cyclic AMP phosphodiesterase, which breaks down cAMP. Dunce flies have an abnormally high level of cAMP and furthermore, normal flies display poor learning after being fed on phosphodiesterase inhibitors. This suggests that the enzyme is necessary for olfactory learning. Another mutant, 'amnesiac', has been isolated which learns normally but forgets very rapidly.

(b) Artificial selection experiments

Selected lines are produced by choosing as parents in each generation those individuals which show the most extreme values of some behaviour character. For example, Aubrey Manning (1961) was able to select for two different mating speeds in the fruit fly, *Drosophila melanogaster*, by selectively breeding from fast and slow maters. Such selection experiments nearly always work, showing that much of the continuous phenotypic variation seen in populations has some genetic basis. Another example is provided by male field crickets, *Gryllus integer*, which either call to attract females or silently intercept females attracted to the callers. Cade (1981) was able to select for males which called a lot and those which called only rarely, thus showing that variation in calling duration had a genetic component. As a final example, Majerus *et al.* (1982) studied mating preferences of the two-spot ladybird, *Adalia bipunctata*, which occurs in two forms, a typical form (*typica*) and a melanic form (*quadrimaculata*). Observation of courting pairs showed that female choice determined the frequency of the different matings. In one selected line, those females which had mated with *quadrimaculata* males were used to lay eggs for the next generation, with the females choosing the *typica* males being discarded. The proportion of females preferring the melanic males increased over successive generations showing again that mating preference has a genetic component which can be changed by artificial selection.

(c) Studying populations with genetic differences

Geographically distinct populations of a species often have a different morphology and behaviour, reflecting adaptations to differing ecological conditions. Stevan Arnold (1981) studied the garter snake, *Thamnophis elegans*, in the SW United States. Inland populations are very aquatic and commonly feed underwater on frogs, fish and leeches. Coastal populations are terrestrial foragers and mainly eat slugs. In laboratory choice experiments it was found that wild-caught inland snakes refused to eat slugs, though the coastal ones readily accepted them. Tests with naive newborn snakes showed that 73 per cent of the coastal individuals attacked and ate slugs while only 35 per cent of the inland snakes did so. Young snakes are incubated inside their mother and so one possibility is that the mother's diet could influence the young directly. Arnold arranged matings between inland and coastal individuals and found that the offspring tended to show an intermediate incidence of slug eating. They did not tend to resemble their mother as opposed to their father and so the influence of maternal diet can be ruled out. These results suggest that differences in food preference are correlated with genetic differences, and that garter snakes in different areas have been selected to respond to different prey types.

Most species of warblers are summer visitors to Europe. If individuals are kept in a cage, they show a period of 'restlessness' in the autumn which corresponds with the time at which they would migrate south to Africa for the winter. Quantitative comparisons of the nocturnal restlessness of caged experimental birds showed that the duration of the restlessness correlated well with the distance which the individuals migrated. Thus, comparing different species of *Sylvia* warblers, the long distance migrant, the garden warbler (*Sylvia borin*) was restless for a longer time than the short distance migrant the blackcap (*Sylvia atricapilla*). When different populations of the blackcap were compared it was found that northerly populations, which have further to migrate to reach winter quarters, were restless for longer periods. Furthermore, when individuals from northern populations (West Germany) were mated to those from the south (Canary Islands), their offspring showed intermediate periods of restlessness. These results suggest that genetic differences underlie the differences in migration behaviour both between closely related species and between different populations within a species (Berthold 1984).

These various examples show that *genetic* differences between individuals can lead to differences in *behaviour* (mating behaviour, learning, singing, foraging and migration). Three points need to be emphasized. First, when we talk about 'genes for' a particular structure or behaviour, we do not imply that one gene alone codes for the trait. Genes work in concert and it is likely that many genes together will influence an individual's mating preference, foraging habits and migration patterns. However, a *difference* in behaviour between two individuals may be due to a *difference* in one gene. A useful analogy is the baking of a cake. A difference in one word of a recipe may mean that the taste of the whole cake is different, but this does not mean that the one word is responsible for the entire cake (Dawkins 1978). Whenever we talk about 'genes for' certain traits this is shorthand for gene differences bringing about differences in behaviour.

Second, genes will often influence behaviour in simple ways.

Gene differences can result in behavioural differences because genes code for enzymes which influence the development of the sensory, nervous and muscle systems of the animal which in turn affect its behaviour. For example, a gene could influence a moth's preference for a particular background by coding for certain visual pigments in the eye. Third, just because it can be shown that genes influence behaviour this does not imply that genes alone produce the behaviour, or even that the behaviour can be usefully divided up into genetic and environmental components. The way in which behaviour develops is the result of a complex interaction between genes and environment. Imagine, for example, that a behavioural ecologist comes across the nest of a long tailed tit (*Aegithalos caudatus*). He or she would be immediately impressed by how well the nest was adapted to the bird's way of life, the wonderful crypsis of the nest, its strength and warmth. There are three different ways in which this adaptation could develop in an individual (Bateson 1983). Individuals could all learn by trial and error how to build good nests. Alternatively they could copy another, more experienced bird. Finally, during evolution genes necessary for the expression of actions leading to the construction of good nests could have spread through the population by natural selection because individuals building the best nests would have left most young. Thus, all individuals may be able to build good nests without practice or observing others. However, even if nest building required learning for its proper development, genetic differences in learning ability may still be involved in its evolution.

Selfish individuals or group advantage?

We now return to our theme of studying the adaptive significance of behaviour, how it contributes to an individual's chances of survival and its reproductive success. We interpreted the behaviour of the lions in relation to individual advantage, reflecting Darwin's emphasis on evolution as a struggle between individuals to outcompete others in the population. Many traits evolve because of their advantage to the individual even though they are disadvantageous to others in the population. For example, it's not to the species' advantage to have a cub killed when a new male takes over a lion pride. It's not to the lionesses' advantage either! However, she is smaller than the male and there is probably not much that she can do about it. Infanticide has evolved simply because of its advantage to the male that practices it.

A few years ago, however, many people thought that animals behaved for the good of the group, or of the species. It was common to read (and sometimes still is) explanations like, 'lions rarely fight to the death because if they did so, this would endanger survival of the species' or, 'salmon migrate thousands of miles from the open ocean into a small stream where they spawn and die, killing themselves with exhaustion to ensure survival of the species.'

Because 'group thinking' is so easy to adopt, it is worth going into a little detail to examine why it is the wrong way to frame evolutionary arguments.

The main proponent of the idea that animals behave for the good of the group is V.C. Wynne-Edwards (1962, 1986). He suggested that if a population over-exploited its food resources it would go extinct and so adaptations have evolved to ensure that each group or species controls its rate of consumption. Wynne-Edwards proposed that individuals restrict their birth rate to prevent over-population, by producing fewer young, not breeding every year, delaying the onset of breeding, and so on. This is an attractive idea because it is what humans ought to do to control their own populations. However there are two reasons for thinking that it is unlikely to work for animal populations.

THEORETICAL OBJECTIONS

Imagine a species of bird in which each pair lays 2 eggs and there is no over-exploitation of the food resources. Suppose the tendency to lay 2 eggs is inherited. Now consider a mutant which lays 6 eggs. Since the population is not over-exploiting its food supplies, there will be plenty of food for the young to survive and the 6 egg genotype will become more common very rapidly.

Will the 6 egg type be replaced by birds that lay 7 eggs? The answer is yes, as long as individuals laying more eggs produce more surviving young. Eventually a point will be reached where the brood is so large that the parents cannot look after it as efficiently as a smaller one. The clutch size we would expect to see in nature will be the one that results in the most surviving young because natural selection will favour individuals that do the best. A system of voluntary birth control for the good of the group will not evolve because it is unstable; there is nothing to stop individuals behaving in their own selfish interests.

Wynne-Edwards realized this and so proposed the idea of 'group selection' to explain the evolution of behaviour that was for the good of the group. He suggested that groups consisting of selfish individuals died out because they over-exploited their food resources. Groups that had individuals who restricted their birth rate did not over-exploit their resources and so survived. By a process of differential survival of groups, behaviour evolved that was for the good of the group.

In theory this might just work, but groups must be selected during evolution; some groups must die out faster than others. In practice, however, groups do not go extinct fast enough for group selection to be an important force in evolution. Individuals will nearly always die more often than groups and so individual selection will be more powerful. In addition, for group selection to work populations must be isolated. Otherwise there would be nothing to stop the migration of selfish individuals into a population of individuals all practising reproductive restraint. Once they had arrived their genotype would soon spread. In nature, groups are rarely isolated sufficiently to prevent such immigration. So group selection is usually going to be a weak force and probably rarely very important (Williams 1966; Maynard Smith 1976a).

EMPIRICAL STUDIES

Apart from these theoretical objections, there is good field evidence that individuals do not restrict their birth rate for the good of the group but in fact reproduce as fast as they can. A good example is David Lack's long term study of the great tit (*Parus major*) in Wytham Woods, near Oxford, England (Perrins 1965; Lack 1966).

In this population the great tits nest in boxes and lay a single clutch of eggs in the spring. All the adults and young are marked individually with small numbered metal rings round their legs. The eggs of each pair are counted, the young are weighed and their survival after they leave the nest is measured by retrapping ringed birds. This intensive field study involves several people working full-time on the great tits throughout the year, and it has been going on for 40 years! Most pairs lay 8 to 9 eggs (Fig. 1.2). The limit is not set by an incubation constraint because when more eggs are added the pair can still incubate them successfully. However, the parents cannot feed larger broods so well. Chicks in larger broods get fed less often, are given smaller caterpillars and consequently weigh less when they leave the nest (Fig. 1.3a). It is not surprising that feeding the young produces a limit for the parents because they have to be out searching for food from dawn to dusk and may deliver over 1000 items per day to the nest at the peak of nestling growth.

The significance of nestling weight is that heavier chicks survive better (Fig. 1.3b). An overambitious parent will leave

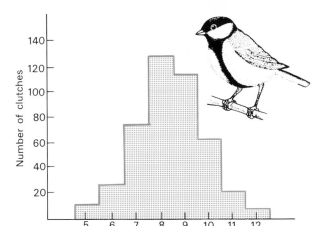

Fig. 1.2. The frequency distribution of the clutch size of great tits in Wytham Woods. Most pairs lay 8−9 eggs (Perrins 1965).

fewer surviving young because it cannot feed its nestlings adequately. By creating broods of different sizes experimentally it can be demonstrated that there is an optimum brood size, one that maximizes the number of surviving young from a selfish individual's point of view (Fig. 1.4). The commonest observed clutch size (Fig. 1.2) is close to the predicted optimum but slightly lower. Why is this? A possible reason is that the optimum in Fig. 1.4 is the one which maximizes the number of surviving young *per brood* whereas, at least in stable populations, we would expect natural selection to design animals to maximize their *lifetime* reproductive output. Experiments involving the addition of extra young to broods of another species of tit, the blue tit (*Parus caeruleus*), have shown that the strain of raising larger broods reduces the chances that the parents will survive to breed another year (Nur 1984a,b). For example, an increase in brood size from 3 to 15 young was associated with a decrease in the chance of female survival from 41 per cent to 19 per cent. Figure 1.5 shows how the mortality costs of raising larger broods can be incorporated into our argument so as to predict the brood size which maximizes overall lifetime reproductive success. In general, the clutch size which maximizes lifetime breeding success will be slightly less than that which maximizes success per breeding attempt.

This model has not yet been tested with the great tit because the costs to adult survival of rearing broods of different sizes are not yet known. Box 1.1 gives a more general model for the optimal trade-off between current and future reproductive effort.

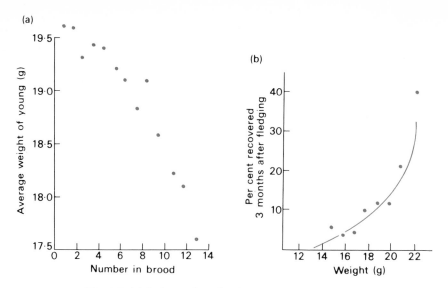

Fig. 1.3. (a) In larger broods of great tits the young weigh less at fledging because the parents cannot feed them so efficiently. (b) The weight of a nestling at fledging determines its chances of survival. Heavier chicks survive better (Perrins 1965).

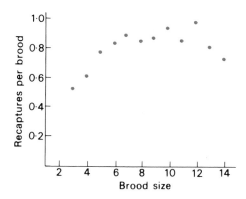

Fig. 1.4. Experimental manipulations of the number of young in a nest show that the optimal brood size for a pair of great tits is between 8 and 12 eggs. This is the brood size which maximizes the number of surviving young (Perrins 1979).

Our conclusion is that reproductive rate in this and other similar studies is close to that which maximizes individual success.

Clutch size is optimal from the selfish individual's point of view. Of course, the exact clutch size may vary a little from year to year and during the season depending on food supplies, so individuals do show some variation. However the variations are in relation to their own selfish optima, not for the good of the group. A good example of this is provided by

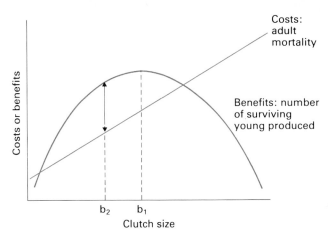

Fig. 1.5. The influence of adult mortality on the optimal clutch size. The number of young produced versus clutch size follows a curve, as in Fig. 1.4, with b_1 being the clutch size which maximizes the number of young produced per brood. Increased clutch size, however, has the cost of increased adult mortality, shown here for simplicity as a straight line. The clutch size which maximizes lifetime reproductive success is b_2, where the distance between the benefit and cost curves is a maximum. This is less than the clutch size b_1, which maximizes reproductive success per brood. From Charnov and Krebs (1974).

Goran Högstedt's study (1980) of magpies, *Pica pica*, breeding in southern Sweden. Observed clutch sizes varied from 5 to 8 depending on feeding conditions in different territories. Our hypothesis would be that some females laid only 5 eggs because this was the maximum number of young they could raise efficiently on their particular territories. Högstedt manipulated clutch sizes experimentally and found that pairs that had produced large clutches did best with large broods, while those which had laid small clutches did best with smaller broods (Fig. 1.6). Variation in clutch size occurred because there was a range of territory quality and each pair raised a brood size appropriate for its own particular territory.

One of the main themes of this book is that different life history strategies will be favoured in different ecological circumstances. For example, continuing our discussion of clutch size, if the European passerine bird species are grouped into two ecological categories, namely those that nest in holes and those that build open nests, it is found that the hole nesters lay larger clutches (Table 1.2). The same relationship occurs in ducks where open nesting species again have smaller clutches than hole nesters (Lack 1968). In holes the young are relatively safe from predators but in the open there is a premium for getting the young out of the vulnerable nest as soon as possible. The same quantity of food could be used to rear a small

Box 1.1. *The optimal trade-off between survival and reproductive effort. From Pianka & Parker (1975) and Bell (1980).*

The more effort an individual puts into reproduction, the lower its chances of survival and so the lower its expectation of future reproductive success. Reproductive costs include allocation of resources to reproduction which would otherwise have been spent on own growth and survival and the increased risks entailed in reproduction, such as exposure to predators. The optimal life history depends on the shape of the curve relating profits in terms of present offspring to costs in terms of future offspring.

The families of straight lines represent fitness isoclines, i.e. equal lifetime production of offspring. In a stable population, present and future offspring will be of equal value and these lines will have slopes of -1. In an expanding population, current offspring are worth more than future offspring (current offspring gain a greater contribution to the gene pool) and the slopes are steeper. In a declining population future offspring are worth more and slopes will be less than -1.

The point of intersection of the curves relating the trade-off between current and future reproductive success, with the fitness isocline furthest from the origin gives the optimal reproductive tactic (indicated by solid dot). When the trade-off curve is convex (a), fitness is maximized by allocating part of the resources to current reproduction and part to survival (i.e. iteroparity, or repeated breeding). When the curve is concave (b), it is best to allocate all resources to current reproduction even at the expense of own survival (semelparity, or 'big bang' suicidal reproduction). If maximal future reproductive success is greater than maximal current reproductive success in case (b), then the optimal tactic is to not breed and save all resources for the future.

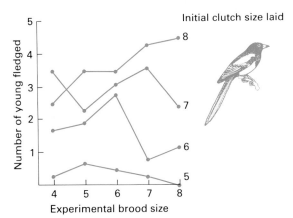

Fig. 1.6. Experiments on clutch size in magpies. Pairs which had initially laid 5, 6, 7 or 8 eggs were given experimentally reduced or enlarged broods. Pairs which had naturally laid large clutches did better with large broods and vice versa. From Högstedt (1980).

brood quickly or a larger brood more slowly. In open nesting species the higher risks of predation have apparently selected for a smaller clutch size and rapid nestling growth.

Table 1.2. A comparison of the clutch size of European passerine birds grouped into two ecological categories. Hole nesting species have larger clutches. After Lack (1968)

Nest type	Predation	Average clutch size	Average length of nestling period (days)
Holes	Low	6.9	17
Open	High	5.1	13

Behaviour, ecology and evolution

We are now in a position to summarize the main principles which underlie the arguments in this book. First, during evolution natural selection will favour individuals who adopt life history strategies which maximize their gene contribution to future generations. Second, the way in which adult survival and reproductive effort are best traded-off in a life history will depend on ecology, the physical environment an individual lives in, its competitors, food, predators, and so on. Third, because an individual's success at surviving and reproducing depends critically on its behaviour, selection will tend to design animals as efficient foragers, efficient predator avoiders, efficient copulators, efficient parents, and so on. What will be 'optimal' will depend on the behavioural alternatives available which will depend on various constraints imposed by phylogeny, physiology and ecology.

Behavioural ecology, therefore, is a meeting point for behaviour, ecology and evolution. We can think of ecology as setting the stage on which animals must perform their behaviour, and evolution as a process which selects individuals whose behaviour results in greatest success in the struggle to contribute genes to the population's gene pool. The aims of the subject are to understand why different species behave in different ways and why, within a species there may also be individual differences in behaviour. To tackle these problems we have to learn about a species' ecology and also about how individuals in a population compete for scarce resources, such as food, mates and territories.

Summary

Behavioural ecology is concerned with functional questions about behaviour, namely how a particular behaviour pattern contributes to an animal's chances of survival and its reproductive success. Experiments show that differences in behaviour can result from differences in genes. Natural selection will favour genes which best promote an individual's chances of passing those genes on to future generations. Individuals are expected to behave in their own selfish interests and not for the good of the species or group. Ecological conditions will determine which behaviour patterns are favoured during evolution.

Further reading

The books by G.C. Williams (1966) and Richard Dawkins (1976, 1982) are excellent discussions of behaviour and evolution. Williams emphasizes the evolution of individual selfishness as opposed to behaviour for the good of the group. Dawkins champions the view that we should think in terms of genes rather than individuals in order to understand the evolution of behaviour.

Partridge (1983) and Bateson (1983) discuss how genes influence behaviour. Southwood (1981) and Horn & Rubenstein (1984) show how different life histories are favoured under different ecological conditions.

In this chapter we have discussed optimal clutch size from the parent's point of view. In fact there will be conflicts of interest between the male and female (see chp. 9), between parents and offspring (Trivers 1974), and between members of the brood. Mock (1984, 1985) shows how chicks within a brood may kill their siblings when competing for food brought by the parents.

Topics for discussion

1 How would you test the causal and functional explanations for lion behaviour in Table 1.1?
2 Under what conditions would the trade-off between present and future reproduction look like (a) or (b) in Box 1.1?
3 What do you understand by the term 'optimal clutch size'?

Chapter 2. Testing Hypotheses in Behavioural Ecology

A rigorous scientific approach to the function of behaviour will involve four stages: observations, hypotheses, predictions and tests. The first two, observation and hypotheses, often go hand in hand. It may take many years getting to know a particular species before it is possible to ask good questions about its behaviour and ecology. Niko Tinbergen's work (1953) on the herring gull, *Larus argentatus*, was the result of over 20 years' painstaking observations of the bird's behavioural repertoire and the environment in which it lives. Having observed some aspect of the behaviour of an animal that we do not understand, how should we proceed?

Let us assume, for example, that we want to discover why our animal lives in a group as opposed to on its own. We may get a strong hint about the function of this simply from observation. If the animal only lived in a group in the breeding season we might suspect that it gained some advantage in terms of increased mating efficiency, for example, whereas if it only lived in a group in winter we may suspect some advantage concerned with improved feeding efficiency or avoiding predation. We can test our ideas in three main ways.

1 *Comparison between individuals within a species.*

Individuals in groups may have greater success at feeding, mating or avoiding predators than solitary individuals. Furthermore, success may vary with group size. The problem is, however, that there may be confounding variables; solitary individuals may be poorer competitors and this, rather than their solitary existence *per se*, may explain their lower success, or individuals in groups may live in better quality habitats, and so on.

2 *Experiments.*

It is often better, therefore, to perform an experiment. With an experiment we can vary one factor at a time; for example we could change group size and see how this influenced success under a particular set of conditions. Niko Tinbergen pioneered the method of elegant field experimentation to answer functional questions. For example, to test the hypothesis that spacing out of gull nests functioned to reduce predation he put out experimental plots of eggs with different spacing patterns and found that those with a clumped distribution suffered greater predation than those that were spaced out as in nature (Tin-bergen *et al.* 1967).

3 *Comparison among species.*

Different species have evolved in relation to different ecological conditions and so comparison among species may help us to understand how differences in feeding ecology or predation pressure, for example, influence the tendency to live in groups or to be solitary. Using the comparative method is rather like looking at the result of experiments done by natural selection over evolutionary time. The results of these 'experiments' are the designs of the various species' behaviour which we now observe. For example, group living may occur in species which experience particular feeding conditions.

In this chapter we will discuss in detail these last two methods for investigating adaptation.

The comparative approach

The idea of comparison lies at the heart of most hypotheses about adaptation. It is the comparative study of different species which gives us a feel for the range of strategies that animals adopt in nature. When we ask functional questions about the behaviour of particular species we are usually asking why it is different from other species. Why does species A live in groups compared with species B which is solitary? Why do males of species B mate monogamously compared with males of species A which are polygynous, and so on? A powerful method for studying adaptation is to compare groups of related species and attempt to find out exactly how differences in their behaviour reflect differences in ecology. We will first describe two examples which pioneered the comparative approach and inspired workers to use the method with other animal groups. Then we will point out some of the methodological difficulties in formulating and testing hypotheses based on comparison. Finally we will describe some recent examples of the comparative method which have attempted to overcome these problems.

SOCIAL ORGANIZATION IN WEAVER BIRDS

The first person to attempt a systematic analysis of this kind was John Crook (1964) who studied about 90 species of weaver birds (Ploceinae). These are small finches which live throughout Africa and Asia, and although they all look rather alike there are some striking differences in their social organization. Some are solitary, some go around in large flocks. Some build cryptic nests in large defended territories while others cluster their nests together in colonies. Some are monogamous, with male and female forming a permanent pair bond; others are polygamous, the males mating with several females and contributing little to care of the offspring. How can we explain the evolution of this great diversity in behaviour?

Crook's approach was to search for correlations between these aspects of social organization and the species' ecology. The ecological variables he considered were the type of food, its distribution and abundance, predators and nest sites. His analysis showed that the weaver birds fell into two broad categories.

1 Species living in the forest tended to be insectivorous, solitary feeders, defend large territories and build cryptic solitary nests. They were monogamous and males and females had similar plumage.

2 Species living in the savannah tended to eat seeds, feed in flocks and nest colonially in bulky conspicuous nests. They were polygamous and there was sexual dimorphism in plumage, the males being brightly coloured and the females rather dull (Fig. 2.1).

Why is the behaviour and morphology of the weaver birds linked to their ecology in such a striking way? Crook invoked predation and food as the main selective pressures that have influenced the evolution of social organization. His argument was as follows.

1 In the forest, insect food is dispersed. Therefore it is best for the birds to feed solitarily and defend their scattered food resources as a territory. Because the food is difficult to find, both parents have to feed the young and therefore stay together as a pair throughout the breeding season. With the male and female visiting the nest, both must be dull coloured to avoid attracting predators. Cryptic nests spaced out from those of neighbours decrease their vulnerability to predation.

2 In the savannah, seeds are patchy in distribution and locally superabundant. It is more efficient to find patches of seeds by being in a group because groups are able to cover a wider area in their search. Furthermore the patches contain so much food that there is little competition within the flock while the birds are feeding.

In open country the birds cannot hide their nests and so they seek safety in protected sites, such as spiny acacia trees. Nests are bulky to provide thermal insulation against the heat of the sun. Because good breeding sites are few and scattered, many birds nest together in the same tree. Within a colony, males compete for nest sites and those that defend the best sites attract several females while males in the poorer parts of the colony fail to breed. In addition, because food is abundant,

Fig. 2.1. Top: some species of weaver birds, like this one (*Malimbus scutatus*) are insectivorous, building cryptic solitary nests in the forest and feeding alone in large territories. Bottom: other species like this one (*Ploceus cucullatus*) feed on seeds out in the open savannah. They build conspicuous nests in colonies and feed in flocks. The males are often brightly coloured.

the female can feed the young by herself and so the male is emancipated from parental care and can spend most of his time trying to attract more females. This has favoured brighter plumage colouration in males and the evolution of polygamy.

Supporting evidence for this interpretation comes from species with intermediate ecology. The grassland seed-eaters have patchy food supplies so group living is favoured for efficient food finding. However, in grassland the nests are vulnerable, so predation favours spacing out. The result is a compromise; these species have an intermediate social organization, nesting in loose colonies and feeding in flocks.

These results show clearly how food and predation may be

important in determining social organization. They also reveal how several different traits such as nests, feeding behaviour, plumage colour and mating system can all be considered together as a result of the same ecological variables. Crook's work with the weaver birds inspired several people to use the comparative method to study social organization in other groups. David Lack (1968) extended the argument to include all bird species and Peter Jarman (1974) used the same approach for the African ungulates.

SOCIAL ORGANIZATION IN AFRICAN UNGULATES

Jarman (1974) considered 74 species of ungulates; all eat plant material but differences in the precise type of food eaten are correlated with differences in movements, mating systems and antipredator behaviour. The species were grouped into five ecological categories (Table 2.1). Just as in the weaver birds, several adaptations seem to go together.

The major correlate of diet and social organization is body size. Small species have a higher metabolic requirement per unit weight and need to select high quality patches of food such as berries and shoots. These tend to occur in the forest and are scattered in distribution, so the small species are forced to live a solitary existence. The best way to avoid predators in the forest is to hide. Because the females are dispersed, the males must also be dispersed and the commonest mating system is for a pair to occupy a territory together.

Table 2.1. The social organization of African ungulates in relation to their ecology. From Jarman (1974)

	Exemplary groups	Body weight (kg)	Habitat	Diet	Group size	Reproductive unit	Antipredator behaviour
Grade I	Dikdik Duiker	3–60	Forest	Selective browsing; fruit, buds	1 or 2	Pair	Hide
Group II	Reedbuck Gerenuk	20–80	Brush, riverine grassland	Selective browsing or grazing	2 to 12	Male with harem	Hide, flee
Group III	Impala Gazelle Kob	20–250	Riverine woodland, dry grassland	Graze or browse	2 to 100	Males territorial in breeding season	Flee, hide in herd
Group IV	Wildebeest Hartebeest	90–270	Grassland	Graze	Up to 150 (thousands on migration)	Defence of females within herd	Hide in herd, flee
Group V	Eland Buffalo	300–900	Grassland	Graze unselectively	Up to 1000	Male dominance hierarchy in herd	Mass defence against predators

At the other extreme, the largest species eat poor quality food in bulk and graze less selectively on the plains. It is not economical to defend such food supplies and these species wander in herds, following the rains and fresh grazing. In these large herds there is potential for the strongest males to monopolize several females by defence of a harem or a dominance hierarchy of mating rights. When predators come along these species cannot hide on the open plains and so either flee or rely on safety in numbers in the herd. Ungulates of intermediate size show aspects of ecology and social organization in between these two extremes (Table 2.1).

Adaptation or story-telling?

This comparative approach to adaptation is persuasive, but there are problems (Clutton-Brock & Harvey 1979; Gould & Lewontin 1979). Many of the following problems are not unique to comparative studies and it is worth bearing them in mind throughout the book.

ALTERNATIVE HYPOTHESES

The explanations for the differences in social organization are plausible, but alternative hypotheses have not been considered in a rigorous manner.

The ecological variables, such as predator pressure and patchy environment, have also been used in rather a vague way. Indeed, in the weaver birds we invoked a patchy food distribution as responsible for the evolution of flocking while in the ungulates we said that high quality patchy foods favour a solitary existence. There is the obvious danger here of explaining things too easily without any rigorous quantification of the ecological factors concerned.

CAUSE AND EFFECT

Consider the observation that weaver birds with a diet of seeds go about in flocks. Our explanation was that seed eating selects for flocking because this is the best way to find a patchy food supply. However we could equally well have suggested that predation selects for flocking and, as a consequence, the birds are forced to select locally abundant food so all the flock can get enough to eat. In this case a diet of seeds is a consequence, or effect, of flocking, not a cause. Maybe predation also selects for flocking in the forest insectivores but because their diet is incompatible with flocking they have to forage singly.

30
*Chapter 2
Testing
Hypotheses in
Behavioural
Ecology*

CONFOUNDING VARIABLES

With the comparative approach there are often confounding variables. For example, we observe the giraffe with its long neck, feeding at the tops of trees and the buffalo with its short neck feeding on the ground. We then say that a long neck is an adaptation for feeding high up in the trees. But a long neck could equally well aid predator detection. How can we control for such confounding variables and decide which selective pressure has favoured the trait? Perhaps it is both?

A particularly important confounding variable in comparative studies is body size. Jarman controlled for this in his analysis of the ungulates by dividing the species up into categories of different body weight (Table 2.1). Most biological traits do not increase in a 1:1 relationship with body size; their relation to body size is said to be allometric (Gould 1966). For example, the brain mass of different bird species increases at about the two-thirds power of body weight. In this case, before we can examine the ecological correlates of brain size we first have to remove the effects of body size. This can be done by calculating the appropriate line of best fit when brain mass is plotted against body weight and then measuring deviations from the line to see whether the size of the trait is greater or less than expected from body weight.

ALTERNATIVE ADAPTIVE PEAKS OR NON-ADAPTIVE DIFFERENCES

It is tempting when comparing between species to assume that differences are always adaptive but some differences may simply be alternative solutions to the same ecological pressures. An ecologist from Mars who visited the Earth would observe that in the United States people drive their cars on the right hand side of the road while in Britain they drive on the left. He would then perhaps make lots of measurements in an attempt to find ecological correlates to explain the adaptive significance of the difference. In fact driving on the right and driving on the left may just be equally good alternatives for preventing accidents (Dawkins 1980).

Some differences between animals may be like this. Sheep use horns for fighting and deer use antlers. Horns are derived from skin while antlers are derived from bone (Modell 1969). The differences between horns and antlers need not necessarily reflect ecological differences; it may simply be a case of evolution working with different raw materials to produce the same functional end (Fig. 2.2). The problem with non-adaptive explanations is that they are hypotheses of the last resort.

Fig. 2.2. The horns of the sheep (left) and antlers of the deer (right) are both used in fighting. Horns are derived from skin and antlers from bone.

Further scientific enquiry is stifled. Maybe there is an adaptive explanation for the difference but we just haven't discovered it yet. For example, antlers are dropped and then renewed each year whereas horns are not. Perhaps this difference is related to the extent of seasonal variation in mating competition and food supply?

These criticisms are important, but they certainly do not mean that the comparative method is a failure. On the contrary, the approach is impressive in the way it brings together such a wide diversity of behavioural and morphological traits within the same ecological framework. Crook's study of the weaver birds and Jarman's work on the antelopes have served as models for ecological work on other groups of species. However the most recent comparative studies have attempted to control for these various problems, and we will now discuss another example, bearing the criticisms in mind, to illustrate how changes in methodology have made comparison between species a more rigorous exercise.

Primate social organization

Early knowledge of primate behaviour came mainly from studies in zoos. In 1932 Lord Zuckerman suggested that primates tend to be social animals because they have continuous sexual activity. It should be clear from the last chapter that this is a causal explanation and leaves the functional significance of sociality unexplained. In the 1950s, the first studies of primates in the field (e.g. Carpenter 1954) revealed that sexual activity was not in fact continuous. It also became clear that

(a)

(b)

(c)

(d)

Fig. 2.3. Four photographs illustrating the variety of primate social organization. (a) Table 2.2, Grade I: A solitary insectivorous prosimian, the lesser bushbaby (*Galago senegalensis*) (photo by Caroline Harcourt). (b) Table 2.2, Grade II: A monogamous pair of folivorous, arboreal, black gibbons (*Hylobates concolor*). The male is on the left. © Ron Tilson/BPS. (c) Table 2.2, Grade IV: Part of a troop of folivorous, arboreal, dusky langurs (*Presbytes obscurus*). © Ron Tilson/BPS. (d) Table 2.2, Grade V: A troop of savannah-dwelling olive baboons (*Papio anubis*). There are two subordinate males in the foreground and the dominant males are in the background near the females and young. Photo by Irven De Vore (Anthro-Photo).

different species have very different social organizations (Fig. 2.3). Tiny tarsiers and lemurs hunt solitarily in the tree tops for insects at night. Some monkeys go around in small groups in the trees by day, feeding on leaves or fruit. Others are terrestrial and live in large troops. Among the apes, the orangutan is solitary, the gibbon lives in pairs and small family groups while the chimpanzee may live in bands of up to 50.

How can we explain the evolution of this bewildering array of social organization? It soon became apparent that ecological factors were important. For example, De Vore (1965) noticed that, compared with other species of primates, anubis baboons live in large groups, the males are large and they have big teeth. He suggested these may all be adaptations to predator defence in a terrestrial environment. By 1966, there were sufficient field data for John Crook and Stephen Gartlan to apply the first comparative approach to a large number of primates.

Like the weaver bird and antelope work, they categorized the species into several groups based on ecology and behaviour. Table 2.2 shows that at one extreme, insectivorous primates are nocturnal, forest animals which are solitary; then there is a variety of fruit and leaf eating species which are diurnal and live in small to large groups; finally, at the other extreme, are the vegetarian browsers of the open country which live in large groups and show intensive competition between males for females and marked sexual dimorphism.

Once again, food and predation were suggested as the main selective pressures responsible for this link between social organization and habitat. Insects are dispersed and difficult to find, so just like the insectivorous weaver birds, these primates are solitary. In open country, predation favours grouping for safety and food is locally abundant so this allows many individuals to congregate at a food source; like the open-country weaver birds, these primates live in groups. In a large group, males compete with each other for mating rights and hence large male body size has been selected for.

Table 2.2. Crook and Gartlan's (1966) division of the primates into five 'adaptive grades'

	Exemplary species	Habitat	Diet	Activity	Group size	Reproductive unit	Sexual dimorphism
Grade I	*Galago* *Lepilemur* *Microcebus*	Forest	Insects	Nocturnal	Solitary	Pairs	Slight
Group II	*Indiri* *Lemur* *Hylobates*	Forest	Fruit or leaves	Crepuscular or diurnal	Very small groups	Single male with family	Slight
Grade III	*Colobus* *Saimiri* *Gorilla*	Forest or forest edge	Fruit or fruit and leaves	Diurnal	Small groups	Multi-male groups	Slight to fairly marked
Grade IV	*Macaca* *Cercopithecus aethiops* *Pan*	Forest edge to tree savannah	Vegetation /omnivore	Diurnal	Medium to large groups	Multi-male groups	Marked
Grade V	*Erythrocebus patas* *Papio hamadryas* *Theropithecus gelada*	Grassland or arid savannah	Vegetation /omnivore	Diurnal	Medium to large groups	One male groups	Marked

Crook and Gartlan's approach was to categorize the primates into a small number of discrete groups. This raises two main problems. First of all, variation in features such as home range size and group size is continuous and so division into hard and fast groups is a bit arbitrary. Because the groups are subjectively defined, it is difficult for subsequent workers to categorize new species in the scheme. Second, different aspects of social organization such as breeding system and group size do not necessarily vary together in the same way. For example, two species of primate could have the same breeding system but live in different sized groups.

The most recent attempt to unravel the complexities of primate social organization has been by Tim Clutton-Brock and Paul Harvey. They have tried to avoid these problems firstly by measuring the various aspects of social behaviour and morphology on a continuous scale. Secondly, they use multivariate statistics to tease out the effects of several different ecological variables on the same traits, and to analyse the influence of ecological factors on each aspect of social organization independently. Their third improvement in approach is a careful consideration about which taxonomic level should be used for analysis, e.g. species, genus, subfamily or family.

This last problem is one about the independence of data points. Imagine plotting all the species of primates on a graph in order to investigate the relationship between body weight and some interesting variable such as home range size, brain size or mating system (e.g. females per male in a breeding group). On our graph we would find that within a genus all the species will be clumped together in a cluster of points. For example, all six species of gibbons are of similar body weight, all are monogamous, arboreal and eat fruit. Our problem is whether we should treat these as six independent points or just one point in any statistical analysis. If we treated them as six independent points our analysis may be biased because it would reflect phylogeny, rather than ecology; all six gibbons may be descended from a single ancestor which was monogamous, arboreal and ate fruit. Because species within a genus tend to have similar characteristics due to phylogenetic constraints, analysis of species data will be statistically biased by those genera containing large numbers of species.

In picking the taxonomic level for our analysis a good rule to use is to choose, within the order (in this case the primates), the taxonomic level where the maximum variance is shown in the variable we are considering. This level can be identified using a nested analysis of variance. As we go up the taxonomic levels the variance will increase. Variance between species within a genus will be small; between genera within a subfamily it will be greater and between families across the whole

order of primates there will be still wider scatter in our points.
For the primates it is found that maximum variance is shown
in most ecological and morphological variables when different
genera are considered as independent points (e.g. all six gibbons would be treated as one point only). No additional variance is shown if families are considered rather than genera.
Therefore, using the genus as the level for analysis means that
there will be no additional variance in the sample which
might obscure the results because it is due to taxonomic relationship rather than ecological differences (Harvey & Mace
1981).

We shall now consider some examples of the comparative
approach to primate social organization and morphology, trea-
ting different genera as independent points for analysis, to
illustrate how comparison has become a more rigorous and
objective exercise.

HOME RANGE SIZE

Larger animals need to eat more food and so, in general, we
would expect them to have larger home ranges. Therefore, if
we want to examine the influence of an ecological variable,
such as diet, on home range size, we have to control for body
weight as a confounding variable. When home range size is
plotted against the total weight of the group that inhabits it, as
expected the larger the group weight the larger the home range
(Fig. 2.4).

The influence of diet on home range size can be seen when
the specialist feeders (insectivores, frugivores) are separated
from the leaf eaters (folivores); the specialist feeders have
larger home ranges for a given group weight. The probable
explanation is that fruit and insects are more widely dispersed
than leaves and so specialist feeders need a larger foraging
area in which to find enough food.

These general trends are confirmed by more detailed studies
of particular species. The red colobus monkey (*Colobus badius*)
is a specialist feeder, eating shoots, fruit and flowers. The food
occurs in scattered in clumps and this species wanders over a
large home range of about 70 hectares. The black and white
colobus (*C. geureza*) is a generalist, eating leaves of all ages.
Its food supply is dense and evenly distributed and its home
range is only 15 hectares (Clutton-Brock 1975).

SEXUAL DIMORPHISM IN BODY WEIGHT

In primates, males are often larger than females. Two hypotheses could explain this observation. Sexual dimorphism
could enable males and females to exploit different food niches

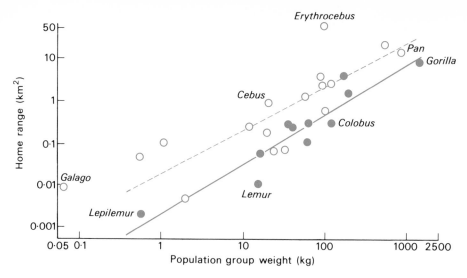

Fig. 2.4. Home range size plotted against the weight of the group that inhabits the home range for different genera of primates. The solid circles (●) are folivores, through which there is a solid regression line. The open circles (○) are specialist feeders (insectivores or frugivores) and the regression line through these points is dashed. Some of the genera are indicated by name. From Clutton-Brock and Harvey (1977).

and thus avoid competition (Selander 1972). If this was true, then we might predict that dimorphism would be greatest in monogamous species where male and female usually associate together and feed in the same areas. Alternatively it could have evolved through sexual selection, large body size in males being favoured because this increases success when competing for females (Darwin 1871). If sexual competition is important then we would predict that dimorphism should be greater in polygamous species, where large male size would be especially advantageous because a male could potentially monopolize several females.

The comparative data show no sign of the trend predicted by the niche separation hypothesis but do support the sexual competition hypothesis; the more females per male in the breeding group, the larger the male is in relation to the female (Fig. 2.5).

SEXUAL DIMORPHISM IN TOOTH SIZE

Males often have larger teeth than females. Again, two hypotheses can be suggested (Harvey et al. 1978). Large teeth may have evolved in males for defence of the group against predators. Alternatively males may have larger teeth for competition with other males over access to females. There is the problem here of body weight as a confounding variable; males

36

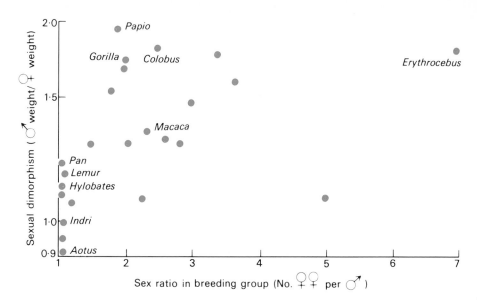

Fig. 2.5. The degree of sexual dimorphism increases with the number of females per male in the breeding group. Each point is a different genus, some of which are indicated by name. From Clutton-Brock and Harvey (1977).

are larger than females and so a difference between the sexes in tooth size could just reflect a difference in body size.

This can be controlled for by calculating the line of best fit when female tooth size is plotted against body weight. If the tooth size of a male is now plotted on the same graph, it can be seen whether its size is greater than expected for a female of the same body weight. The results show that in monogamous species male tooth size is as expected for a female of equivalent body weight. However it is larger than expected in harem-forming species. These data support the sexual competition hypothesis for the evolution of larger teeth in males. However we cannot exclude the predator defence hypothesis because maybe the harem-forming species are the ones most vulnerable to predation.

The analysis can be taken a step further by considering species where several males live together in a group (multimale troops). It is found that, within this type of social organization, the males of terrestrial species have larger teeth for their body size than arboreal species. Therefore even within the same mating system there is a difference in tooth size in different habitats. The terrestrial environment is usually thought to present greater risks of predation and so predation pressure may have been responsible for the evolution of larger teeth in terrestrial species.

Our conclusion is that both sexual competition and predation may have influenced the evolution of sexual dimorphism in tooth size. There is also the further possibility that differences in tooth size are important in reducing diet overlap between the sexes and so preventing competition for food. This example shows that, even with careful analysis, it may be difficult to tease out the effect of several variables on the evolution of a trait.

TESTIS SIZE AND BREEDING SYSTEM

The heaviest primates, the gorilla (*Gorilla gorilla*) and orangutan (*Pongo pygmaeus*) have breeding systems that involve one male monopolizing mating with several females, and have testes that weigh 30g and 35g respectively (average weight of both testes). The smaller chimpanzee (*Pan troglodytes*), by contrast, has a breeding system where several males copulate with each oestrus female and this species has testes weighing 120g! It seems likely that the marked differences in testes weights are related to differences in breeding system. In single-male breeding systems (gorilla and orangutan) each male need ejaculate only enough sperm to ensure fertilization. In multi-male systems (chimpanzee), however, a male's sperm has to compete with sperm from other males. Selection should, therefore, favour increased sperm production and hence larger testes.

Harcourt *et al.* (1981) tested this hypothesis by comparing 20 genera of primates, varying in body size from the 320g marmosets (*Callithrix*) to the 170 kg gorilla. Figure 2.6 shows that, as expected, testes weight increases with body weight. For a given body weight, however, it is clear that genera with multi-male breeding systems have heavier testes than genera with single-male or monogamous breeding systems. The data points for the former group lie above the line, and those for the latter lie below ('single-male' indicates that there is only one breeding male although, as in the gorilla, there may be more than one male in the social group; 'monogamous' indicates that there is just one male and one female in a group). These data therefore support the sperm competition hypothesis.

The comparative approach reviewed

The statistical approach we have described for the primates is certainly a major improvement on the first applications of the comparative method. To summarize, the main improvements are:

1 Different aspects of social organization are treated independently and as continuous variables.

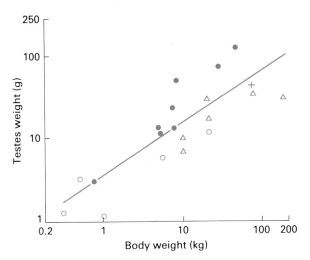

Fig. 2.6. Log combined testes weight (g) versus log body weight (kg) for different primate genera. ● Multi-male breeding system; ○ monogamous; △ single-male; + *Homo*. From Harcourt *et al.* (1981).

2 Confounding variables are dealt with in a rigorous manner.
3 Care is taken to choose the most appropriate taxonomic level for analysis.
4 The data are used wherever possible to discriminate between alternative hypotheses such as predation or sexual competition.

The end result of many of the analyses is a plausible interpretation which may be treated as an hypothesis for further testing. In conclusion, the comparative approach is very useful for looking at broad trends in evolution and the general relationship between social organization and ecology. It generates hypotheses which can be used as predictions for other groups of animals. It can also be used to test hypotheses which are not amenable to experimentation, such as the effect of polygamy on sexual dimorphism. Furthermore, it is impressive in the way it shows how diet, predation, social behaviour and body size, for example, can all be interrelated.

However, we need a different approach to understand in detail why animals adopt particular strategies in relation to their ecology. Can we actually measure patch structure and predation risk and then come up with precise predictions as to how an animal will behave? Can we explain why a monkey goes round in a group of 20 rather than one of 16 or 25, why its home range is 10 hectares rather than 8 or 12 hectares, and why it spends 1 hour in a patch of fruit? Indeed we can attempt to answer precise questions like these. No one has yet attempted this for anything as complicated as primate social behaviour. However a start has been made for simpler kinds of behaviour using optimality theory and an experimental approach.

Experimental studies of adaptation

We now turn to a different, and complementary, way of looking at how selection moulds behaviour. Instead of broad scale comparisons between species, the emphasis will be on the behaviour of individuals of the same species and analysing their behaviour in terms of *costs* and *benefits*.

The idea of trying to measure costs and benefits grew out of Niko Tinbergen's experimental approach to studying the survival value of behaviour. For example, Tinbergen observed that in a colony of blackheaded gulls (*Larus ridibundus*) nesting on sand dunes in north-western England, incubating parents always pick up the broken eggshell after a chick has hatched and carry it away from the nest (Fig. 2.7). Although carrying the shell takes only a few minutes each year it is crucial for the survival of the young. The eggs and young of the blackheaded gull are well camouflaged against the grass, sand and twigs around the nest. The inside of the broken shell, however, is white and highly conspicuous. Tinbergen carried out an experiment to test the hypothesis that the conspicuous white broken shell reduces the camouflage of the nest. He painted hens' eggs to resemble cryptic gull eggs and laid them out at regular intervals in the gull colony. Next to some he placed a broken shell. The results confirmed his prediction that the cryptic eggs were much more likely to be discovered and eaten by predators such as crows if they were close to a broken shell. So it is easy to visualize why the parent benefits by removing the conspicuous empty shell soon after the chick has hatched: the camouflage of the brood is preserved and the likelihood of the parent perpetuating its genes is increased.

But there is more to the story than this. The parent does not remove the eggshell immediately; it stays with the newly hatched chick for an hour or more and then goes off with the shell. In order to explain the delay in removing the shell we have to introduce the idea of a trade-off between costs and benefits. If the parent flies off with the shell at once, it has to leave the newly hatched chick unattended (the second parent is away at the feeding grounds fuelling up for its next stint at the nest). Tinbergen observed that the new chick, with its plumage still wet and matted, is easily swallowed and therefore makes a tempting meal for a cannibalistic neighbouring adult. However, when the chick's down has dried out and become fluffy it is much harder for a gull to swallow, and is therefore less vulnerable to attacks from neighbours. The

Fig. 2.7. A blackheaded gull removing an eggshell from its nest. Photo by N. Tinbergen.

parent's delay before removing the shell therefore probably reflects a balance between the benefits of maintaining the camouflage of the brood and the costs associated with leaving a newly hatched chick at its most vulnerable moment.

When the balance between costs and benefits is changed, the length of the parent's delay might also be expected to change. This is borne out by observations of the oystercatcher, another ground nesting bird with camouflaged eggs and young. The oystercatcher (*Haematopus ostralegus*) is a solitary nester and cannibalism by neighbours is therefore not a risk associated with leaving the newly hatched chicks. The parents benefit by restoring camouflage of the nest as soon as possible after hatching and as expected the parent removes broken eggshells more or less as soon as a chick has hatched and before its down is dry.

OPTIMALITY MODELS

Tinbergen's study of eggshell removal illustrates how experimental studies of costs and benefits can be used to unravel behavioural adaptations, but it has an important limitation.

The hypothesis about the trade-off between camouflage and chick vulnerability made only a qualitative prediction. The idea would be consistent with observations of gulls removing eggshells 1, 2, 3 or perhaps even 4 hours after the chick has hatched, so that it is hard to test whether the hypothesis is right or wrong. One way of trying to make an hypothesis more easily testable is to try to generate quantitative predictions. If one could predict that the parent gull should remove its egg-shell after 73.5 mins then one would have produced a very testable model indeed. This is an approach which has been developed by using *optimality models* to study adaptations. An optimality model seeks to predict which particular trade-off between costs and benefits will give the maximum net benefit to the individual.

Thinking back to the gulls, if one could measure exactly how much the survival of the brood is reduced by the conspicuous broken eggshell next to the nest and exactly how the risk of cannibalism by neighbours changes with time since the chick hatched, one could start to calculate the optimum time for the parent to delay removal of the shell. In this case the optimum might well be defined as the time that maximizes total reproductive success for the season. But the currency of an optimality model does not have to be survival or production of young. The overall success of an individual at passing on its genes may depend on finding enough food, choosing a good place to nest, attracting many mates, and so on. In solving any of these problems an animal makes decisions, and the decisions can be analysed in terms of an optimal trade-off between appropriate costs and benefits. For a foraging animal, for example, currencies might be energy and time.

Gathering food

CROWS AND WHELKS

On the west coast of Canada, as in many coastal areas, crows feed on shellfish. They hunt for whelks at low tide, and having found one they carry it to a nearby rock, hover and drop it from the air to smash the shell on the rock and expose the meat inside. Reto Zach (1979) observed the behaviour of north-western crows in detail and he noted that they take only the largest whelks and on average drop the shell from a height of about 5 m. Zach carried out experiments in which he dropped whelks of different sizes from various heights. This, together with data on the energetic costs of flying and searching gave him the information to carry out calculations of the costs and benefits associated with foraging. The benefit obtained by the crow and the cost paid could both be measured in calories, and Zach's calculations revealed that only the largest whelks (which contain the most calories and break open most readily) give enough energy for the crow to make a net profit while foraging. As predicted from these calculations, the crows ignored all but the very largest whelks even when different sizes were laid out in a dish on the beach.

Usually the crow has to drop each whelk twice or more in order to break it open. Since ascending flight is very costly, Zach thought that the crow might have chosen the dropping height which would minimize the total expenditure of energy in upward flight. If each drop is made from close to the ground, a very large number of drops is required to break open the shell, while at greater and greater heights the shell becomes more and more likely to break open on the first drop (Fig. 2.8a). The experiment of dropping shells from different heights allowed Zach to calculate the total vertical flight needed to break an average shell from different dropping heights (Fig. 2.8b). The crows fly to an average height of 5.2 m and at this height the total vertical flight per whelk is close to its minimum. However, the crow would have to undertake almost the same total upward flight even if each drop was made from a height somewhat greater than 5.2 m (this is indicated by the very shallow U-shaped curve of Fig. 2.8b) because slightly fewer drops would be needed. Zach suggests that there may be an additional penalty for dropping from too great a height: the whelk may bounce away and be lost from view or may break into so many fragments that the pieces are too small to retrieve.

The story of crows and whelks shows how calculations of costs and benefits can be used to produce a quantitative prediction. The crow seems to be programmed to choose a dropping height that comes close to minimizing the total vertical flight per whelk. Other currencies, such as maximizing net rate of energy gain, predict much greater drop heights (Plowright *et al.* in press).

Summary

Throughout this book we will be using both the comparative method and the experimental analysis of costs and benefits to individuals to test hypotheses about behavioural adaptations. The comparative method involves comparing different species to see whether differences in behaviour are correlated with differences in ecology. In weaver birds, antelopes and primates the main factors determining the evolution of social

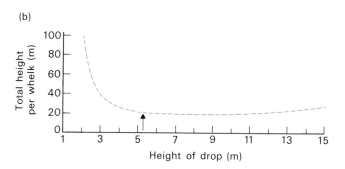

Fig. 2.8. Results obtained by dropping whelk shells from different heights. (a) Fewer drops are needed to break the shell if it is dropped from a greater height. (b) The total upward flight needed to break a shell (no. of drops × height of each drop) is close to its minimum at the height most commonly used by the crows (shown by arrow). (Zach 1979.)

behaviour are the distribution and abundance of food, predators and mates. These have been shown to influence group size, home range size, mating behaviour and sexual dimorphism. Two of the main problems in comparative studies are those of confounding variables and the choice of taxonomic level for comparison.

The experimental approach involves a detailed study of the costs and benefits of a behaviour pattern to an individual of a particular species. Behaviour can be viewed as having costs and benefits and animals should be designed by natural selection to maximize net benefit. Ultimately the net benefit must be measured in terms of gene contribution to future generations. This will depend on various shorter-term goals such as foraging efficiency, mating efficiency and efficiency of avoiding predators. Optimality models can be used to predict which particular trade-off between costs and benefits gives maximum net benefit.

Further reading

Clutton-Brock and Harvey (1979) provide a good discussion of some of the problems in using the comparative approach. David Lack's (1968) book applies the comparative method to the breeding biology of birds. Mark Ridley (1983) discusses the methodology of the comparative approach, particularly the problem of what to use as independent data points. Niko Tinbergen's book on gulls (1953) and his book *Curious Naturalists* (1974) give a marvellous sense of the excitement of observing animals in the field and show the power of performing simple field experiments to help us understand why animals behave the way they do.

Topics for discussion

1 What are the relative merits of the comparative method and the experimental approach for studying the adaptive significance of behaviour?

2 How can we decide what are the independent units of observation in comparative studies? (see chp. 1 in Ridley 1983).

3 How could Jarman's (1974) analysis of ungulate social organization be improved by applying Clutton-Brock and Harvey's methods?

Chapter 3. Economic Decisions and the Individual

In this chapter we will describe in more detail how the idea of economic analysis of costs and benefits can be used to understand behaviour. Most of our examples will refer to foraging, but the same principles apply to other aspects of behaviour.

The economics of carrying a load

STARLINGS

Starlings feed their young mainly on leatherjackets (*Tipula* larvae) and other soil invertebrates. A busy parent at the height of the breeding season makes up to 400 round trips from its nest to feeding sites every day, ferrying a load of food to its insatiable nestlings on each trip (Fig. 3.1). In this section we are going to focus our miscroscope of economic analysis on one aspect of the parent starling's behaviour and ask: how

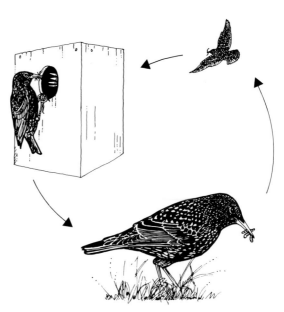

Fig. 3.1. Starlings fly from their nest to a feeding site, search for a beak-full of leatherjackets by probing in the grass, then take them home to the nestlings. The question examined in the first part of this chapter is how many items the parent should bring on each trip in order to maximize the rate of delivery of food to the nestlings.

many leatherjackets should the parent bring home on each trip? This may seem like an inconsequential question, but the size of load has a critical effect on the parent's overall rate of delivering food to the nest, which in turn determines whether or not the chicks survive to become healthy fledglings. As we saw in chapter 1, the reproductive success of small birds is often limited by their ability to feed their young. There is therefore strong selective pressure on the parents to perform as effective food deliverers.

It is not hard to imagine why it would be a poor strategy to bring one leatherjacket on each trip, given that the parent is in fact capable of carrying several at a time: the amount of travelling back and forth for the parent would be unecessarily high. It would be as if you did your week's shopping at the supermarket by making a separate trip for each individual item instead of collecting a whole bag full of groceries on one trip. Should the starling, therefore, bring back the biggest load it can carry on every trip? The answer is no, because of a crucial fact about starling foraging behaviour. A starling hunts for food by probing with its bill in a special way. It thrusts its closed bill into the turf of a meadow and then spreads its mandibles to part the vegetation and expose the leatherjackets just below the soil surface. The bird is very efficient at probing, but when it has a load of leather jackets already in its bill it becomes less efficient. For this reason it is not necessarily the best thing for the starling to fully load its beak before flying home to unload and start again. A similar kind of problem faces many animals that carry a load back to their nest or some other central place such as a cache of food.

The starling's problem of load size can be summarized as a graph (Fig. 3.2a). The graph shows time along the horizontal axis and load (measured in leatherjackets) on the vertical axis. Consider a starling at the nest about to embark on a round trip. It has to fly to (and eventually from) the feeding site; the times of these two trips are added together and plotted on the graph as 'travelling time'. When it arrives at the part of the meadow where the leatherjackets are abundant it starts to load up with food. The first couple of leatherjackets are found quickly and easily, but because of the encumbrance of the prey in its beak, the bird takes longer and longer to find each successive prey. The result is a 'loading curve' (or 'gain curve' as it is sometimes called) that rises steeply at first but then flattens off. This is a curve of diminishing returns — the longer the starling has been foraging, the less likely it is to find another leatherjacket in the next few moments — and the starling's problem, as summarized in the graph, is when to give up on this curve. If it gives up too early it spends a lot of

(a)

(b)

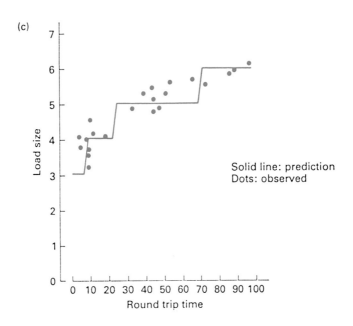

(c)

time travelling for a small load; if it struggles on too long it spends time in ineffective search which could be better spent by going home to dump its load and starting again at the beginning of the loading curve. Somewhere in between these extremes is the starling's 'best' option. A reasonable hypothesis (but at the moment it is no more than this), is that for the starling 'best' means 'providing the maximum net rate of delivery of food to the chicks'. Any starling that is slightly better at producing chicks than its rivals will be at a selective advantage, so selection should in the long run favour behaviour that maximizes chick production.

The best load can be found by drawing the line AB in Fig. 3.2a. The slope of this line is (load/[travel-time + foraging-time]) or in other words rate of delivery of food; this can be seen by the fact that it forms the hypotenuse of a right angled triangle with a base measured in 'time' and a vertical corresponding to 'load'. Now, travel time and the loading curve are constraints — fixed properties of the environment (or more precisely of the interaction between the starling and its environment), so any line such as AB representing the starling's rate of delivery of food has to start at the beginning of the round trip travel time and touch the loading curve somewhere. Any other line you could draw with these constraints will have a shallower slope (that is, give a lower rate of delivery) than the line AB. A couple examples are shown in Fig. 3.2a.

Fig. 3.2. Opposite. (a) The starling's problem of load size. The horizontal axis shows 'time' and the vertical axis shows 'load'. The curve represents the cumulative number of leatherjackets found as a function of time spent searching. The line AB represents the starling's maximum rate of delivery of food to the nestlings. This rate is achieved by taking a load of 7 leatherjackets on each trip. Two other lines, corresponding to loads bigger (8) and smaller (1) than 7 are shown to make the point that these loads result in lower rates of delivery (shallower slopes). Note that although the cumulative load is shown here as a smooth curve, in reality it is a stepped line since each food item is a discrete package. (b) When the round trip travel time is increased from short to long the load size that maximizes delivery rate increases from b to b'. (c) When starlings were trained to collect mealworms from a feeder, they brought bigger loads from greater distances. Each dot is the mean of a large number of observations of loads brought from a particular distance. The predicted line goes up in steps because the bird is predicted to change its load size in steps of one worm (of course the mean loads do not have to be integers). The prediction shown here is one based on the model of Fig. 3.2b, but it also includes the refinement of taking into account the energetic costs to the parent of foraging and to the chicks of begging (Kacelnik 1984).

Figure 3.2b develops the argument a little further. Suppose that the starling now switches to feeding at a closer site with a short travel time, how should its load per trip change? Using the same method as before, we can now draw two lines (Fig. 3.2b): when the travel time is shorter, the load that maximizes rate is smaller. One way to think of this is to imagine the starling at its moment of decision to go home. If it goes it loses the opportunity to continue foraging, if it stays it loses the opportunity to go home and start afresh. When it is far away the expected returns from going home are relatively low, since there is a long way to fly before the next chance to forage. It therefore pays to persist a little longer on the present trip, until current gains drop to a slightly lower level.

. Alex Kacelnik (1984) tested this prediction of the model of load size in the following way. He trained parent starlings in the field to collect mealworms for their young from a wooden tray onto which he could drop mealworms through a long piece of plastic pipe. Rather than letting the birds generate their own loading curve by diminishing search efficiency, Kacelnik generated the curve for them by dropping mealworms at successively longer and longer intervals. The trained bird would simply wait on the wooden tray for the next worm to arrive, until eventually it flew home with a beak-load for its chicks. The beauty of this experimental method is that Kacelnik knew the shape of the loading curve precisely and was hence able to present exactly the same loading curve at randomly varying distances (ranging from 8 to 600 m) from the nest on different days. The results were striking (Fig. 3.2c): not only did the load size increase with distance from the feeder to the nest, but there was a close quantitative correspondence between the observed load sizes and those predicted by the model of maximizing delivery rate.

Let us briefly summarize what the results of the starling study show. We started off by considering load size from the point of view of costs and benefits. We formulated a specific hypothesis about how costs and benefits might influence load size in the form of a model (Fig. 3.2a), and then used the model to generate a quantitative prediction (Fig. 3.2b). In making the model we did three important things. First we expressed a general conviction that starlings are designed by natural selection to be good at their job of parenting. This is not something that we aimed to test, but it is our general background assumption to justify thinking in terms of maximizing pay-off in relation to costs and benefits. Second, we made a guess about the *currency* of costs and benefits; we suggested that for a parent starling the crucial feature of doing a good job is maximizing net rate of delivery of food to the nestlings (rather than, say,

(a)   (b)

Fig. 3.3. (a) The model in Fig. 3.2 can be used to predict the duration of copula in dungflies, Box 3.1 and (b) to make predictions about many other examples of animals exploiting patchily distributed resources, such as these goldfinches (*Carduelis carduelis*) feeding on teazel heads.

waking them up as little as possible which is what human parents of young babies might favour!). Third we specified certain *contraints* on the starling's behaviour. Some of these constraints are to do with features of the environment (the time required to travel, the shape of the loading curve) and in the experiment of Kacelnik, they were clearly identifiable. Another important assumption about constraints is that the starling is assumed to 'know', or at least to behave as if it knows, the travel time and the shape of the loading curve. When we worked out the optimum load size we assumed that these were known. The experimental results supported the predictions of the model, and in so doing they supported the hypotheses about the currency and the constraints that were used to construct the model. Kacelnik did actually compare the predictions of models based on several different currencies, and he found for example, that one based on energetic efficiency (energy gained/energy spent) as opposed to rate gave a rather poor fit to the data.

The model of load size (usually called the 'marginal value theorem' (Charnov 1976) is applicable to lots of situations in which an animal exploits a resource that occurs in discrete sites or patches, and within a patch it experiences diminishing returns (see Box 3.1 and Fig. 3.3).

BEES

A similar problem to that of the starling is faced by a worker honey bee as it flies from flower to flower filling its honey crop with nectar to take back to the hive. Bees, like the starlings, often return to the hive with less than the maximum load they could carry and their behaviour can be explained by a model similar to that used for the starling. There is, however an important difference: the bee experiences a curve of diminishing returns neither because the nectar in its crop makes it less able to suck more flowers nor because of resource depression (Box 3.1) but because the weight of nectar in the crop adds an appreciable energetic cost to flight. The more the bee loads up its crop the more of its load it will burn up as fuel before it gets home. As a consequence, while the gross quantity of nectar harvested increases at a constant rate, the net yield of energy for the hive increases at a diminishing rate as the crop fills (producing, in effect, a loading curve like that of the starling).

Paul Schmid-Hempel (Schmid-Hempel et al. 1985) tested whether these diminishing net returns influence the bee's decision about when to go home and empty its crop. He trained bees to fly from the hive to a cluster of artificial flowers, each containing 0.6 mg of nectar. By varying the amount of flight the bee had to do between each flower in the cluster, he could alter the total cost of carrying the crop load and therefore the extent to which the bee experienced diminishing net returns. If, for example the bee could collect a load of 10 flower's-worth of nectar while flying for a total of 5 s it would experience little decrease in returns as it loaded up, while a bee collecting the same load by flying for a total of 50 s would suffer sharply diminishing returns. As predicted, the bees went home with smaller loads when they were forced to fly a greater distance between flowers (Fig. 3.4a). Figure 3.4a also shows two predicted lines based on maximizing two different currencies. One is based on the currency used for the starlings, net rate of energy delivery, while the other is based on a currency that did not work for the starlings, energetic efficiency. In contrast to the starlings, the second currency but not the first accounts for the bees' behaviour.

Why should there be this difference between bees and starlings? A simple example shows why the 'starling currency' is normally a sensible one to consider. Compare a starling that forages for one hour, spending 1 Kcal and gaining 9 Kcal with one that spends 10 and gains 90. Both have the same efficiency (9) but the former has 8 Kcal to spend on its chicks and the latter has 80. In other words net rate ((gain − cost)/time) tells us how much the animal has left at the end of the day to spend on reproduction or survival while efficiency does not. On the other hand efficiency may be a sensible currency when the crucial variable for the animal is not just the amount of

Fig. 3.4. (a) The relationship between load size (expressed as number of flowers visited) carried home by worker bees and flight time between flowers in a patch. Each dot is the mean of an individual bee and the two lines are predictions based on maximizing efficiency (e) and maximizing rate (r). (Schmid-Hempel *et al.* 1985). (b) By placing tiny weights on the bee's back while it is foraging Schmid-Hempel was able to study the bee's rule of thumb for departure from a patch to go home to the hive with a load of nectar. The weights, in the form of brass nuts, are placed on a fine rod that is permanently glued to the bee's back. They can be added or removed to simulate loading and unloading. (Schmid-Hempel 1986.)

gain, but also the amount spent. If, for example you have to drive from A to B on a fixed amount of fuel, efficiency might be very important indeed. It turns out that bees may be in this position. Bees seem to 'wear out' after a certain amount of flying so that the longevity of a worker is inversely related to the distance it flies (Neukirch 1982). Therefore a bee that delivered food to the hive at a high net rate but that was a profligate spender of flight energy would die before a more efficiency-oriented individual and may, as a result, contribute less to the total nectar store of the hive.

The contrast between bees and starlings serves to underline the point that one of the aims of economic cost benefit analyses is to compare alternative currencies and to try to understand why a particular currency is appropriate in each case. In each study one of the major advantages of the quantitative analysis was that it allowed us to see when there was a *discrepancy* between observed and predicted results. Without this potential for discrepancy it would have been impossible, for example, to tell whether bees were maximizing rate or efficiency, or nothing at all.

The bee example also illustrates another important point. We have been thinking of animals as well-designed problem solvers making decisions that maximize an appropriate currency, but of course we do not believe that bees and other animals calculate their solutions in the same way as the behavioural ecologist. Instead the animals are programmed to follow rules of thumb which give more or less the right answer. The bees, for example might use a rule that involved a threshold body weight ('if weight greater than x then go home'). Schmid-Hempel (1986) investigated this by adding tiny (7 mg) weights to the bee's back while it was foraging (Fig. 3.4b). He found that when he added five weights at intervals during a foraging bout the bees went home with a smaller load, as predicted if they were using a threshold weight rule. However another experiment showed that the rule is not this simple. Instead of adding five weights gradually, Schmid-Hempel added five weights at the start of a foraging bout and then took them off gradually as the bee filled its crop. These bees also went home with smaller loads than unmanipulated bees (or than controls where the weights were placed on the bee's back for a brief moment). The most reasonable interpretation of these results is that the bee in some way integrates the total weight it has carried since arriving at the foraging site.

Box 3.1. *The marginal value theorem and reproductive decisions.*
The model of load carrying for starlings is applicable to many other situations in which animals experience diminishing returns within a patch. It has been used to predict how much time an animal foraging for itself (as opposed to carrying loads) will spend in each site before moving on (Cowie 1977). Diminishing returns in each patch (generally referred to as 'resource depression') might arise, for example, simply because of depletion, or because prey in the patch take evasive action and become harder to catch, or because the predator

becomes less likely to search new areas in the patch (it crosses its own path more) as time goes by, or because the predator starts with the easy prey and then goes on to hunt for those that are more difficult to catch or are less rewarding. An example of the last of these is when bumblebees or other nectar-feeders visit the biggest and most rewarding flowers on an inflorescence first, and then go on to the smaller flowers which hold less nectar (Hodges & Wolf 1981).

Reproductive decisions can be analysed as though they are foraging decisions with the same model. An example is Geoff Parker's (1978) analysis of how male dungflies search for mates. Males compete with one another for the chance to mate with females arriving at cowpats to lay their eggs. Often one male will succeed in kicking another off a female during copulation and take her over. When two males mate with the same female the second one is the individual whose sperm fertilizes most of the eggs. Parker (1978) showed this by the clever technique of irradiating males with cobalt[60], which sterilizes them but does not alter sperm activity (the sperm can still fertilize an egg but the egg does not develop). If a normal male is allowed to mate after a sterile one about 80 per cent of the eggs hatch, whereas if the sterile male mates second only 20 per cent of them hatch. The conclusion from these 'sperm competition' experiments is clear: the second male's sperm fertilizes about 80 per cent of the eggs. It is not surprising, therefore, that after a male has copulated he sits on top of the female and guards her until the eggs are laid, only relinquishing his position to a rival male after a severe struggle.

When a second male takes over (or when a male encounters a virgin) how long should he spend copulating? Parker carried out sperm competition experiments in which he interrupted the second male's copulation after different times; this showed that the longer the second male mates the more eggs he fertilizes, but the returns for extra copulation time diminish rapidly (see Fig. on p. 56). There is a cost associated with a long copulation: the male misses the chance to go and search for a new female. After the male has copulated for long enough to fertilize about 80 per cent of the eggs, the returns for further copulating are rather small and the male might do better by searching elsewhere for a new mate.

The analogue of travel time in the starling model is the time the male dungfly must spend guarding the present female until she has laid her eggs plus the time he spends searching for a new female. This total is 156 min on average. As shown below, this estimate of travel time can be used to predict with reasonable accuracy how long the male spends copulating with a female.

Some other reproductive decisions that can be analysed in a similar way are discussed by Charnov and Skinner (1984).

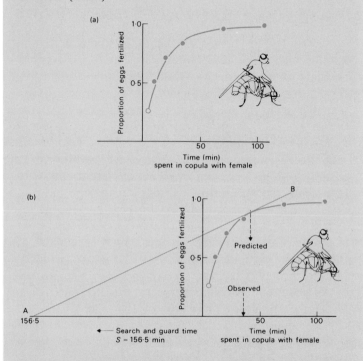

(a) The proportion of eggs fertilized by a male dungfly (*Scatophaga stercoraria*) as a function of copulation time: results from sperm competition experiments. (b) The optimal copulation time (that which maximizes the proportion of eggs fertilized per minute), given the shape of the fertilization curve and the fact that it takes 156 min to search for and guard a female, is 41 min. The optimal time is found by drawing the line AB (Parker 1978).

The economics of prey choice

The same kind of economic approach that we have used for bees and starlings can also be used to account for the kinds of prey items that predators decide to eat.

When shore crabs are given a choice of different sized mussels they prefer the size which gives them the highest rate of energy return (Fig. 3.5). Very large mussels take so long for the crab to crack open in its chelae that they are less profitable in terms of energy yield per unit breaking time (E/h) than the preferred, intermediate sized, shells. Very small mussels are easy to crack open, but contain so little flesh that they are hardly worth the trouble. However the story cannot be as simple as this, because the crabs eat a range of sizes centred around the most profitable ones. Why should they sometimes eat smaller and larger mussels? One possible hypothesis to explain why several sizes are eaten is that the time taken to search for the most profitable sizes influences the choice. If it takes a long time to find a profitable mussel, the crab might be

Fig. 3.5. Shore crabs (*Carcinus maenas*) prefer to eat the size of mussel which gives the highest rate of energy return. (a) The curve shows the calorie yield per second of time used by the crab in breaking open the shell and (b) the histogram shows the sizes eaten by crabs when offered a choice of equal numbers of each size in an aquarium (Elner & Hughes 1978).

able to obtain a higher overall rate of energy intake by eating some of the less profitable sizes.

In order to calculate exactly how many different sizes should be eaten we need to develop a more precise argument based on handling time, searching time, and the energy values of the various prey (Box 3.2). The equations in Box 3.2 show the following for the simple example of a predator faced with a choice of two sizes of prey. First, when the more profitable

---

Box. 3.2. *A model of choice between big and small prey.*

Consider a predator searching for T_s seconds and encountering two types of prey at rates λ_1 and λ_2 prey per sec. The two kinds of prey contain E_1 and E_2 calories and take h_1 and h_2 seconds to handle. They therefore have profitabilities E_1/h_1 and E_2/h_2.

If the predator eats both kinds of prey it will obtain the following in T_s seconds:

$$E = T_s(\lambda_1 E_1 + \lambda_2 E_2) \tag{3.1}$$

and this will take a total time:

$$T = T_s + T_s(\lambda_1 h_1 + \lambda_2 h_2) \tag{3.2}$$
$$= \text{searching time} + \text{handling time}.$$

Therefore the predator's rate of intake is (3.1) divided by (3.2):

$$E/T = \frac{\lambda_1 E_1 + \lambda_2 E_2}{1 + \lambda_1 h_1 + \lambda_2 h_2} \tag{3.3}$$

(note that T_s has cancelled out).

Let prey type 1 be the more profitable type. If the predator is to maximize E/T it should specialize on type 1 if:

$$\frac{\lambda_1 E_1}{1 + \lambda_1 h_1} > \frac{\lambda_1 E_1 + \lambda_2 E_2}{1 + \lambda_1 h_1 + \lambda_2 h_2} \tag{3.4}$$

(energy gain from just eating type 1 > energy gain from eating both).

This can be rearranged to give:

$$\frac{1}{\lambda_1} < \frac{E_1}{E_2} h_2 - h_1 \tag{3.5}$$

(note that λ_2 has cancelled out).

This is a useful form of the equation. $1/\lambda_1$ is the expected time needed to find a type 1 prey.

type (higher E/h) is very abundant the predator should special-ize on this alone. This is intuitively obvious: if something giving a high rate of return is readily available, an efficient predator should not bother with less profitable items. Second-ly the availability of the less profitable prey should have no effect on the decision to specialize on the better prey (this is because the term λ_2 does not appear in equation 3.5 in Box 3.2). This also makes sense; if good prey are encountered sufficiently often to make it worthwhile to ignore the bad ones, it is never worth taking time out to handle bad prey regardless of how common they are. The third conclusion from Box 3.2 is that as the availability of the good prey in-creases, there should be a sudden change from no preference (the predator eats both types when encountered) to complete preference (the predator eats only the good prey and always ignores the bad ones). Only when the two sides of equation 3.5 in the box are exactly equal will it make no difference to the predator whether it eats one or both types of prey.

An experiment which tried to test these predictions is illus-trated in Fig. 3.6. The predators were small birds (great tits) and the prey were large and small pieces of mealworm. In order to control precisely the predator's encounter rate with the large and small worms the experiment involved the un-usual step of making the prey move past the predator rather than vice versa (Fig. 3.6a). The big worms in the experiment were twice as large as the small ones ($E_1/E_2 = 2$) and h_1 and h_2 could be accurately measured as the time needed for the bird to pick up a worm and eat it. During the experiment the bird's encounter rate with large worms was varied so as to cross the predicted threshold from non-selective to selective foraging (equation 3.5) in Box 3.2. The results were roughly as pre-dicted, the main difference between observed and expected results was that the switch was not a step but a gradual change (Fig. 3.6b). When big worms were abundant the birds, as predicted, were selective even if small worms were extremely common.

Sampling and information

The discussion so far has referred to animals that know their environment. In Fig. 3.2a, the model assumes that the animal knows the quality of each patch and the travel time before it starts. Sometimes this may be a reasonable assumption, but at other times it may be more realistic to assume that the animal learns as it goes along. Steve Lima (1984) studied this kind of problem in downy woodpeckers. He trained woodpeckers in the field to hunt for seeds hidden in holes drilled in hanging

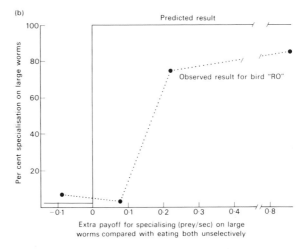

Fig. 3.6. (a) The apparatus used to test a model of choice between big and small worms in great tits (*Parus major*). The bird sits in a cage by a long conveyor belt on which the worms pass by. The worms are visible for half a second as they pass a gap in the cover over the top of the belt and the bird makes its choice in this brief period. If it picks up a worm it misses the opportunity to choose ones that go by while it is eating. (b) An example of the results obtained. As the rate of encounter with large worms increases the birds become more selective. The x-axis of the graph is the extra benefit obtained from selective predation. As shown in Box 3.1 the benefit becomes positive at a critical value of $1/\lambda_1$. The bird becomes more selective about the predicted point, but in contrast with the model's prediction this change is not a step function (Krebs *et al.* 1977).

logs. Each log had 24 holes and in each experiment some logs were quite empty and others had seeds hidden in some or all of the the holes. The woodpeckers could not tell in advance which were the empty logs so they had to use information gathered at the start of foraging on each log to decide whether or not it was likely to be empty and therefore should be abandoned. When the logs contained 0 or 24 seeds the task was easy: looking in a single hole in theory gave sufficient information to decide and the woodpeckers in fact took an average of 1.7 looks. The task was more complicated when the two kinds of log contained 0 and 6 or 0 and 12 seeds: finding one empty hole is no longer enough to reject a log, but there must be some point at which the information gained from seeing a succession of empty holes makes it worthwhile giving up. Lima calculated how many empty holes the woodpeckers ought to check before giving up on a log in order to maximize their rate of food intake. The calculated values were 6 and 3 while the observed means were 6.3 and 3.5; thus, the wood-peckers use information gleaned while foraging in a way that comes close to maximizing their overall rate of intake.

This is one simple illustration of how animals might learn about their environment while foraging. A rather different way of thinking about the problem is illustrated in Box 3.3.

The risk of starvation

Two kinds of currency for foraging animals — rate of food intake and efficiency — have come up so far. Another curr-ency that may be important for foraging animals is the risk of starvation. This is especially likely to be important when the animal lives in an environment that is unpredictable; the exact amount of food the animal will obtain is uncertain.

For example imagine you are offered the choice of two daily food rations: one is fixed at 10 sausages per day, the other is uncertain; on half the days you get 5 sausages and on the other half, 20 sausages. Although the *average* of the second diet is higher than that of the first, it is a riskier option since there is no way of telling whether you will get 5 or 20 on any parti-cular day. Which is the better option? The answer depends on the benefit (or 'utility' in economic jargon) of eating different numbers of sausages per day. If a diet of 10 is enough to survive on while 5 is not, then nothing is to be gained by choosing the risky option. If, on the other hand, 10 is not quite enough to survive on, the only viable option may be to take the risk and hope for 20 sausages. This option offers a 50 per cent chance of survival while the certain option offers no chance.

This idea has been tested in an ingenious experiment by Tom Caraco *et al.* (1980b). They offered yellow-eyed juncos (*Junco phaeonotus*) (small birds) in an aviary a choice between two feeding options: either a risky one or one with a certain pay-off. Both feeding stations offered the same mean reward rate, but at one the reward was always the same (e.g. two seeds) while at the other the pay-off was variable (e.g. half the time no seeds, half the time four seeds). They tested the birds after two lengths of starvation period: after the short starvation period of one hour the certain option provided just enough food for daily survival, but after a period of four hours starvation the birds had to take the risk of the uncertain choice to stand a chance of getting enough to eat. As predicted by the argument above, the birds changed their preference from the certain to the risky option as the starvation period increased. The conclusion is that *variation* in rate of food intake may sometimes have a more important effect than the *average* rate.

The hypothesis that animals minimize the risk of starvation can be applied to other foraging decisions. For example it provides a possible explanation of the phenomenon of 'impulsiveness' of hungry animals. Hungry animals are often less selective in their choice of prey and are more likely to take unprofitable food items than are animals close to satiation (e.g. Snyderman 1983). One explanation for this is that for the hungry animal the risk of starvation is high and by eating any item which has a positive effect on its energy reserves, the animal moves away from the risk of death. When it is less

Box 3.3. *Psychology and foraging decisions.*
While behavioural ecologists have tended to study foraging and other decisions in terms of their costs and benefits, psychologists often focus on 'how' questions about decision making: they try to analyse the rules by which animals make decisions.

The study of learning and choice in animals by psychologists is relevant to the analysis of foraging decisions. Two examples will serve to make the point. The first concerns the way in which an animal allocates its foraging behaviour to two options (which might be thought of as equivalent to patches). A common set-up for such experiments is to train a pigeon in a box to peck a key to get food rewards. The pigeon (or whatever species is under study) is then offered a choice of two keys that yield rewards at different rates. Sometimes the rate is determined by the number of pecks the pigeon makes (say one in every 10 pecks is on average rewarded) and sometimes it is determined by the amount of time that has passed (say a reward is available on average once every 10 s). The difference between these various 'schedules of reinforcement' is important for some purposes but does not matter for the moment. One of the many aspects of behaviour studied in this kind of experiment is the pigeon's rule for choosing at any moment in time which key to peck next, in other words its rule of thumb for foraging decisions. One rule that accounts for much of the observed behaviour is called momentary maximizing: 'peck the key with the higher probability of reinforcement on the next peck' (Hinson & Staddon 1983). There is a related idea called melioration in which the pigeon responds to the option with the higher local rate of reinforcement (Herrnstein & Vaughan 1980). One question of interest, as yet unanswered, is how well this kind of rule describes what animals do in a wider a range of foraging situations; another is how close the rule comes to maximizing overall rate of obtaining food, or other likely currencies for natural foraging situations.

A second example concerns the way in which animals use their past experience to estimate the rate of reward obtained from pecking a key (the rule of thumb referred to above has mainly been analysed in experiments where the animal has had a lot of experience of the reward rates used and is not learning about them). One simple learning rule that has been proposed is called the linear operator rule. The idea is that the animal updates its estimate of reward rate at frequent intervals using a rule of the following form:

$$R_{t+1} = Q_t (1 - a) + a (R_t), (1 > a > 0)$$

R is the animal's estimate of reward rate at time intervals t and $t + 1$ (the time intervals could be measured in pecks or in seconds), Q is the reward obtained in interval t and a is the animal's memory of the past. If a is nearly 1 the animal places a lot of weight on past experience, if it is zero no weight is given to the past. The equation says that at each moment in time the animal's estimate of reward probability or rate is a weighting average of its previous estimate and its present experience. This simple description (and modifications of it) of how animals might learn about changes in rate of reward in their environment accounts for a range of different experimental results obtained by psychologists (e.g. Killeen 1984).

close to starvation, the animal can afford to be choosy and maximize its rate of intake like the shore crabs and great tits we described earlier (Houston & McNamara 1985).

Feeding and danger: a trade-off

If you watch a squirrel eating chocolate chip cookies in the park, as Steve Lima and colleagues did (Lima *et al.* 1985) you will notice that the squirrel generally comes to your picnic table, grabs a cookie and retreats to a tree to eat it. If you put out small fragments of cookie the squirrel will often make repeated sorties to the table and take each morsel back to the tree to eat it. This is obviously not a very efficient way to eat food: if maximizing net rate of energy intake or efficiency was the only important factor for a squirrel it would simply sit on the table and eat pieces of cookie until it was full. One interpretation of the squirrel's behaviour is that it is balancing the demands of feeding and safety from predators. It could feed at maximum rate and run a good chance of being killed by a cat by staying on the table or it could be completely safe from cats but die of starvation in the trees. Neither of these is the best solution to maximize survival, so the squirrel does a mixture of the two. Lima *et al.* argued that the squirrel should be more prone to seek safety in the trees while feeding when this involves a smaller sacrifice in terms of feeding rate. Consistent with this they found that when the feeding table was close to the trees the squirrels were more likely to take each item to cover. Big pieces of cookie were more likely to be taken to cover than small ones; they take a long time to eat and are therefore more dangerous to handle out in the open and when handling time is long the relative cost of travelling back and forth is reduced.

The balance between the benefits of feeding and of avoiding danger is also influenced by an animal's hunger. On a very cold day in winter normally shy birds become quite tame at the garden bird table, presumably because their increased need for food overrides the danger of coming into the open. Manfred Milinski and Rolf Heller (1978, 1979) studied a similar problem with sticklebacks (*Gasterosteus aculeatus*). They placed hungry fish in a small tank and offered them a simultaneous choice of different densities of water fleas, a favourite food. When the fish were very hungry they went for the highest density of prey where the potential feeding rate was high, but when they were less hungry the fish preferred lower densities of prey. Milinski and Heller hypothesized that when the fish feeds in a high density area it has to concentrate hard to pick out water fleas from the swarm darting around in its field of vision, so it is less able to keep watch for predators (Milinski 1984). A very hungry fish runs a relatively high chance of dying from starvation and so is willing to sacrifice vigilance in order to reduce its food deficit quickly. When the stickleback is not so hungry it places a higher premium on vigilance than on feeding quickly, so it prefers the low density of prey. The balance of costs and benefits shifts from feeding to vigilance as the stickleback becomes less hungry.

Consistent with this hypothesis Milinski and Heller found that predation risk influences choice of feeding rate. When they flew a model kingfisher (*Alecedo atthis*) (a predator on sticklebacks) over a tank containing hungry fish they found

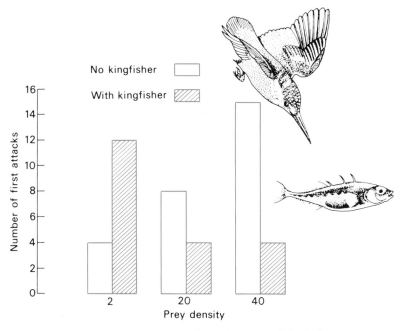

Fig. 3.7. Hungry sticklebacks normally prefer to attack high density areas of prey but after a model kingfisher was flown over the tank they preferred to attack low density areas (Milinski & Heller 1978).

that the sticklebacks preferred to attack low rather than high prey densities (Fig. 3.7). This is to be expected if the hungry fish, in spite of its high chance of starvation, places a very high premium on vigilance when a predator is in the vicinity.

An important difference between Milinski and Heller's analysis of foraging and those described earlier is that the cost-benefit calculations include the animal's hunger state. An optimization model in which the animal's state changes as a result of its behaviour (the fish becomes less hungry as a result of feeding) is referred to as a dynamic, as opposed to static model. In fact the traditional view that an animal's internal state controls its behaviour can be turned on its head and the animal can be seen as using its behavioural repertoire to control the internal state in an optimal way. The influence of the kingfisher on the stickleback is to alter the optimal allocation of time to feeding and vigilance, so that the fish decreases its hunger at a slower rate.

The idea of a dynamic feedback between foraging, body reserves and danger of predation has been used by Jim Gilliam (1982) to predict how an individual should shift between habitats as it grows up. His analysis applies to the bluegill sunfish (*Lepomis macrochirus*). In experimental ponds in Michigan, Earl Werner found that these fish could obtain a higher rate of food intake by foraging on benthic invertebrates such as chironomid larvae than they could by foraging either on the plankton or near the emergent vegetation at the edge of the pond. As might be expected the fish spend most of their time (more than 75%) foraging on the benthos. However, when predators in the form of largemouth bass (*Micropterus salmoides*) were added to the pond, a significant change in habitat use by the sunfish was seen. The bass could eat only the smallest sunfish (the others were too big) and these fish now spent more than half their foraging time in the reeds feeding on plankton where they were relatively safe, even though as a result their food intake was reduced by about one third and their seasonal growth rate by 27%. The bigger sunfish continued to forage with equanimity on the benthos (Werner *et al.* 1983). The little fish thus face a trade-off: is it better to stay in the relative safety of the reeds and grow slowly, prolonging the period of vulnerability to predators, or is it better to gamble on rapid growth to a safe size in the benthos? Gilliam was able to show that the best thing (to maximize its total chance of survival) for the fish to do is to stay in the safety of the reeds until a certain size is reached and then to go for the benthic prey. This accords with observation: the young fish in the presence of predators tend to feed in safe places, and as they get bigger they shift to the better feeding areas.

Nutrient constraints: herbivores and plants

The examples we have described so far have illustrated the variety of currencies that might be important for foraging animals, but we have said little about the range of constraints that might be important. The diet of herbivores is a good example to illustrate the importance of constraints.

As a general rule nutrient quality of food is more important to herbivores than to carnivores and insectivores. This is because plants often lack essential dietary components and only by careful selection of plant species can a herbivore obtain a balanced intake. For example, the diet of moose (*Alces alces*) on the shores of Lake Superior in Michigan is strongly influenced by sodium requirements. The moose feed in two habitats: forest, where they browse on deciduous leaves, and small lakes, where they crop plants growing under water. The aquatic plants are rich in sodium but relatively poor in energy, while terrestrial plants have little sodium but a high energy content. The moose need both energy and sodium to survive and therefore have to eat a mixed diet, but to predict the exact mixture involves making an optimality model.

Since the diet of the moose contains two components we can plot it as a point on a graph, the axes of which are intake of terrestrial and aquatic plants (Fig. 3.8a). If, for example, the moose ate largely terrestrial plants with just an occasional aquatic, its diet would be represented by a point at the lower right-hand side of the graph. Now as we have already said, the daily diet has to contain a certain minimum amount of sodium. This is represented in the graph as a constraint line: the horizontal dot-dash line shows the minimum intake of aquatic plants needed to satisfy sodium requirements. But this is not the only constraint on the animal's diet. It also needs a certain amount of energy per day. This could be obtained by eating a pure diet of y grams of aquatic plants of x grams of terrestrial plants or a mixture of the two, as shown by the solid line in Fig. 3.8a. This line shows the mixture of plants that would provide just enough energy to survive the day. Finally, the diet is constrained by the size of the moose's rumen. The moose has a specially modified stomach, the rumen, in which food is slowly fermented by micro-organisms prior to digestion. The size of the rumen sets an upper limit on the amount of food that can be processed at any one time, and therefore limits the total daily intake. The broken line on Fig. 3.8a shows the maximum amount of food that could be eaten per day with different combinations of terrestrial and aquatic plants.

The total effect of these constraints can now be assessed. Only the diets inside the small shaded triangle on Fig. 3.8a satisfy all three constraints. The diet has to be above the sodium line, above the energy line, and below the rumen line. But where inside the triangle is the optimal diet? This depends on the goal or optimality criterion. If the moose is designed, for example, to maximize its daily sodium intake, the diet should include as much aquatic material as possible and lie in the top left corner of the triangle. If the moose was designed to minimize the time spent in water each day, its diet might be near the bottom right of the triangle. Gary Belovsky (1978) carried out a detailed study of the diet and found that the mixture of plants eaten was at the point within the

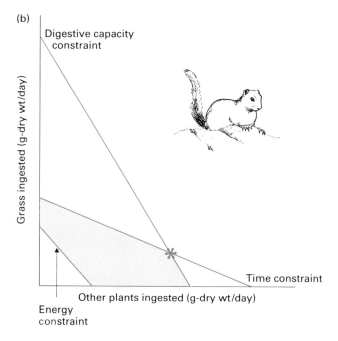

Fig. 3.8. (a) The diet of moose is constrained by the requirements for sodium and energy: the daily requirements are shown as the dot-dash and solid lines respectively and the moose has to eat a mixture of plants which lies in the space above these two lines. The third constraint is the size of the moose's rumen (broken line). Aquatic plants are bulkier than terrestrial ones so fewer grams of these can be fitted into the rumen. The mooses' diet was found to lie at the point inside the triangle that maximizes daily energy intake (indicated by a star) (Belovsky 1978). (b) The Columbian ground squirrel (*Spermophilus columbianus*) also maximizes daily energy intake subject to contraints. In this case digestive capacity (equivalent to the moose's rumen constraint), time available and minimum energy requirements define the feasible set of diets (Belovsky 1986).

triangle which would be predicted if the moose maximizes its daily energy intake subject to the constraints of sodium need and rumen size. This point is at the right hand corner of the triangle (marked with a star). To see why, think of lines parallel to the energy constraint line (energy isoclines) at different distances from the origin. Any point along one of these lines is a point of equal energy intake per day, while lines further from the origin represent higher energy intake. The highest energy isocline that the moose could achieve is at the bottom right hand corner of the triangle of feasible diets.

The conclusion from the moose study, that generalist herbivores choose foraging habitats so as to maximize their daily energy intake subject to constraints seems to be generally applicable (Belovsky 1984). The constraints vary from species to species, but they often include time available for foraging and rate of digestion (Fig. 3.8b).

Optimality models and behaviour: an overview

In this chapter we have seen how optimality models can be used to analyse decisions about foraging and mating. This approach is an extension of the idea of interpreting behaviour in terms of costs and benefits that we introduced in the last chapter. Let us now try to summarize some of the advantages and limitations of optimality modelling. Three main advantages that are illustrated by this chapter are:

1 Optimality models often make testable, quantitative predictions so that it is relatively easy to tell whether the hypotheses that are represented in the model are right or wrong. For example the honeybee workers were shown not to be maximizing net rate of energy delivery to the hive, but were maximizing efficiency in their foraging. The hypotheses that were tested in the bee study, and in all optimality studies, were hypotheses about the *currency* (net rate or efficiency) and about the *constraints* on the animal's performance (energy costs, handling times and so on). The currency is a hypothesis about the costs and benefits impinging on the animal, for example for bees energetic costs and benefits seem to be much more important than, say, predation and other dangers. The constraints are hypotheses about the mechanisms of behaviour and the physiological limitations of the animal, whether it is able to recognize differences in nectar concentration, how fast it can fly and so on.

2 A second advantage is that the assumptions underlying the currency and constraint hypotheses are made explicit. In the model used to analyse load size of starlings, for example, we had to make explicit assumptions about the loading curve, about the fact that the bird could encounter only one patch at a time, about the time taken to fly to the nest, and so on. By making these things explicit in the model one is forced to think clearly about the problem.

3 Finally, optimality models emphasize the generality of simple decisions facing animals. The starling model applied equally to dungflies, and we could have illustrated it with reference to many other animals and other decisions.

Now for a difficulty with the optimality approach: deciding what to do when the model fails to predict what the animal does. Take the dungflies as an example: the model predicts reasonably well, but not exactly, the duration of copula. What should be done about the discrepancy? Should we ignore it, assuming it is within the acceptable range of error, or should we try to analyse it further? Assume for the moment that we wanted to take the latter course. One possibility is that the currency of the model was incorrect; dungflies may trade-off feeding and mating, or danger and mating, rather than simply maximizing rate of fertilization. A second possibility is that the currency is correct but that the constraints have not been identified correctly; perhaps males run out of energy reserves while in copula. Finally, the whole idea of dungflies or other animals maximizing a currency may be incorrect. Animals may simply not be that well tuned by the process of natural selection or they may be lagging behind when some aspect of the environment changes. This kind of argument, although it is often put forward, should really be saved as a last resort, since it is not very fruitful in leading to new experiments or observations. There is, however, no straightforward recipe for distinguishing between the first two possibilities, although one useful step is, as we saw in the honeybee and moose examples, to compare different currencies (or equally, different constraints).

Summary

Behaviour can be viewed as having costs and benefits and animals should be designed by natural selection to maximize net benefit. This idea can be used as a basis for formulating optimality models in which the criterion of maximum benefit, the constraints on the animal, and the currency for measuring benefit are specified. Different kinds of currency might be appropriate for measuring benefits and costs of different behaviours, for example with feeding behaviour rate of intake might be a good currency and with male mating behaviour, rate of fertilizing eggs seems reasonable (Table 3.1).

The emphasis of this approach is on quantitative testable

Table 3.1. A summary of the decisions, currencies and constraints discussed in this chapter

Animal	Decision	Currency	Some constraints	Test
Starling	Load size	Maximize net rate of gain	Travel time, loading curve energetic costs	Load versus distance
Bee	Crop load	Maximize efficiency	Travel time, sucking time energetic costs	Load versus flight time
Dungfly	Copulation time	Maximize fertilization rate	Travel time, guarding time fertilization curve	Predict copula duration
Downy woodpecker	Patch time	Maximize rate of gain	Travel time, recognition time	Number of holes inspected
Yellow-eyed junco	Where to feed	Minimize risk of starvation	Handling time, daily energy budget	Choice of risky or certain place
Squirrel	Where to eat	Maximize survival	Travel time, handling time	Vary size of food and distance
Stickleback	Where to feed	Minimize danger and starvation	Vigilance and foraging incompatible	Vary hunger and danger
Bluegill sunfish	Habitat choice	Maximize survival	Growth depends on food intake, danger related to size	Habitats used at different ages
Moose	Food choice	Maximize daily energy intake	Sodium need, digestion limit energy limit	Per cent composition of diet

predictions. Often the results of experiments deviate slightly from the predictions of simple models; these deviations can be just as valuable as successful predictions in helping to understand how behaviour is designed.

Further reading

The paper by Maynard Smith (1978) discusses some of the pros and cons of optimality modelling. Three books which contain useful reviews of foraging economics are the monograph by Stephens and Krebs (1986) for a summary of mathematical models; the volume edited by Kamil et al. (1987) for empirical studies and two more general reviews by Russell Grey (who is very critical of the approach) and by Kacelnik and Cuthill; the volume edited by Commons et al. (1986) incudes several papers giving a psychologists' perspective on foraging behaviour. Heinrich (1979) and Heinrich & Collins

(1983) describe how caterpillars balance foraging and predation danger. Palatable caterpillars forage in ways which reduce their conspicuousness to birds. For example they eat leaves in a way that reduces the appearance of damage and they often move a long way from their resting site to where they feed. Birds hunting for caterpillars use leaf damage as a cue, so the caterpillars' strategies reduce the danger of predation. Unpalatable caterpillar species do not take these precautions.

Topics for discussion

1 Do animals have to be clever to forage optimally?
2 Are laboratory experiments on decision-making in simple environments useful for understanding behaviour in the field?
3 Is average net rate of intake a sensible currency for foraging animals?

Chapter 4. Predators versus Prey: Evolutionary Arms Races

In chapter 3 we examined the ways in which predators search for and select their prey. We reguarded various parameters (e.g. handling time) as more or less constant. Over the evolutionary time, however, these may vary. For example, during evolution we expect natural selection to increase the efficiency with which predators detect and capture prey. On the other hand, we would also expect selection to improve the prey's ability to avoid detection and to escape. The complex adaptations and counter-adaptations we see between predators and their prey are testament to their long coexistence and reflect the result of an arms race over evolutionary time (Table 4.1; Fig. 4.1). In this chapter we will consider three questions about predator-prey arms races.

1 *Adaptation or story telling?* Do the proposed adaptations of one party make functional sense given the adaptations by the other party in the arms race? This is not a trivial question. One temptation when presenting functional explanations in biology is to invent clever stories which are difficult to test. For example we might suggest that flamingoes are pink so as to blend in with the setting sun, so that lions find them difficult to see in the evening when out hunting. You may think this is an unlikely idea! But for all we know, the proposal that brown moths are so coloured to blend in with the tree trunks on which they rest may be equally unlikely. We need to perform experiments to test our functional hypotheses.

2 *How can an arms race begin?* Prey cannot suddenly evolve perfect counter-adaptations any more than a vertebrate could instantly evolve a perfect complex structure, such as an eye. Presumably at the start of an arms race even slight and very crude counter-adaptations could confer a selective advantage. Improvements in the predators then selected for improvement in the prey, resulting in better and better counter-adaptations. Our second question is, therefore, can even crude counter-adaptations by prey decrease predation pressure and thus serve as a starting point for an evolutionary arms race?

3 *How do arms races end?* Why don't predators become so efficient that the prey are driven to extinction? Alternatively, why don't prey evolve to be so good at escaping that the predators become extinct?

(a)

(b)

(c)

Fig. 4.1. Examples of adaptations by prey. (a) The syrphid fly, *Metasyrphus americanus* (Dipteran) is harmless to predators but gains protection by mimicry of the wasp's yellow and black pattern on the abdomen. (b) When the Brazilian toad, *Physalaemus nattereri*, is attacked it rears up to display two large eyespots on its sides. Many predators show avoidance responses to patterns resembling eyes. (c) A cryptic lizard, which blends in beautifully with its background.

Predators versus cryptic prey

We begin by attempting to answer the first two questions for one example of an arms race, namely birds as predators on cryptic prey. Pietrewicz and Kamil (1981) and Sargent (1981) studied underwing moths (*Catocala* spp.) in the deciduous woods of North America. There are up to 40 species living in a particular locality and they are hunted extensively by birds, including blue jays and flycatchers.

74

Table 4.1. Examples of predator adaptations and counter-adaptations by prey

Predator activity	Predator adaptations	Counter-adaptations by prey
Searching for prey	Improved visual acuity	Crypsis
	Search image	Polymorphism
	Search limited area where prey abundant	Space out
Recognition of prey	Learning	Mimicry
Catching prey	Motor skills (speed, agility)	Escape flights, 'startle' response,
	Weapons of offence	weapons of defence
Handling prey	Subduing skills	Active defence, spines, tough integument
	Detoxification ability	Toxins

TESTING FUNCTIONAL HYPOTHESES ABOUT ADAPTATION

The forewings of the moths appear cryptic, looking very much like the bark of the trees on which the moths rest (Fig. 4.2). The hindwings, on the other hand, are often strikingly coloured yellow, orange, red or pink (Fig. 4.2). The moths rest with their forewings covering the hindwings but when they are disturbed the hindwings are suddenly exposed. Our hypotheses, therefore, are that the forewings are coloured so as to decrease detection by a predator and the hindwings may have a 'startle' effect on a predator that has succeeded in detecting the moth, causing the bird to stop momentarily and thus giving the moth time to escape.

In support of the crypsis hypothesis for the forewings, different species of underwing moths select different backgrounds which match their own reflectance and thus maximize the cryptic effect. Furthermore, they orient themselves in particular ways so that the wing patterns merge in with the fissure patterns on the bark. Pietrewicz and Kamil tested the importance of crypsis by giving a slideshow to blue jays in an aviary (Fig. 4.3a). Slides were projected on a screen and in some cases there was a moth present in the picture while in other cases there was no moth. If there was a moth, the jay was rewarded with a mealworm for pecking at the slide, and the next slide was then shown after a short interval. If there was no moth in the picture then the jay could peck a second smaller 'advance' key to produce the next slide almost immediately. If the jay made either of two errors, namely pecking at

Fig. 4.2. Underwing moths. *Catocala* spp. have cryptic forewings and conspicuously coloured hind wings. Left: *Catocala neogama*; Right: *C. retecta*. Photos by Ted Sargent.

the slide when no moth was present or pecking the advance key when a moth was present, it was 'punished' in the form of a delay to the next slide presentation.

This experimental procedure is ingenious for two reasons. First, the predator is faced with a perceptual problem only; there is no complication caused by other factors which might influence predation such as prey taste, activity or escape efficiency. Second, because the predator is stationary and the prey are, in effect, moving past in front of it (in the form of a sequence of slides), it is easy to control the frequency and order in which the predator encounters prey. This would be more or less impossible if the predator was moving about the cage searching for real moths. It was found that the jay made many more mistakes if the moth was presented on a cryptic background than if presented on a conspicuous background (Fig. 4.3b) This provides direct support for the hypothesis about crypsis.

In many species of underwing moths the forewings are polymorphic, that is there are different colour forms coexisting within the same population. One hypothesis for this is that when a predator discovers a moth it may form a 'search image' for that particular colour pattern and concentrate on looking for another which looks the same (Box 4.1). If all the population were of exactly the same colour then all would be at risk, but if there was a polymorphism then a predator which had a search image for one morph may be more likely to miss the other morphs. Pietrewicz and Kamil were able to test this idea by presenting the jay with different sequences of slides. When, for example, runs were presented of 'all morph A' or 'all morph B', the jay quickly got its eye in and improved its

(a)

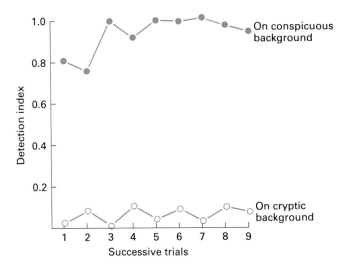

(b)

Fig. 4.3. (a) A blue jay in the testing apparatus. The slides are back projected on a screen in front of the bird. The advance key (see text) is to the right. A mealworm is delivered through the circular hole if the jay makes a correct response. Photo by Alan Kamil. (b) The jays were more likely to detect *Catocala* moths on a conspicuous background. A jay pecking indiscriminately at all slides gets a low score on the detection index. From Pietrewicz and Kamil (1981).

success at pecking at the moth as the trial proceeded. When, however, morphs A and B were presented in random order, the jay did not improve its detection success with successive slides (Fig. 4.4). This shows that encountering polymorphic prey does indeed seem to prevent effective search image formation by the predator.

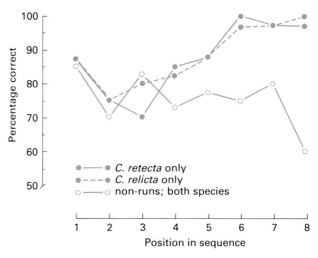

Fig. 4.4. The mean percentage of correct responses by jays when moths are presented in sequences of the same species (runs of either *Catocala retecta* or *Catocala relicta*) or in a sequence containing both species in random order. The jays improved their performance when runs of the same species were presented but not when the two species were presented in a mixed sequence. From Pietrewicz and Kamil (1981).

Box 4.1. *Search images.*

Luc Tinbergen (1960) studied the feeding behaviour of birds in Dutch pinewoods. He discovered that they did not eat certain insects when they first appeared in the spring but then suddenly started to include them in their diet. He suggested that the sudden change was due to an improvement in the birds' ability to see the cryptic insects, a process he called 'adopting a specific searching image'. There are, however, other hypotheses which could explain Tinbergen's observations. For example, the birds could have seen the insects all along but only decided to include them in their diet when their abundance increased sufficiently to make it profitable to search for them (Royama 1970). Alternatively, the birds may have been reluctant at first to eat novel prey or have improved in their ability to capture the prey.

Experiments by Marian Dawkins (1971) eliminated these alternative explanations and showed that predators can indeed undergo changes in their ability to see cryptic prey. Her predator was the convenient, if unspectacular, domestic chick and her prey were coloured rice grains. The clever design of the experiment was to keep the prey the same (therefore handling, acceptability constant) and to vary the background. Two examples are shown below. Chick (a) was presented with orange grains on a green background (solid line) and on an orange background (dashed line). Chick (b) had green grains on an orange background (solid line) and on a green background (dashed line). The two tests were run separately for each chick. In both cases the chicks found the prey quicker on a conspicuous background. On the cryptic background the chicks mainly pecked at background stones at first but after 3−4 min they eventually started to find the grain and by the end of the trial they were eating the cryptic prey at the same rate as when it was on the conspicuous background.

What about the bright-coloured hind wings? Debra Schlenoff (1985) tested the responses of blue jays to models which had variously patterned 'hindwings' concealed behind cardboard 'forewings'. The model moths were attached to a board and the jays were trained to remove them to get a food reward underneath. When the models were removed, the hindwings suddenly expanded from behind the forewings to mimic the reaction of the real moths. Jays which had been trained on models with grey hindwings showed a startle response when they were exposed to the brightly patterned hindwings typical of *Catocala*, whereas subjects trained on brightly-patterned models did not startle to a novel grey hindwing. After repeated presentations the birds habituated to a particular *Catocala* pattern but a novel bright pattern elicited another startle response. These results provide good evidence for the startle hypothesis, and the habituation effect suggests an adaptive advantage for the great diversity in hindwing patterns of different sympatric species of *Catocala*. Other experiments have shown that eye spots are particularly effective in causing a startle response in the predator (Fig. 4.5).

Fig. 4.5. A great tit about to pick up a mealworm (left). As it does so the light in the box is switched on to show a slide with an artificial 'wing display' on it. Some of the slides used are shown on the right; those towards the bottom were found to be most frightening to the bird. From Tinbergen (1974).

DOES EVEN CRUDE CRYPSIS CONFER AN ADVANTAGE?

The profitability of a prey to a predator is the value of the prey (e.g. in terms of energy) divided by the recognition time plus handling time (chp. 3). The evolution of crypsis can be thought of as the way in which the prey decreases its profitability by increasing recognition time. Can even a small increase in recognition time, caused by crude crypsis, still bring a selective advantage? Erichsen *et al.* (1980) presented a great tit in a cage with a 'cafeteria' of prey moving past on a conveyor belt like the one in Fig. 3.6. As with the jays, the clever design of this experiment is that is enables the observer to control precisely the order in which the predator encounters prey and the rate of encounter. Three items came past the great tit on the belt.
1 *Inedible twigs.* In fact these were opaque pieces of drinking straw containing brown string. They had zero energy value, a handling time h_t (time for the great tit to pick up the item and reject it as inedible) and the encounter rate was λ_t (items per second).
2 *Large cryptic prey.* These were opaque pieces of drinking straw with a mealworm inside; energy value E_1, handling time h_1 (time to pick up and eat the prey) and encounter rate λ_1.
3 *Small conspicuous prey.* These were clear pieces of drinking straw with half a mealworm inside, clearly visible; energy value E_2, handling time h_2 (time to pick up and consume the prey), and encounter rate λ_2.

The large prey were worth more energy per unit handling time than the small prey ($E_1/h_1 > E_2/h_2$). However, the problem of selecting large prey is that of recognizing them from the inedible twigs; the opaque straw which came past on the belt had to be picked up and examined to see whether it contained a mealworm or just inedible string. The experimental design therefore mimics the problem faced by a predator searching for a profitable, but cryptic prey item.

The equations describing the predator's decisions are similar to the ones in Box 3.2, but include the extra factor of recognition time. If the predator was a *generalist*, picking up every item that came past, then in T_s seconds, its energy intake would be:

$$\bar{E} = T_s \left(\lambda_1 E_1 + \lambda_2 E_2 \right)$$

in a total time

$$T = T_s + T_s \left(\lambda_1 h_1 + \lambda_2 h_2 + \lambda_t h_t \right).$$
$$(\text{search time} + \text{handling time})$$

The rate of energy intake by a generalist would therefore be:

$$\frac{E}{T} = \frac{\lambda_1 E_1 + \lambda_2 E_2}{1 + \lambda_1 h_1 + \lambda_2 h_2 + \lambda_t h_t}. \tag{4.1}$$

(Note that T_s has cancelled out in the division).

If the predator ignored the cryptic prey altogether (i.e. all the opaque straws) and just played safe and ate the conspicuous small prey (i.e. picked up all the clear straws, all of which have a food reward inside), its rate of energy intake would be:

$$\frac{E}{T} = \frac{\lambda_2 E_2}{1 + \lambda_2 h_2} \tag{4.2}$$

The predator should, therefore, specialize on the less profitable but conspicuous prey provided that the rate in equation 4.2 is greater than the rate in equation 4.1.

By similar reasoning, it will pay the predator to specialize on large, cryptic prey if:

$$\frac{\lambda_1 E_1}{1 + \lambda_1 h_1 + \lambda_t h_t} > \frac{\lambda_1 E_1 + \lambda_2 E_2}{1 + \lambda_1 h_1 + \lambda_2 h_2 + \lambda_t h_t} \tag{4.3}$$

The encounter rates were arranged so that in one treatment the birds were predicted to specialize on the small conspicuous prey and ignore the large cryptic prey (i.e. rate in equation 4.2 greater than rate in equation 4.1), and in another treatment they were predicted to specialize on the large cryptic prey (equation 4.3 holds). The tits behaved more or less as predicted (Fig. 4.6). Now for the important point: in the experiment h_t, the discrimination time for an inedible twig, was only 3 to 4 s. Therefore the tit, given time, could easily tell a twig from a large prey. Nevertheless, provided conspicuous prey were en-

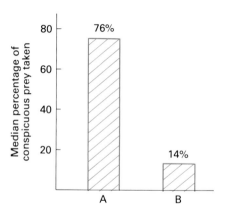

Fig. 4.6. Great tits foraging for artificial cryptic prey. The cost of recognizing cryptic profitable prey may cause the predator to specialize on other more conspicuous prey. In treatment A, twigs were 4 times as common as the large prey that resembled them; in B the large prey were 4 times as common as the twigs. The abundance of conspicuous small prey was constant in A and B. According to equations 4.1, 4.2 and 4.3, it pays the predator to specialize on conspicuous prey in treatment A, and on cryptic prey in B. From Erichsen *et al.* (1980).

countered sufficiently frequently, or provided many of the large items were twigs rather than real prey, it payed the tit to ignore the large prey altogether. The conclusion is that even slight crypsis, enough to impose only a second or two extra discrimination time, can still be of selective advantage to a prey. To gain an advantage the crypsis only has to be sufficient to make another prey item more profitable for the predator. The results of this experiment support the idea that crude crypsis can indeed provide a starting point for an evolutionary arms race.

The advantage and evolution of warning colouration

Some prey are very brightly coloured, rather than cryptic. Fruit often becomes brightly coloured when ripe to increase the probability that it is eaten and so the seed is dispersed. This is an example of a prey which is in fact selected to be eaten by predators! On the other hand, many insects are also brightly coloured yet they are presumably selected to *avoid* predation (Fig. 4.7). Bright colouration is often associated with distastefulness, and A.R. Wallace suggested that 'some outward sign of distastefulness is necessary to indicate to its would-be destroyer that the prey is a disgusting morsel'. The hypothesis is, therefore, that gaudy colours are more easily learned by predators as representing noxious prey.

Are brightly coloured prey really more easily recognized? Gittleman and Harvey (1980) tested this idea by presenting chicks with different coloured breadcrumbs. The chicks showed

Fig. 4.7. Cinnabar moth caterpillars (*Callimorpha jacobaenae*) are distasteful and warningly coloured with orange and black bands.

equal preference for blue or green crumbs. In the experiment, the crumbs were made distasteful by dipping them in quinine sulphate and mustard powder. Four groups of chicks were used: (a) blue crumbs or (b) green crumbs on a blue background, and (c) blue crumbs or (d) green crumbs on a green background. Whatever the background colour, the chicks took more of the conspicuous prey early on in the experiment. However, overall the cryptic prey suffered the greatest predation (Fig. 4.8). This suggests that it does indeed pay a distasteful prey to be conspicuous. Chicks may learn to avoid conspicuous prey more readily simply because they are more easily recognized on account of their bright colour, or the eating of a large number of prey in a short time may be a more powerful aversive experience than eating an even greater number of prey over a longer time period. Other experiments have shown that learning to avoid bright noxious prey can occur after just one trial and last for a long time. Miriam Rothschild had a pet starling which had just one experience of a brightly coloured distasteful caterpillar and it still refused to pick this species up a year later even though it had no further experience in the meantime. Sometimes predators will avoid brightly coloured and dangerous prey even though they have had no experience of them. For example, hand-reared 'naive' flycatchers (the great Kiskadee *Pitangus sulphuratus*) will avoid coral snakes (*Micrurus* spp.) which are coloured with red and yellow bands (Smith 1977).

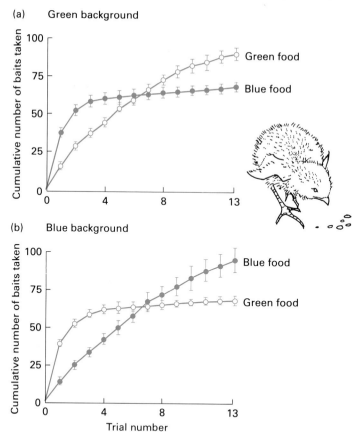

Fig. 4.8. Cumulative number of conspicuous and cryptic distasteful prey taken in successive trials by chicks. In (a) the green food is cryptic, in (b) the blue food is cryptic. In both experiments, the distasteful prey is eaten less when it is conspicuous. From Gittleman and Harvey (1980).

There may be another advantage for the conspicuousness of warning colouration. Not only can predators learn to avoid conspicuous colours more easily, but having learned to avoid a prey type they are probably less likely to attack it by mistake if it is conspicuous (Guilford 1986). A bird, for example, often pecks at an item before it is certain that it is a prey. This behaviour often saves time and may be advantageous when prey are mobile. Given that predators may make hasty decisions, it may pay unpalatable prey to be conspicuous to reduce recognition errors.

How did warning colouration evolve in the first place? Imagine a population of unpalatable but cryptic larvae. A mutation arises in an adult which causes its larvae to be brightly coloured. These larvae will be conspicuous and may be destroyed by predators so that the mutant form goes extinct before it has a chance to spread. R.A. Fisher was the first to propose a solution to this problem. He suggested that if the distasteful prey are clumped in family groups then the predator may learn to avoid the distasteful families. Although the predator may kill one individual, its experience would then cause it to avoid the others and so siblings, which also possessed copies of the mutation by inheritance, would have increased survival chances. The importance of family grouping is that it produces a local abundance of similar phenotypes (Guilford 1985). The conclusion is that warning colouration (aposematism) may be more likely to evolve if the species is dispersed in family groups. Warningly coloured insects are, in fact, more often aggregated in family groups than are cryptic species (e.g. Table 4.2). There are, however, still many examples of aposematic species which are solitary. One possibility is that their predators have large ranges, so that what looks like solitary prey is, in fact, clumped prey as far as the predator is concerned; even solitary individuals may be protected if predators, have learned to avoid the same colouration elsewhere.

Another idea is that there may be a direct advantage to an *individual* from being brightly coloured. For example, a predator may attack a bright caterpillar and then release it unharmed (Jarvi *et al.* 1981). The individual may gain an advantage if the predator is then less likely to attack it again. Many aposematic insects have tough integuments compared to cryptic forms perhaps to decrease the costs of attacks from predators. Not all warning colouration may signal distastefulness. Some froghopper bugs (Hemiptera) are very brightly coloured and are difficult for predators to catch because they can suddenly leap away. The predators may learn that their particular bright pattern signals an 'unprofitable prey'.

Table 4.2. Brightly coloured species of caterpillars of British butterflies are more likely to be aggregated in family groups than cryptic species (Harvey *et al.* 1983)

Dispersion	No. species of caterpillars	
	Aposematic	Cryptic
Large family groups	9	0
Solitary	11	44

Brood parasites and their hosts

Some species of birds, fish and insects are brood parasites, laying their eggs in the nests of other species. As with the arms race between a predator and its prey, one party (the brood parasite) gets a benefit and the other (the host) suffers a cost, in this case rearing someone else's young for them! Among birds, brood parasites show various adaptations for increasing their efficiency of host exploitation, including mimicry of the host's eggs or young and killing host young or ejecting host eggs (Fig. 4.9). Insect brood parasites often gain entrance to their host nests by mimicry of the chemical communication system of their hosts (Fig. 4.10).

Stephen Rothstein (1975a,b, 1982) has made an extensive experimental study of the brown-headed cowbird (*Molothrus ater*) which is a brood parasite widely distributed in North America. It is a generalist, having been recorded laying in the nests of 216 different host species, 139 of which have been known to successfully raise a young cowbird! The female usually lays one egg in the host's nest and removes one host egg. At the very least, therefore, the host suffers the cost of one lost egg and the time and energy costs of rearing a cowbird chick for no genetic profit. Although in many cases the host rears some of its own young together with the cowbird young, sometimes the cowbird egg may hatch first, well before the host's own eggs. In these cases the hosts stop incubating in order to feed the early hatching parasite and may therefore fail to produce any of their own young at all. For example, parasitized nests of the eastern phoebe (*Sayornis phoebe*) produce on average only 0.32 phoebe young compared with the 4.45 young produced by unparasitized nests. In the phoebes, 18.6 per cent of the population are parasitized by cowbirds and so the costs suffered are very heavy. In other species over 50 per cent of the host population may suffer parasitism from cowbirds! Indeed, several species appear to be in danger of being driven to extinction by cowbird parasitism (May & Robinson 1985).

Given these costs, we would expect counter-adaptations by the hosts. The most obvious host defence would be to eject the cowbird egg. If, however, hosts rapidly ejected parasitic eggs, soon after they were deposited, most would disappear before observers saw them and so field observations on naturally parasitized nests would tend to under-estimate rejection rates. Rothstein therefore adopted an experimental approach. He made model cowbird eggs from plaster of Paris and, adopting the procedure of real cowbirds, he deposited one model in the host's nest during the laying sequence and also removed one host egg. The results showed, surprisingly, that in some species of hosts all individuals, or almost all, were 'acceptors' of the cowbird egg while in other species, all individuals, or again almost all, were 'rejectors' and ejected the model (Fig. 4.11).

How do rejectors recognize the parasitic egg? One possibility is that they could reject eggs which were unlike their own. Alternatively they could reject the egg type which was in the

Fig. 4.9. Some adaptations of avian brood parasites. (a) A female cuckoo, *Cuculus canorus*, laying an egg in the nest of a reed warbler, *Acrocephalus scirpaceus*. The cuckoo first removes one host egg, which she is holding in her bill. (b) The cuckoo egg (bottom) is a little larger than the two reed warbler's eggs (above) but it matches them in shade and colour. (c) When the young cuckoo hatches, it ejects the reed warbler's eggs or young from the nest and so gets all the food which the host parents bring. (d) In other species the parasite's young may be raised alongside those of the host, in which case there is often mimicry of the host young. In this example, the mouthparts of the young parasite, *Vidua macroura* (a weaverbird, Ploceinae) on the left, mimic those of the young finches, *Estrilda astrild*, on the right, with which it is reared. From photos by Ian Wyllie and drawings in Lack (1968).

minority in their nest. Rothstein's experiments favoured the first hypothesis. For example for two hosts, the American robin (*Turdus migratorius*) and the grey catbird (*Dumetella carolinensis*), he replaced all, or all but one, of their eggs with cowbird models, so that the host egg, if present, was now in

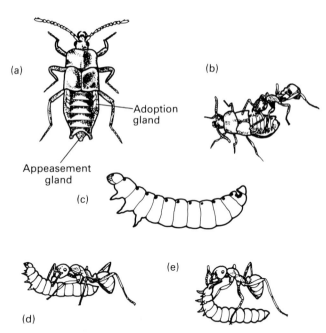

Fig. 4.10. The rove beetle, *Atemeles* gains entrance to the nests of its host, *Myrmica* ants, by mimicking the ant's chemical communication system. Secretion from the beetle's appeasement gland (a) suppresses the ant's aggressive behaviour towards intruders, and secretion from the adoption gland stimulates the ant to carry the beetle into its nest (b), where the beetle lays eggs. The beetle's developing larvae have rows of glands (c), the secretions from which stimulate the ant to regurgitate droplets of food to it (d, e). From Hölldobler (1971).

the minority. The results showed that the hosts ejected only the foreign eggs and left their own egg if one was present. Further experiments showed that rejectors learn the characteristics of their own eggs when they breed for the first time. What about the accepter species? Experiments showed that not only will these hosts accept one cowbird egg, they even accept a clutch of 'all cowbird eggs' or 'all but one cowbird eggs'! Acceptor species are not, however, oblivious to all experimental manipulations of their clutch. Like rejector species, they will remove damaged eggs and will desert if their clutch is severely reduced in size. Their 'mental block' appears to be limited to that of failing to recognize a foreign egg in their nest.

Rothstein considered various hypotheses for why there was a dichotomy between rejector versus acceptor hosts.

1 *Costs of parasitism.* It might be suspected that rejector species were the ones most heavily parasitized in nature while acceptors were rarely, if ever, parasitized so that there has been no selection during evolution for ability to recognize

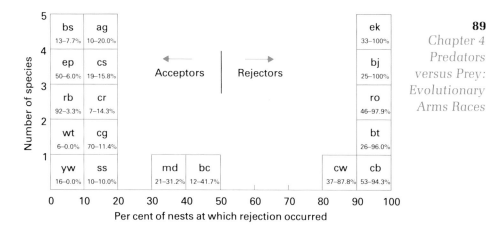

Fig. 4.11. Frequency distribution of the percentage rejection of artificial and real cowbird eggs experimentally placed in the nests of various species. The numbers under each species' abbreviation are the number of nests experimented on and the percentage rejection for that species. Species abbreviations are as follows. Acceptors: *yw* = yellow warbler (*Dendroica petechia*); *wt* = wood thrush (*Hylocichla mustelina*); *rb* = red-winged blackbird (*Agelaius phoeniceus*); *ep* = eastern phoebe (*Sayornis phoebe*); *bs* = barn swallow (*Hirundo rustica*); *ss* = song sparrow (*Melospiza melodia*); *cg* = common grackle (*Quiscalus quiscula*); *cr* = cardinal (*Cardinalis cardinalis*); *cs* = chipping sparrow (*Spizella passerina*); *ag* = American goldfinch (*Spinus tristis*); *md* = mourning dove (*Zenaida macroura*); and *bc* = black-billed cuckoo (*Coccyzus erythropthalmus*). Rejectors: *cw* = cedar waxwing (*Bombycilla cedrorum*); *cb* = grey catbird (*Dumetella carolinensis*); *bt* = brown thrasher (*Toxostoma rufum*); *ro* = robin (*Turdus migratorius*); *bj* = blue jay (*Cyanocitta cristata*); *ek* = eastern kingbird (*Tyrannus tyrannus*). From Rothstein (1975a).

cowbird eggs. This was not true; acceptors were just as heavily parasitized as rejectors.

2   Acceptors are recent hosts. There has not been sufficient time for rejection behaviour to evolve. The cowbird inhabits short grasslands in the Western States. When eastern North America was settled by immigrants, they cut down the forests and cowbirds spread east. We could propose, therefore, that acceptors are species which have only come into recent contact with cowbirds. However, acceptor species are not more 'forest' species than rejectors and many rejector and acceptor hosts have probably had long histories of sympatry with the cowbird.

3   Egg mimicry. Perhaps rejector species have eggs most unlike those of the cowbird, and hence more easily recognizable? We can also reject this idea, because most acceptor and rejector hosts have eggs which are very unlike those of the cowbird.

4   Taxonomy. If some feature of a host species biology increased the likelihood that it could evolve rejection behaviour (e.g. larger size made it better able to pick up and reject the cowbird egg), then rejectors and acceptors should sort out along taxonomic lines. They don't. For example, four passerine families had both rejector and acceptor species.

Because no factor was a good predictor of a species' status as a rejector or acceptor, Rothstein suggested that acceptors simply lacked the ability to discriminate against foreign eggs and this constraint led them to suffer a cost from cowbirds, even though there would be a huge selective advantage for rejection behaviour if it occurred. The suggestion is, therefore, that some species have, by chance, had suitable mutations or recombinations that enable rejection and that selection could act on. These species have evolved into rejectors. Other species have remained acceptors simply because suitable genetic variants have not yet appeared. Knowing the cost of parasitism and the proportion of hosts parasitized, Rothstein was able to calculate the time needed for a hypothetical 'rejector genotype' to spread through the population, making some simple assumptions about the genetics of rejection behaviour (the details of which are not known). These calculations showed that, once rejector genes occur, hosts will change from acceptors to rejectors within as little as 20 to 100 years! For example, given the costs of parasitism suffered by the eastern phoebe (see above), an acceptor species, the population would be expected to go from 80 per cent acceptance to 80 per cent rejection in less than 100 years if rejector variants occurred.

These calculations show why the various hosts fall into two such distinct classes of 'all accept' versus 'all reject' (Fig. 4.11). Because the selective advantage of rejection is so strong, the time span over which any species would show an intermediate rejection rate is of very short duration, once rejectors occur in the population. The evolution of rejection behaviour may therefore be a difficult first step, which once realized will rapidly sweep through the population to fixation. Rothstein's study provides a good example where the adaptations of a species appear to be limited simply by genetic variation. It would clearly be adaptive for acceptors to reject, but if there is no variation for selection to work on then rejection can clearly not evolve.

Evolutionary arms races

Why don't predators and brood parasites become so efficient that they drive their prey and hosts to extinction? Or why don't prey and hosts evolve such excellent counter-adaptations that the predators and brood parasites become extinct? In other words why do we see both parties in the arms race coexisting together? In many cases the complexity of their adaptations and counter-adaptations is such that they must have coexisted for a long time. We consider three hypotheses (Dawkins & Krebs 1979; Slatkin & Maynard Smith 1979).

1   *Prudent predation.* Man has the capacity to be a prudent predator and avoid over-exploiting his food supplies to extinction. Should animal predators likewise be prudent? The problem with this idea is that it relies on group selection. In a population of prudent predators any individual that cheated and ate more than its 'fair share' would pass on more genes to future generations than prudent individuals (see chp. 1). Prudent predation could, however, evolve where an individual had exclusive use of a resource (e.g. by defending a territory) and so was saving food for its own future use rather than for the good of the population.

2   *Group extinctions.* If group extinctions were common then the reason we see stable predator-prey systems in nature may be because all the unstable systems have indeed gone extinct.

3   *Prey are ahead in the arms race.* Predator-prey systems may be stable because the prey are always one step ahead in the arms race. One hypothesis for why this might be so can be described as the 'life-dinner principle'; rabbits run faster than foxes because the rabbit is running for its life while the fox is only running for its dinner. The cost of a mistake is clearly greater for the rabbit. As Dawkins and Krebs (1979) put it, 'A fox may reproduce after losing a race against a rabbit. No rabbit has ever reproduced after losing a race against a fox. Foxes who often fail to catch prey eventually starve to death, but they may get some reproduction in first'. Therefore, selection pressure will have been stronger on improving the ability of rabbits to escape than the ability of foxes to make successful captures.

In many cases the prey may also be ahead in the arms race because they have a shorter generation time than the predator and can therefore evolve at a faster rate. This will apply to examples such as weasels versus mice, and birds versus insects, but in some cases it is the predator which has the capacity to reproduce at a faster rate (aphids versus roses). We may also ask why prey do not become so efficient at escaping so as to drive their predators extinct. One hypothesis is that as predators become rare, because of increased prey efficiency, they exert little selection on the prey for further improvement.

Finally a predator may often be unlikely to drive a prey to extinction because if a prey species became rare, due for example to excessive predation, then the predator is likely to seek other prey species. Because adaptations for feeding on different prey are usually different, a predator is unlikely to be sufficiently specialized on any one species to drive it to extinction.

Summary

The complex adaptations of predators and counter-adaptations by their prey reflect the result of an evolutionary arms race. Experiments on predation by birds on cryptic prey show that:
1 Crypsis does indeed reduce predation, and polymorphism in the prey population is effective in reducing the efficiency of search image formation by the predator (blue jays versus underwing moths).
2 Even slight crypsis, enough to impose just a few seconds extra discrimination time, can bring an advantage to prey and thus serve as the starting point for an arms race (great tits and conveyor belt experiment).

Noxious prey are often brightly coloured. Experiments show that predators learn to reject noxious prey quicker if they are conspicuous rather than cryptic (chicks versus coloured breadcrumbs). Adaptations and counter-adaptations of brood parasites and their hosts are described. Hosts of cowbirds are either 'rejectors' or 'acceptors'. Acceptor species may be those which simply lack the genetic variation to allow rejection behaviour to evolve. Ideas are discussed concerning why prey may often be 'ahead' in an arms race.

Further reading

We have not attempted to be at all comprehensive in this chapter, rather to take a few well defined problems and show how these can be analysed by field and laboratory experiments. For general discussions of predator-prey adaptations see the books by Wickler (1968) and Edmunds (1974). Payne (1977) reviews brood parasitism. For theoretical accounts of coevolution and arms races, see Slatkin & Maynard Smith (1979) and Dawkins & Krebs (1979). Baker and Parker (1979) discuss bird colouration in relation to the unprofitable prey hypothesis.

Topics for discussion

1 Why are warning colours conspicuous?
2 What influences the outcome of an evolutionary arms race between (a) cuckoos and their hosts, and (b) pathogens and their hosts.
3 Design a field experiment to test whether prey choice is influenced by searching images.

Chapter 5. Competing for Resources

Our discussion in chapter 3 of how individuals exploit resources such as food omitted a crucial factor: competition. When many individuals exploit the same limited resources, they are competitors, and the decisions made by one competitor may be influenced by what others are doing. This is true, for example, of the male dungflies introduced in chapter 3. Female dungflies (the limited resource for which males compete) occur at higher densites on cowpats than in the surrounding grass, but nonetheless it pays some males to search in the grass where competition is less intense.

In this chapter we will start by discussing the simplest form of competition, *exploitation*, which simply means 'using up resources', and then go on to describe another form of competition, *resource defence*, in which animals keep others away from resources by fighting or by aggressive displays.

Competition by exploitation: the ideal free distribution

Let us start with a model of competition. To keep things simple, imagine there are two places ('habitats'), a rich one containing a lot of resources and a poor one containing few, and that each individual chooses to exploit the habitat in which it can achieve the higher pay-off, measured as rate of consumption of resource. With no competitors, an individual would simply go to the better of the two habitats and this is what we assume the first arrivals will do. But what about the later arrivals? There is no territoriality or fighting and so there is no limit to the number of individuals who can go there. However the more competitors that occupy the rich habitat the more the resource will be depleted, and so the less profitable it will be for further newcomers. Eventually a point will be reached where the next arrivals will do better by occupying the poorer quality habitat where, although the resource is in shorter supply, there will be less competition (Fig. 5.1). Thereafter the two habitats will be filled so that the profitability for an individual is the same in each one.

In other words, competitors adjust their distribution in relation to habitat quality so that each individual enjoys the same rate of acquisition of resources. This theoretical pattern of distribution of competitors between resources has been

93

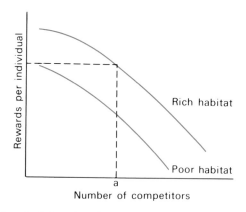

Fig. 5.1. The ideal free distribution. There is no limit to the number of competitors who can exploit the resource. Every individual is free to choose where to go. The first arrivals will go to the rich habitat. Because of resource depletion, the more competitors the lower the rewards per individual so at point *a* the poor habitat will be equally attractive. Thereafter the habitats should be filled so that the rewards per individual are the same in both. After Fretwell (1972).

termed the 'ideal free' distribution by Stephen Fretwell (1972) because it assumes that animals are free to go where they will do best and that the animals are ideal in having complete information about the availability of resources.

We can see a good example of this in action when people stand in line at the counters of a supermarket. If all the serving clerks are equally efficient and all the customers are equal in the time they require for service, then the lengths of all the lines should end up being equal. If one line gets shorter, then customers would profit by joining it until its length becomes the same as the others. Because everyone is free to join which-ever line they like, each person goes to the best place at the time, and the lines fill up in an ideal free way with the result that every customer should have the same waiting time for service.

An equivalent in animals is Manfred Milinski's (1979) ex-periment with sticklebacks. Six fish were put in a tank, and prey (*Daphnia*) were dropped into the water from a pipette at either end. At one end prey were dropped into the tank at twice the rate of the other end. The best place for one fish to go depends on where all the others go. There was no resource defence and Milinski found that the fish distributed them-selves in the ratio of the patch profitabilities, with four fish at the fast-rate end and two at the slow-rate end. When the feeding regimes were reversed, the fish quickly redistributed themselves so that four were again at the fast end (Fig. 5.2). This is the only stable distribution under ideal free condi-tions. With any other distribution it would pay an individual

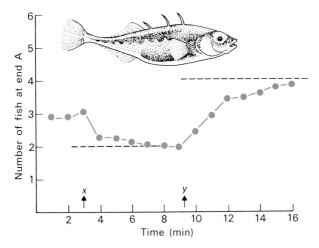

Fig. 5.2. Milinski's (1979) feeding experiment with six sticklebacks. At point x, end B of the tank had twice the amount of food as end A. At point y the profitabilities were reversed. The dashed lines indicate the number of fish predicted at end A according to ideal free theory, and the solid line is the observed numbers (mean of several experiments).

to move. For example, if there were three fish at each end then one fish would profit by moving from the slow to the fast-rate end. Once it had done so, it would not pay any of the other fish to move. In our supermarket analogy, this experiment is equivalent to what should happen if one clerk is twice as efficient at serving customers as another; the stable distribution would be for this line to be twice as long.

The stable distribution of searching individuals could be achieved in two ways. For example, if one habitat was twice as profitable as another then stability could come about by:
1 Competitor numbers adjusting so that twice as many individuals go to the good habitat as to the poor one.
2 All individuals visiting both habitats, but each spending twice as much time in the good habitat as the poor one.

Competition by resource defence

Consider the same situation as before: two habitats, one rich and one poor, but this time the first competitors to settle in the rich habitat defend resources by establishing territories (pieces of ground containing the resource) so that later arrivals are forced to occupy the poor habitat even though they do less well there than the individuals in the rich area. When the poor habitat fills up with territory-defending individuals the latest arrivals of all may end up being excluded from the resource altogether (Fig. 5.3). This kind of situation is very common in nature. In Wytham Woods, near Oxford, England, the best breeding habitat for great tits is in oak woodland. This is quickly occupied in the spring and becomes completely filled with territories. Some individuals are excluded from the oak wood and have to occupy the hedgerows nearby where there is less food and consequently have lower breeding success. If great tits are removed from the best habitat then birds rapidly move in from the hedgerows to fill the vacancies (Krebs 1971). Similarly, in red grouse (*Lagopus lagopus scoticus*) territorial birds defend the richest areas of the heather moors as breeding and feeding territories. Excluded birds have to go about in flocks and exploit poor habitats where their chances of survival are low. Once again, if a territory owner is removed its place is quickly taken by a bird from the flock (Watson 1967).

In these examples the strongest individuals are despots, grabbing the best quality resources and forcing others into low quality areas or excluding them from the resource altogether.

Resource defence is one kind of competition that comes under the general heading of *interference*. Interference occurs whenever competitors interact with one another in such a way as to reduce their efficiency at searching for or exploiting resources. In addition to resource defence, interference might occur, for example, competitors bump into each other and waste time which would otherwise be spent searching, or as happens in some insect parasitoids, scent marks left by one competitor confuse another individual (Hassell 1971).

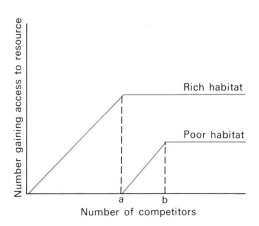

Fig. 5.3. Resource defence. Competitors occupy the rich habitat first of all. At point *a* this becomes full and newcomers are now forced to occupy the poor habitat. When this is also full (point *b*), further competitors are excluded from the resource altogether and become 'floaters'. After Brown (1969).

Most examples in nature will have features of both the simple models we have discussed above. Perhaps the commonest situation will be where the best place to search depends on where all the other competitors are, but within a habitat some individuals get more of the resource than others. In the fish experiment, for example, our population counts may show a stable, ideal free distribution of individuals but the chances are that some fish will be better competitors than others. At each end of the tank there could be one or two large fish grabbing most of the prey. The ideal free distribution could come about because of the way the subordinates distribute themselves in relation to the despots. In effect, the despots are part of the habitat to which the subordinates respond when deciding where to search (Milinski 1984).

It is unlikely that there will be any populations where all individuals are of equal competitive ability. Even though male dungflies obey an ideal free distribution around cow pats, the larger males get more females than small males (Borgia 1979). In our supermarket there is probably no fighting between the customers but some will have more items of shopping than others and will require more time to be served. In a sense, they are greater competitors because standing in line behind them will impose greater waiting times.

A good example which shows features of both the resource defence and ideal free models is the study by Thomas Whitham (1978, 1979, 1980) of habitat selection in the aphid (*Pemphigus betae*). In the spring, females known as 'stem mothers', settle on leaves of narrowleaf cottonwood (*Populus angustifolia*) to feed and they become entombed by expanding leaf tissue, so forming a gall. A stem mother reproduces parthenogenetically and the number of progeny she produces depends on the quantity and quality of the juices she can tap from the leaf. The largest leaves provide the richest supplies of vascular sap and result in the greatest reproductive success, with up to seven times the number of progeny that are produced by settling on a small leaf. As we would expect, all the large leaves are quickly occupied and so additional settlers have the problem of whether to settle on large leaves and share the resources or occupy smaller leaves alone.

Whitham made measurements of reproductive success which enabled him to plot a family of fitness curves for habitats of varying quality (leaves of different sizes) and with different densities of competitors (number of galls per leaf). Figure 5.4 shows the results, which enable us to draw three conclu-

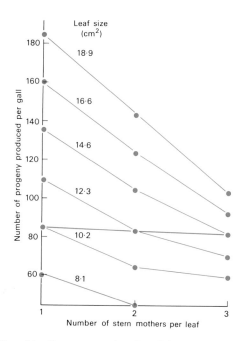

Fig. 5.4. The thin lines are a family of fitness curves for habitats of varying quality (leaf size) and competitor density (no. stem mothers per leaf) in the aphid *Pemphigus betae*. The solid horizonal line is the average success for one, two and three stem mothers per leaf. See text for explanation. From Whitham (1980).

sions. First, for any competitor density, the average reproductive success increases with habitat quality. Second, within a habitat of a certain quality, success decreases as the number of competitors increases. This shows that stem mothers settling on the same leaf must compete with each other for resources. Third, if the *average* reproductive success is calculated for aphids who are alone on a leaf, those who share a leaf with one other and those who share with two others, no significant differences are found. There was also no significant difference in average success on leaves with different numbers of competitors when other fitness measures were used such as body weight of the stem mother, abortion rate, development rate or predation. The results support the predictions of the ideal free model. The conclusion, therefore, is that the stem mothers settle on leaves of different sizes such that the average success in good habitats with a high density of competitors is the same as in poor habitats with fewer competitors.

However, although the results of average success on different sized leaves are in accord with ideal free predictions, within a habitat not all individuals get equal rewards. This is because a leaf is not a homogeneous habitat. The best place to be is on the mid-rib at the base of the leaf blade because

Fig. 5.5. Stem mother aphids *Pemphigus betae* fight for prime positions on a leaf by kicking and pushing. The winner will settle at the base of the mid-rib where food is richest (Whitham 1979).

everything translocated into and out of the leaf must flow past this point. Basal galls on a leaf give rise to more young than distal galls and the stem mothers spar with each other, like boxers in a ring, for occupancy of these prime positions (Fig. 5.5). As we would predict from the defence model, if a basal individual is removed her place is quickly occupied by another aphid from a distal site.

The economics of resource defence

Some animals, as we have seen, compete for resources by exploitation, others by territoriality. Is it possible to predict when the latter form of competition should be adopted instead of the former? The first attempts to analyse this question did so by asking 'What is the function of territorial defence?' (Hinde 1956; Tinbergen 1957). This approach led to long lists of different functions for different species, emphasizing the diversity of resources defended. Hole nesting birds defend territories to secure nest sites, barnacles defend areas of rock (by chemical signals) to guarantee enough space in which to grow, bullfrogs defend bits of habitat which are attractive to females and so on. The 'functions' approach did not, however, answer the question of when it pays to compete by means of territorial defence. A more fruitful approach stems from the concept of economic defendability.

(a) Economic defendability

Jerram Brown (1964) first introduced the idea of economic defendability. He pointed out that defence of a resource has costs (energy expenditure, risk of injury and so on) as well as

99

the benefits of priority of access to the resource. Territorial behaviour should be favoured by selection whenever the benefits are greater than the costs. This may seem a rather obvious conclusion and indeed as stated so far it is. However it led field workers to look in more detail at the time budgets of territorial animals, in particular the feeding territories of nectivorous birds. While Brown's idea is in principle applicable to any kind of territory, it has been most useful in looking at birds such as hummingbirds, sunbirds, and honeycreepers in which the costs and benefits can be measured in calories in the field. Frank Gill and Larry Wolf (1975), for example, were able to measure the nectar content of territories of the golden-winged sunbird (*Nectarinia reichenowi*) in East Africa, where it defends patches of *Leonotis* flowers outside the breeding season. They also calculated from time budget studies and laboratory measurements of the energetic costs of different activities such as flight, sitting and fighting, how much energy a sunbird expends in a day. When the daily costs were compared with the extra nectar gained by defending a territory and excluding competitors, it turned out that the territorial birds were making a slight net energetic profit. The resource was economically defendable (Box 5.1).

Box 5.1. *The economics of territory defence in the golden-winged sunbird (Gill & Wolf 1975).*
(a) The metabolic cost of various activities was measured in the lab:

Foraging for nectar	1000 cal/h
Sitting on a perch	400 cal/h
Territory defence	3000 cal/h

(b) Field studies showed that territorial birds need to spend less time per day collecting enough energy in the form of nectar to survive when the flowers contain more nectar:

Nectar per flower (µl)	Time to get energy (h)
1	8
2	4
3	2.7

(c) By defending a territory a bird excludes other nectar consumers and therefore increases the amount of nectar available in each flower. The bird therefore saves foraging time because it can satisfy its energy demands more rapidly. It spends the spare time sitting on a perch, which uses less energy than foraging. For example, if defence results in an increase in the nectar level from 2μl to 3μl per flower, the bird saves 1.3h per day foraging time [from (b)]. It therefore saves:

$$(1000 \times 1.3) - (400 \times 1.3) = 780 \text{ cal}$$
foraging   resting

(d) But this saving has to be weighed against the cost of defence. Measurements in the field show that the birds spend about 0.28 h per day on defence. This time could otherwise be spent sitting, so the cost of defence is:

$$(3000 \times 0.28) - (400 \times 0.28) = 728 \text{ cal}$$

In other words, the flowers are just economically defendable when the nectar levels are raised from 2 to 3μl as a result of defence. Gill and Wolf found that most of their sunbirds were territorial when the flowers were economically defendable.

The idea of economic defendability has also been used to predict the levels of resource availability which could lead to territorial defence. If resources are very scarce, the gains from excluding others may not be sufficient to pay for the cost of territorial defence. Instead the animal might abandon its territory and move elsewhere. There may also be an upper threshold of resource availability beyond which defence is not economical. This upper boundary could arise for a number of reasons.

1 There may be so many intruders trying to invade rich areas that defence costs would be prohibitively high.

2 There may be no advantage of territoriality at high resource levels if the owner cannot make use of the additional resources made available by defence. In Gill and Wolf's sunbirds, one advantage of territorial defence was that it raised the amount of nectar per flower (by exclusion of nectar thieves) and hence saved foraging time (Box 5.1). But if nectar levels are already high, the extra increment resulting from territorial defence saves hardly any foraging time. This is because the bird's rate of food intake at high nectar levels is limited by the time it takes to probe its beak into the flower (the handling time). For example Gill and Wolf calculate that an increase from 4μl to 6μl per flower would save the birds less than 0.5h of foraging time, while as shown in Box 5.1 an increase of from 1μl to 2μl saves 4h. Thus when nectar levels are high, territorial exclusion of nectar thieves does not pay for itself in savings of foraging time.

3 A third hypothesis which predicts an upper boundary was proposed by Carpenter and McMillen (1976). They suggested that territory defence has associated risks such as increased conspicuousness to predators, so that whenever resource levels are high enough to allow an animal to satisfy its needs without excluding others, territories should be abandoned. As yet there is no critical evidence to test this idea.

(b) Optimal territory size

Although the use of Brown's concept to predict the range of resource levels over which animals should defend territories is a step forward from simply showing that territories are economically defendable, a more quantitative and powerful development is to predict the optimal amount of resource for an individual to defend. Gill and Wolf found that their sunbirds always defended about 1600 flowers, even though the territory area varied by 300 fold! Graham Pyke (1979b) has tried to find out why sunbirds defend 1600 flowers by building an economic model of optimal territory size. He used Gill and Wolf's data to calculate the time budgets and energy expenditures of birds defending different numbers of flowers, assuming that a territory with more flowers and therefore more nectar would have more intruders. As we have seen before (chp. 3), an optimality model involves choosing a criterion of benefit which is to be maximized. Pyke considered four possibilities. The birds might choose a territory size which maximizes daily net energy gain: this could be a good measure of benefit for a bird that is short of food or one that is storing food for future hard times. The second idea was that the birds might maximize the ratio of gains to costs. A third hypothesis was that the birds might maximize the daily sitting time, a reasonable goal if a sitting bird is much safer from predators than one flying around in the open. The fourth hypothesis was that the birds would minimize their daily energy costs. The rationale for this is that metabolizing energy wears out the machinery of the body, so that an animal intent on lasting out until the next breeding season should minimize costs. Pyke's calculations showed that the fourth hypothesis accurately predicts both the number of flowers defended and the daily time budget of the sunbirds (Box 5.2). The third hypothesis is good at predicting the number of flowers defended but not the time budget.

In contrast to the sunbirds intent on surviving from one breeding season to the next, the rufous hummingbirds (*Selasphorus rufus*) studied by Lynn Carpenter. (Carpenter *et al.* 1983) appear to defend territories of a size that maximizes

Box 5.2. *Pyke's model of optimal territory size in sun-birds.*

1 The bird's time budget consists of four components:

sleeping + feeding + sitting + defence

$$Z \qquad F \qquad S \qquad D \qquad \text{(i)}$$

2 The total daily energy costs (C) are made up of the 'fixed cost' of sleeping (Z) and the variable costs of F, S and D. These are considered to be variable because the bird can alter its time budget to change the total cost.

3 The total daily gain (G) from the territory is:

standing crop of energy/flower × rate of visiting

$$\times \text{ feeding time/day} = erF \qquad \text{(ii)}$$

Pyke assumes that the food supply is in equilibrium, i.e. the birds eat as much as is produced each day. The daily production is given by:

number of flowers × daily production/flower

$$= np$$

$$G = erF = np$$

$$e = \frac{np}{rF}. \qquad \text{(iii)}$$

4 The defence time is assumed to be related to the quality of the territory. More invaders try to get into better quality territories.

$$D = kne^{\alpha} = \frac{k'n^{\alpha+1}}{F^{\alpha}} \text{ (by substitution of iii)} \qquad \text{(iv)}$$

(k and k' are constants, α determines the form of the relationship between D and the standing crop (ne). If $\alpha > 1$, the curve of D against ne is accelerating.)

5 Pyke's hypotheses are that the birds might choose n and F (size of territory, measured as number of flowers defended, and feeding time) so as to maximize $E = G - C$, or minimize C, or maximize S, or maximize G/C. All of these have to be subject to the constraint that $E \geqslant 0$, otherwise the bird would die of starvation! Note that the choice of n and F completely determines the time budget: n and F determine D (equation iv) and hence S (by equation i). This in turn determines the value of C (see point 2 above).

6 Pyke used laboratory measurements of the costs per minute of F, S, D and Z, and field estimates of e and n. He assumed that the value of α lies between 1 and 3 and calculated k' from equation (iv).

7 With α set equal to 2 the various hypotheses made the following predictions:

	Four different hypotheses				
	Max E	Max G/C	Max S	Min C	Observed
Number of flowers	7070	6722	1653	1595	1600
Foraging time (h)	7.45	8.18	1.72	2.41	2.42
Defence time (h)	2.55	1.82	0.61	0.28	0.28
Sitting time (h)	0	0	7.67	7.31	7.30

Both maximizing S and minimizing C predict approximately the correct number of flowers but the latter is better at predicting the time budget. The reason is that defence costs are very high relative to other activities, therefore minimizing C produces a low value of defence time. However maximizing S takes no account of the cost of defence, so it predicts a higher value of D.

One way to test Pyke's model further would be to measure the relationship between territory quality and defence time. Pyke predicts that α should be about 2.

energy gain (Fig. 5.6). These birds were studied on temporary territories around patches of Indian Paintbrush (*Castilleja linariaefolia*), defended during pauses on their southward migration through the Sierra Nevada mountains in California, so it is not surprising that their currency is different from that of the sunbirds. Rather than trying to survive a long period at minimum expenditure, they are trying to put on weight as rapidly as possible in order to continue with their migration. By the ingenious method of putting out perches attached to a small balance, Carpenter et al. were able to measure the weight gain of birds defending territories of different sizes. At least some individuals appeared to adjust their territory size over a series of days towards the size that maximized rate of weight gain (Fig. 5.6).

One feature of hummingbird behaviour that is at first sight puzzling in view of their goal of maximizing weight gain, is the fact they spend about 75 per cent of their time perching and only 20 per cent of their time foraging. If maximizing energy gain is so important why do the birds not forage more often? A neat answer has been provided by Jared Diamond and colleagues (Diamond et al. 1986). They used isotope labelled glucose and polyethylene glycol to measure the time that it takes the hummingbird to absorb energy and the time required for a meal to pass through the digestive tract. They found that hummingbirds have an extremely rapid rate of active absorption of glucose through the intestinal wall, but that the limiting step in digestion is the time taken for

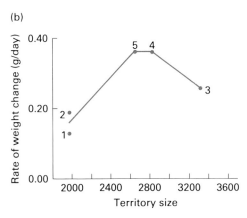

Fig. 5.6. (a) A rufous hummingbird weighing itself by sitting on a perch attached to a balance. Photo by Mark Hixon. (b) The daily weight gain of one territorial bird plotted against its territory size for five successive days. The bird started with a small territory on the first day, enlarged its territory on the second day and then gradually moved towards an intermediate size at which daily weight gain was maximal.

food to pass from the crop, where it is stored when the bird first takes it in, to the stomach. It takes about four minutes for the crop to half empty after a 100 ml meal, which corresponds quite closely to the average interval between meals in the wild. In other words, while the bird is sitting it is in fact hard at work making room for the next meal! (Hixon 1982).

(c) Graphical models of optimal territory size: a caveat

Brown's concept of economic defendability is often represented in the form of a graph of costs and benefits which can be used to predict the effects of changes in resource density and number of competitors on the amount of resource defended (Fig. 5.7). In a cautionary note, Tom Schoener (1983) has pointed out that the graphical model can lead to diametrically opposing predictions, depending upon the exact shape of the curves (Fig. 5.7). The message is that graphical models, in this book and elsewhere, should be handled with care!

SHARED RESOURCE DEFENCE

So far we have assumed that territories are defended by single individuals, but very often two or more competitors share the

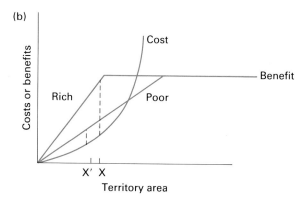

Fig. 5.7. (a) The idea of economic defendability. As the amount of resource defended (or territory size) increases so do the costs of defence.
The benefits (e.g. amount of food available) are assumed to increase at
first but level off as the resource becomes superabundant and the animal
is limited in its capacity to process the resource. Two benefit curves are
shown, one for a rich environment and one for a poor environment: the
benefit curve rises more steeply in the former because the density of
resources is higher. The resource is economically defendable between A
and B. Within this range the optimal territory size depends on the currency: for maximizing net gain the optimal size is smaller for the rich
environment (X) than for the poor one (X') (note that this is where the
slopes of the cost and benefit curves are equal). (b) The same model, but
with slightly different shaped curves. In this case the optimal territory
size to maximize net gain is predicted to *increase* in a rich environment
(X now greater than X'). After Schoener (1983).

same territory. Typically, sharing individuals are a mated pair,
as in most territorial birds for example, and here the reasons
for shared defence may be unrelated to economic considerations of costs and benefits of resource defence. The fact that
European robins (*Erithacus rubecula*) share a territory in the
breeding season but not at other times of year is more related
to the necessity of the two birds to be together for mating and

parental duties than because resource defence is more efficient when done by two birds. However, sometimes shared defence may occur because of economic considerations. An example which has been analysed in some detail is Nick Davies and Alasdair Houston's (1981) study of the pied wagtail (*Motacilla alba*).

In winter, these birds defend territories along stretches of the river Thames near Oxford. They feed on insects washed up by the river onto the bank. After a bird has foraged in a particular spot and depleted the insects, the numbers gradually build up as new insects are washed ashore. Eventually the rate of deposition and washing away is about equal and the insect abundance remains at a plateau. An efficient way to exploit this kind of renewing resource is to visit each place only after the food has had a chance to renew itself, and this is exactly what the wagtails do. The territory owner works systematically around its territory and revisits each spot on average about once every 40 min after the insect supply has been replenished.

It is not hard to see why territorial defence pays: without exclusive use of the river bank, a wagtail's strategy for harvesting the renewing food could easily collapse. It might turn up at a spot it has not visited for 40 min only to find that another bird has just been there, rather like the mushroom picker who shows up too late in the morning. One might expect, therefore, that wagtails would always defend exclusive territories, but this is not so. Sometimes the territory owner tolerates a second bird, a so-called 'satellite'. The two sharers move around the territory out of phase with one another, so that the average return time to a site is halved to 20 min, resulting in a lower feeding rate for the owner. To counterbalance this cost there is a benefit of sharing because the satellite chases away intruders. The owner therefore saves defence time leaving more time to feed. The net effect of these costs and benefits on maximizing feeding rate, the currency hypothesized by Davies and Houston, depends on the food supply. On days when the rate of renewal of food is high the cost of sharing is relatively low, and on days when intrusion rate is high, the benefit of sharing is relatively large. Intrusion rate is closely related to asymptotic food levels, so sharing will tend to pay on days with a high rate of food renewal and a high asymptotic level. By calculating the exact values of the costs and benefits in terms of feeding rate Davies and Houston were able to predict on which days it would pay a territory owner to share and on which days it would not. Their predictions were right for 34 out of 40 days (Fig. 5.8).

The wagtail study illustrates two general points: first it is an example of how apparently different kinds of costs and bene-fits (defence and feeding) can sometimes be reduced to a single currency — feeding rate in this case. Second, it shows that one advantage of group living is shared resource defence. The wagtail groups were never larger than two, but the same argument could be generalized to larger groups (Brown 1982). In chapter 6 we will come back to the costs and benefits of group living.

INTERSPECIFIC TERRITORIALITY

Although competition for resources is usually more intense within than between species, individuals of different species with similar ecological requirements may compete and some-times they do so by defending interspecific territories. In the past there has been some debate over whether interspecific territoriality arises because individuals of one species mis-identify individuals of another species and treat them as com-petitors even though they are not, or whether it is a genuine case of economic resource defence. The instances that have been studied in detail favour the latter view. A nice example is Tim Reed's work (1981) on interspecific territoriality between chaffinches (*Fringilla coelebs*) and great tits (*Parus major*). The species have very different songs and are quite distinct in appearance. Over most of the British Isles they occur in the same habitats and do not interact at all, but on the small islands of Arisaig and Eigg off the west coast of Scotland they

Fig. 5.8. Wagtails were predicted to share their territories when the rate of renewal of food and the asymptotic abundance of food were above the curve. These combinations represent instances where the costs of sharing are outweighed by the benefits. The observed outcomes are shown as dots, each one representing a single day: solid dots — shared territories, open dots — not shared.

defend interspecific territories. Reed showed this in two ways. First he carried out playback experiments on the islands and on the nearby mainland; these showed that neither species responds at all to the other's song on the mainland, but that they respond by aggressive approach and counter-singing on the islands. Secondly, Reed did a removal experiment on the islands: he captured chaffinches and removed them from their territories. Great tits soon moved in to occupy the empty spaces, indicating that before the removals they have been kept at bay by chaffinch territorial defence.

Why do the two species defend interspecific territories on the islands but not on the mainland? The likeliest explanation is that on the islands the wooded habitats which both species occupy are structurally much simpler than on the mainland. Because of the simplicity of the habitat the two species are forced to use the same resources (especially food) which they would not normally do.

Summary

Animals may compete for resources by pure exploitation or by resource defence, or by a mixture of both. A simple model of exploitation is the 'ideal free' distribution of competitors between resources. This pattern is seen in a number of examples. A useful concept in thinking of when it pays to compete by resource defence is that of economic defendability. This can be used in combination with studies of time budgets to predict conditions under which defence should occur and to formulate hypotheses about the optimal size of territories.

Further reading

Myers *et al.* (1981) provide a detailed review of the economics of resource defence and optimal territory size. Hixon *et al.* (1983) describe experiments on optimal territory size in hummingbirds. Verner (1977) develops the idea that animals might defend more resources than they actually need to maximize their own success, simply in order to reduce the success of others (superterritories). Appealing though it is, Verner's idea has been criticized on theoretical grounds by several authors (e.g. Rothstein 1978; Parker & Knowlton 1980). The essence of the criticism is that defending a superterritory might pay when the trait is rare, but it will not pay when most of the population adopts this strategy. Harper (1982) shows how duck flocks distribute themselves between patches of different profitability.

this strategy. Harper (1982) shows how duck flocks distribute themselves between patches of different profitability.

Topics for discussion

1 Do animals ever defend more resources than they need?
2 How should individuals of unequal competitive ability be distributed among resources? (see Sutherland & Parker 1985).
3 How would you apply the idea of economic defendability to resources other than food (e.g. nest sites, mates)?

Chapter 6. Living in Groups

Show anyone 10 000 flamingeos nesting beak by jowl in a colony and the chances are that sooner or later they will ask 'Why on earth are they all nesting so close together?'. In this chapter we will look at why animals live in groups: why flamingeos flock, horses herd and sardines shoal. Using the methods described in chapter 2; comparison between or within species and experimental studies of costs and benefits, we will show how ecological pressures might favour living in groups.

Comparisons between species suggest that the two main environmental influences on group size are food and predators (chapter 2) and comparisons between populations within a species also emphasize their importance (Fig. 6.1). In many studies either costs or benefits related to feeding and predation have been measured and we will describe some of these in the first part of the chapter, before moving on to consider whether different kinds of cost and benefit can be combined to predict optimal group size. Animals which do not live in groups (and also some which do) often defend resources from which they exclude other members of the same species. Therefore the question 'Why live in a group?' is a natural complement to the question 'Why defend resources?' that we discussed in chapter 5.

Living in groups and avoiding predation

The guppies in Fig. 6.1a live in groups when they are in streams where predators are common, which suggests that being in a group might help an individual to avoid becoming a meal. This could happen in several different ways.

INCREASED VIGILANCE

For many predators success depends on surprise: if the victim is alerted too soon during an attack, the predator's chance of success is low. This is true, for example, of goshawks hunting for pigeon flocks (Fig. 6.2). The hawks are less successful in attacks on large flocks of pigeons mainly because the birds in a large flock take to the air when the hawk is still some distance away. If each pigeon in the flock occasionally looks up to scan for a hawk, the bigger the flock the more likely it is that one

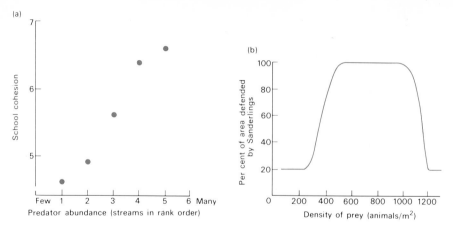

Fig. 6.1. Intraspecific variation in group size may be related to predators and food. (a) Guppies (*Poecilia reticulata*) from different streams in Trinidad: guppies from streams with many predators live in tighter schools than those from streams with few predators. Each dot is a different stream and 'cohesion' was measured by counting the number of fish in grid squares on the bottom of a tank (Seghers 1974). (b) Sanderlings (*Calidris alba*) in Bodega Bay, California: The birds defend stretches of beach in some parts of the intertidal zone and feed in roving flocks on other parts of the beach. Whether or not the birds defend territories depends on the density of the major prey, an isopod called *Excirolana linguifrons*. Territories are mainly defended in areas of intermediate prey density. At very low densities there are not enough prey to make defence worthwhile and at very high densities there are so many sanderlings trying to feed that defence would not be feasible because of high intruder pressure. In the area where birds defend territories there is an inverse correlation between territory size and food density (Myers *et al.* 1979).

bird will be alert when the hawk looms over the horizon. Once one pigeon takes off the others follow at once.

The precise way in which vigilance changes with flock size depends on how individuals in the group spend their time. In ostrich flocks, for example, Brian Bertram (1980) found that each individual spends a smaller proportion of its time scanning than when alone but that the overall vigilance of the group (proportion of time with at least one bird scanning) increases slightly with group size (Fig. 6.3). Therefore each bird in the flock has more time to feed and enjoys greater awareness of approaching lions (a potential predator of ostriches). The increase in vigilance with group size is as predicted if each bird raises its head independently of the others. The ostriches also raise their heads at random time intervals which makes it impossible for a stalking lion to predict how much time it has to creep forward undetected between look-ups by its victim. Any predictable pattern of looking could be exploited by the lion in its tactics of approach.

The problem of how individuals in a group scan is complicated by the fact that in a large group, where overall vigilance

(a)

(b)

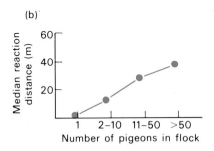

Number of pigeons in flock

Fig. 6.2. (a) Goshawks (*Accipiter gentilis*) are less successful when they attack larger flocks of wood pigeons (*Columba palumbus*). (b) This is largely because bigger flocks take flight at greater distances from the hawk. The experiments involved releasing a trained hawk from a standard distance (Kenward 1978).

is at the maximum value of 100 per cent, it would pay an individual to 'cheat' and spend all its time with its head down feeding. The cheater loses nothing in terms of vigilance because others are busy scanning and it gains extra time to feed. It is not known how this kind of cheating is prevented from evolving, but one suggestion is the following. Although the 'innocent' strategy of scanning regularly regardless of what others do is susceptible to cheating, a flock made up of more canny individuals, which do not scan unless they have seen their neighbours doing the same thing, might be resistant to cheaters (Pulliam *et al.* 1982). The general point is that even when there is an overall benefit of being in a group, each individual will be expected to try to get more benefit than the others.

DILUTION AND COVER

Although there is only a slight increase a vigilance with increasing group size in ostriches, the chances that any one individual will be eaten during an attack by lions decreases

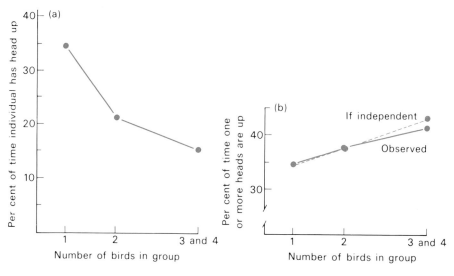

Fig. 6.3. Vigilance in groups. (a) An ostrich (*Struthio camelus*) spends a smaller proportion of its time scanning for predators when it is in a group. (b) The overall vigilance of the group increases slightly with group size, (solid line) as predicted if each individual looks up independently of the others (broken line). (Bertram 1980.)

rapidly with group size, because the lions can kill only one ostrich per successful attack. By living in a group the ostrich dilutes the impact of a successful attack because there is a good chance that another bird will be the victim. To some extent this dilution effect may be offset by the increased number of attacks on larger and more conspicuous groups, but usually the net effect probably favours living in a group, as the following hypothetical example illustrates. An individual antelope in a herd of a hundred has (all things being equal) only a one in a hundred chance of being the victim in a single attack and the herd is not likely to attract more than a hundred times as many attacks as a solitary antelope (see also Fig. 6.4). In fact if the herd is more vigilant it may pay the predator to concentrate its attacks on small groups and solitary individuals.

One study in which the survival rate of individuals in different sized groups was measured showed an overall benefit of group living from dilution. The monarch butterfly (*Danaius plexippus*) migrates from North America to spend the winter in warmer places such as Mexico. They assemble into enormous communal roosts in which the trees over an area of up to 3.0 ha may be clothed in resting butterflies. The monarch is not a very palatable butterfly, but some birds attack them in the winter roosts. Counts of the remains of predated butterflies showed that predation rate is inversely related to colony size,

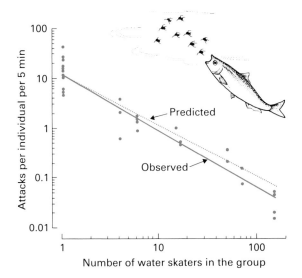

Fig. 6.4. An example of the dilution effect. The prey are insects called water skaters (*Halobates robustus*) that sit on the water surface, their predators are small fish (*Sardinops sagax*). The fish snap the insects from below, so there is little possibility that vigilance increases with group size. The attack rate by the fish was similar for groups of different sizes, so the attack rate for individual varies only because of dilution. The 'predicted' line is what would be expected if the decline in attack rate with group size is entirely caused by dilution; this line is very close to the observed (Foster & Treherne 1981).

so the advantage of dilution seems to outweigh any disadvantage of greater conspicuousness in a large roost (Calvert *et al.* 1979).

The dilution effect is probably a very widespread advantage of being in a group and it might explain the strange behaviour of birds such as ostriches and goosanders when they have young. When two females meet, each appears to try and steal the other's young and incorporate them into its own brood. Usually caring for someone else's young doesn't pay, but if predation pressure is severe it might, because of dilution. A more concrete example of the dilution effect comes from a study of semi-wild horses in the Carmargue, a marshy delta in the South of France. In the summer months the horses are plagued by biting tabanid flies and during this period they are more likely to cluster together in large groups. Measurements of the number of flies per horse in large and small groups showed that horses in a large group are less likely to be attacked. An experiment in which horses were transferred from large to small groups and vice versa confirmed that living in a group gives protection by the dilution effect (Duncan & Vigne 1979).

In some animals dilution is achieved by synchrony in time as well as in space, and this might explain the remarkable 13 and 17 year life cycles of certain species of cicada. These insects live as nymphs underground and the adults emerge after 13 or 17 years depending on the species and location. In the 17 year cicada studied by Dybas and Lloyd (1974) millions of adults (of three species) emerge in synchrony over a wide area, effectively 'flooding the market' so that the chances of any one individual falling victim to a predator is reduced. Lloyd and Dybas (1966) and others have speculated on why the cycle should be 13 or 17 years long and not, for example, 15 or 18. The advantage of a very long dormant stage between emergence periods is that it forces specialist predators and parasites out of business. When there are no cicadas around for 13 or 17 years the predators have either to die or to switch to other prey or to become dormant themselves. The very long cycle could have evolved as a result of an 'evolutionary race' (chp. 4) in which both cicadas and their predators gradually extended their life cycles until the cicadas eventually 'won'. The significance of the 13 and 17 year periods is that these are prime numbers which means that a predator could not regularly fall into synchrony with the cicadas if it had a short life cycle of which the cicada cycle is a multiple. If, for example cicadas had a 15 year cycle, predators with 3 or 5 year life cycles would fall into step with their prey every fifth or third generation.

This idea remains an interesting speculation, but synchrony is certainly an advantage. Field evidence shows that cicadas emerging at the peak of the cycle have a lower chance of succumbing to predators than those emerging early or late (Simon 1979). Selection therefore acts to maintain synchrony once it is established.

Just as a cicada in the middle of the emergence period is safer than one at either end, individuals in the middle of a flock, school or herd may enjoy greater security than those at the edge. If the predators pick off victims from the edge, each member of the group should jockey for a central position and, in effect, seek cover behind the others (Hamilton 1971). This may explain why starling flocks, for example, bunch together in a tight group when a predator approaches. Why should predators attack the edge of the group? The old trick of throwing three tennis balls to a friend at the same time shows how difficult it is to track one of a number of rapidly moving objects in the visual field for long enough to catch one. There is some evidence that predators suffer from the same type of

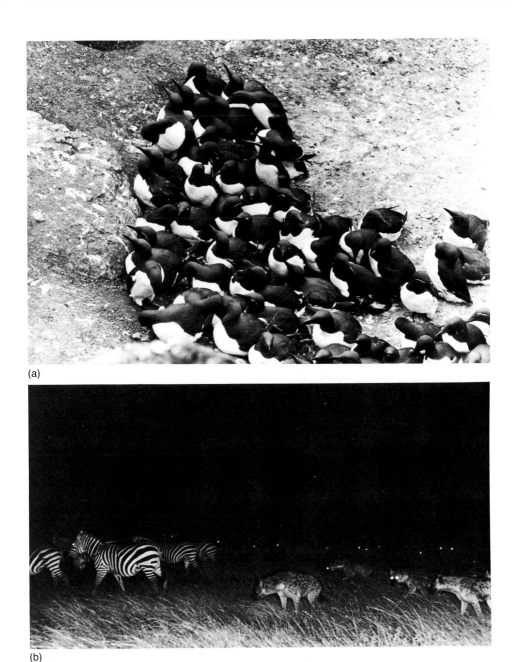

(a)

(b)

Fig. 6.5. (a) Group living and avoiding predators. In dense colonies of guillemots (*Uria aalge*) like this one, breeding success is higher than in sparse colonies because of more effective defence against nest predators such as gulls (Birkhead 1977), photo by T. R. Birkhead. (b) Group living and hunting for food: Hyenas (*Crocuta crocuta*) can successfully attack prey which are larger than themselves because they hunt in a group (Kruuk 1972). Photo by Hans Kruuk.

confusion when attacking a dense group of prey (Neill & Cullen 1974) and this may provide an explanation of why attacks should be directed at the edge of a group.

GROUP DEFENCE

Prey animals are often not just passive victims and by living in a group they may be able to defend themselves against the unwelcome attentions of a predator. In colonies of blackheaded gulls nesting pairs will mob a crow when it flies near their nest and in the centre of a dense colony many gulls mob the crow at the same time because it is close to many nests. The effect of this is to reduce the success of the crows in hunting for gulls' eggs (Kruuk 1964) (see also Fig. 6.5a).

COSTS OF BEING IN A GROUP

As we mentioned earlier, one of the costs of group living might be increased conspicuousness. This cost was studied experimentally by Malte Andersson (Andersson & Wicklund 1978) using artificial nests of the fieldfare, a thrush-like bird which breeds colonially in Scandinavian boreal forests. The bulky nests are quite conspicuous and a colony of artificial nests attracted more predators than did solitary nests. However, fieldfares vigorously mob and defecate on crows and other predators, and Andersson and Wicklund found that artificial nests placed near a colony of fieldfares survived better than those placed near solitary fieldfare nests. They concluded therefore that the benefit of group mobbing by members of a colony more than offsets the disadvantage of being conspicuous. This is supported by Volker Haas's (1985) observation that nesting success is higher for colonial than for solitary fieldfares.

Living in groups and getting food

FINDING GOOD SITES

The comparative studies described in chapter 2 revealed that species which feed on large ephemeral clumps of food such as seeds or fruits often live in groups. For these animals, the limiting stage in feeding is the problem of finding a good site: once the patch has been found there is usually plenty of food, at least for a short while. Peter Ward and Amotz Zahavi (1973) developed the idea that communal roosts and nesting colonies of birds may act as 'information centres' in which individuals find out about the location of good feeding sites by following others. The idea is that unsuccessful birds return to the colony or roost and wait for the chance to follow others who have had more success on their last feeding trip. Unsuccessful birds might recognize successful ones by, for example, the speed with which they fly out from the colony on their next trip.

It is perhaps unfortunate that Ward and Zahavi used the phrase 'information centre' which carries with it a connotation of mutual co-operation in the transfer of information, as for example in a honeybee or ant colony. As we shall see in chapter 13 there are special reasons to expect co-operation in social hymenopteran colonies, but these do not apply to bird roosts or nesting colonies. 'Mutual parasitism' might be a more appropriate label here, since the successful foragers are in effect parasitized by unsuccessful birds. Each individual is out to maximize its own success and not the success of the colony as a whole. In some species the 'informer' might be unable to avoid being followed because it is conspicuous when it leaves the nest, for example seabirds leaving a colony on a cliff. The 'informer' might, however, benefit from being followed. The benefit could be long term: on a later trip the leader becomes a follower, or short term: there may be an advantage to feeding in a group because, for example, of the reduced risk of predation. If these benefits do not outweigh the disadvantage of competition at the feeding grounds which arises from being followed, a successful bird should conceal as much as possible information about its success.

The most direct experimental test of Ward and Zahavi's idea is a study of communally roosting weaver birds (*Quelea quelea*) by Peter de Groot (1980). *Quelea* nest in a colony and roost in groups sometimes estimated to contain over a million birds. They are serious agricultural pests in parts of central Africa and can devastate a grain field in a few hours. De Groot's experiments were done on a somewhat more modest scale (Fig. 6.6). Two groups of birds roosted together in the large aviary labelled X and had access to foraging areas in the small compartments labelled 1−4. The birds could not see into the foraging compartments from the roosting area but had to pass through small entrance funnels to explore them for food or water. In one experiment one of the groups (A) was trained to find water in one of the four compartments and the other group (B) was trained separately to find food in another compartment. The two groups were allowed to roost together and were deprived of food or water. When they were thirsty, the birds of group B followed A to the drinking site, and when

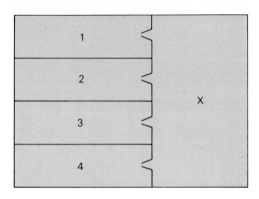

Fig. 6.6. An experiment to test the 'information centre' hypothesis with *Quelea*. The birds roost in the large area labelled X and feed in the smaller compartments labelled 1 – 4.

they were hungry A followed B to the feeding site. Somehow the 'naive' birds assessed that the other group was knowledgeable and followed it to the resource supply.

In a second experiment group A was trained to forage on a good food supply (pure seed) in one of the compartments while group B was separately trained to fly to another compartment for a poor food supply (seed in a bed of sand). When the two groups roosted together, the members of the second group followed the first group when they left the roost at dawn. It is not yet known how the birds recognize which individuals to follow.

In another study, however, the mode of information transfer has been identified. Geoff Galef and Stephen Wigmore (1983) trained rats (*Rattus norvegicus*) to search for food in a three-arm maze. Each arm had food with a different flavour, cocoa in one, cinnamon in another, and cheese in the third. In the first part of the experiment the rats learned that on any particular day only one of the three sites contained food, but the site was unpredictable. Then on the days of the actual experiment each of the seven test rats was allowed to sniff a 'demonstrator' rat in a neighbouring cage. The demonstrator had been allowed to feed on whatever randomly chosen food was available for that day, and four of the seven test rats, having sniffed the demonstrator, went to the correct site on their first choice of the day. 'Sniff' is the operative word, because other experiments showed that the cue the test rat picks up from the demonstrator is the smell of the food it has eaten, just as you can tell when your friend has had a garlic pizza.

Learning about potential food sources in a more direct way, by seeing others exploiting them, is important in both flocks of birds and shoals of fish (Krebs *et al.* 1972; Pitcher *et al.* 1982).

Individuals in groups may be able to capture prey which are difficult for a single individual to overcome either because the prey is too large for one predator to handle (e.g. lions hunting adult buffalo) or because it is too elusive for one predator to catch (e.g. killer whales hunting porpoise). When the prey themselves are in a group the predators may, by hunting in a group, succeed in separating a victim from its companions and subsequently chasing it until they overtake it. This is how predatory fish such as the jack (*Caranx ignobilis*) hunt for schooling prey. Individuals in a group are more successful than single fish when hunting for schools of Hawaiian anchovy (*Stolephorus purpureus*) (Fig. 6.7). However the benefit is not shared equally between members of the hunting group: fish at the front of the school when it is chasing a prey get more than those at the back (Fig. 6.7). In fact the fourth and fifth fish could do better by hunting alone, but it may be that different individuals occupy the lead position during different chases. This is a reminder of the general point that benefits of being in a group may not be shared equally.

HARVESTING RENEWING FOOD

Suppose an animal eats food which renews itself continuously, for example growing vegetation. The amount of food available in a site increases with time since the last visit, so an individual could get the maximum possible foraging returns by coming back to the same site after the appropriate time interval. Returning too soon means not finding enough food and returning too late means missed opportunities to eat a plentiful food supply. As we saw with the wagtails in chapter 5, the problem with harvesting a renewing food supply in this way is that it only works if there is no interference by others with the renewal pattern. Individual A's strategy of returning after ten days would fail if B visited the site after, for example, nine or eight days. One way to prevent interference by others is to defend a territory (Charnov et al. 1976) and another way is to visit sites in a group so that everyone returns at the same time. Wintering flocks of Brent geese (*Branta bernicla*) feeding on salt marshes in Holland seem to do the latter, where territory defence would not be feasible because the marsh is frequently inundated at high tide. Continuous observation of 40 one hectare plots from dawn to dusk for 24 days during the spring showed that the flocks return to exactly the same site on the marsh at regular 4 day intervals. No only does this allow the sea plantain (*Plantago maritima*) to recover between

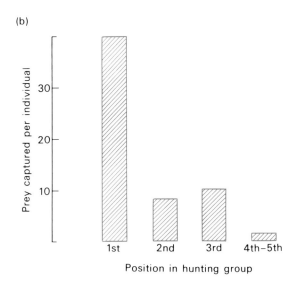

Fig. 6.7. The jack is a predatory fish which hunts in schools. (a) Each fish on average captures more prey per experiment when hunting in a group. (b) But fish at the front of the group benefit the most (Major 1978).

visits by the geese, but also the regular cropping pattern actually stimulates the growth of young leaves which are rich in nitrogen. Experiments in which sea plantain was cut with scissors to simulate goose grazing at different time intervals suggested that given the average bite size, the geese may even return after a time interval which maximizes the growth of young shoots (Prins *et al.* 1980).

Do individuals at the back of the group fare less well than those at the front, as in the hunting schools of fish described earlier? The answer is not yet known, but it is possible that the overall benefit in terms of food intake is similar for birds in different parts of the flock. The front birds eat most of the vegetation, but the youngest and most nutritious parts of the *Plantago* plants are the base of the leaves close to the ground. These parts are exposed only after the older, taller, parts of the leaves have been cropped off, so it is reasonable to hypothesize that the first birds eat larger mouthfuls while the later birds eat food of higher quality. The overall effect may be that all birds obtain the same quantity of nutrients.

COSTS ASSOCIATED WITH FEEDING

The goose study suggests an important potential cost of feeding in a group: competition for food. Competition may take the form of direct exploitation as in the jack where fish at the front catch the prey and deprive those at the back of the school, or it may arise as a result of interference in which the availability of food to a group member is reduced as a result of the behaviour of nearby companions. This occurs in the redshank (*Tringa totanus*), a shorebird studied by John Goss-Custard on the coastal mudflats of Britain. Redshank feed in tight flocks at night and more loosely scattered or solitarily during the day: this difference seems to be related to interference. In the daytime, redshank feed by sight on small shrimps (*Corophium*) which live with their tails just sticking out of the mud surface, while at night, when visual search is impossible, the birds turn to feeding by touch on snails (*Hydrobia*), sweeping their long beaks through the mud. When feeding on shrimps, the birds are likely to interfere with one another because the shrimps retreat into the mud and become unavailable as soon as they detect the heavy clump of redshank feet (this was shown in experiments with captive birds). Feeding rate is therefore greater with increasing neighbour distance and the birds tend to space out. At night there is no feeding interference because the birds do not depend on seeing the prey, and the snails in any case do not react quickly to disturbance. Feeding rate in these conditions is not related to flock density and the birds crowd into tight groups (Goss-Custard 1976). One interpretation of these results is that there is an advantage to tight clumping in redshank (perhaps as a consequence of seeking cover from predators in the middle of the flock), and that during the night there is little cost to feeding close together. During the day, however, feeding close together causes interference and the birds spread out. The nearest neighbour distance under these conditions appears to reflect a balance between the costs and benefits of group living.

Table 6.1. Examples of studies in which possible costs and benefits of group living other than those mentioned in the text have been measured

Hypothesis	Test	Reference
1. Warm-blooded animals save energy because of thermal advantage of being close together	Pallid bats (*Antrozous pallidus*) roosting in groups use less energy than solitary roosters	Trune & Slobodchikoff (1976)
2. Inferior competitor can overcome competitive advantage of another species by group foraging	Blue tang surgeon fish (*Acanthurus caeruleus*) are excluded from algal mats by dusky damsel fish (*Stegastes dorsopunicans*) when solitary but not when in a group	Foster (1985)
3. Hydrodynamic advantage for fish swimming in a school. They save energy by positioning themselves to take advantage of vortices created by others in the group	Measurements of distances and angles between individuals show that they are *not* correctly positioned to benefit according to the predictions of the theory	Weihs (1973) Partridge & Pitcher (1979)
4. Increased incidence of disease as a result of close proximity of others	Measure number of ectoparasites in burrows of prairie dogs (*Cynomys* spp.). There are more parasites per burrow in larger colonies	Hoogland (1979b)
5. Risk of cuckoldry by neighbours	In colonial nesting red-winged blackbirds (*Agelaius phoeniceus*) the mates of vasectomized males laid fertile eggs. They must have been fertilized by males other than their mates	Bray *et al.* (1975)

Table 6.1. *Continued*

125

Chapter 6
Living in
Groups

Hypothesis	Test	Reference
6. Risk of predation on young by cannibalistic neighbours	In colonies of Belding's ground squirrels (*Spermophilus beldingi*) females with small territories are more likely to lose their young to cannibalistic neighbours than are females with large territories around their burrows	Sherman (1981)

Another study which illustrates how competition for food influences group size is Mark Elgar's (1986) work on house sparrow (*Passer domesticus*) flocks. The first sparrow to arrive at a feeding site on the ground gives a 'chirrup' call from a perch before going down to feed. The chirrup draws in other sparrows and the group goes to the food source together. It may seem maladaptive for a sparrow, having found a good source of food, to attract other competitors to it, but it is known that sparrows spend less time scanning in larger groups (Elgar & Catterall 1981) so there is some benefit to the first sparrow in generating a flock in which to feed. However, and this is the most interesting point, the first sparrow does not give a chirrup to recruit others if the food source is indivisible. The cost of recruiting in terms of competition for food now outweighs the benefits derived from reduced scanning time. Elgar's experiment was simply to provide the same amount of food — a piece of bread — either in a lump (indivisible) or in crumbs (divisible). The first sparrow to arrive was more likely to chirrup when the food was in crumbs.

Weighing up costs and benefits — optimal group size

The message of the chapter so far is that there are many different costs and benefits of living in a group, some or all of which might be relevant to a particular species. Pigeons, horses ostriches and cicadas do not necessarily get together for the same reasons, but they could do so for any or all of the reasons discussed. Our list has been by no means comprehensive; we could have gone on to describe costs of group living such as transmission of diseases, cannibalism, and cuckoldry, or benefits such as protection from the elements, co-operative territorial defence, and increased efficiency of locomotion. (Some of these are summarized in Table 6.1). But rather than try to extend the list of possible costs and benefits to its exhaustive and exhausting conclusion, we will return to the more interesting question of whether different kinds of cost and benefit can be combined to predict an optimal group size.

COMPARATIVE STUDIES

A qualitative picture of how costs and benefits interact can be drawn from comparisons between species. For example, some species of shorebirds such as the knot feed in dense flocks while others such as the ringed plover feed in loose flocks or as solitary birds (Fig. 6.8) It is known that living in a flock confers protection on shorebirds against attacks by birds of prey (Page & Whitacre 1975) so why do not all species feed in tight flocks? The species which feed in dense flocks hunt by touch and walk slowly, probing or swinging their beaks through the mud, while solitary and loose flock feeders hunt by sight and move rapidly, picking prey off the surface of the mud or water. Perhaps, as with the redshank, the costs of feeding interference are so big in the latter species that the net benefit for an individual is higher when alone, even though the risk of predation is greater.

TIME BUDGETS

In order to predict more precisely how different costs and benefits combine to determine group size we will return to the approach of chapter 3. Ultimately the costs and benefits influence survival and reproduction, but as with the optimality models described in chapters 2 and 3 it is more useful to think in terms of a proximate currency that ultimately relates to fitness.

Ron Pulliam (1976) and Tom Caraco (1979a) have used time as a currency and developed a model of optimal group size based on time budgets. The model is meant to illustrate the factors influencing winter flocks of small birds. The survival of birds in a flock is considered to be dependent on two main risks, starvation and predation, and the birds' time budget is divided into three types of behaviour associated with these risks: scanning (for predators), feeding, and fighting (for food). Based on their observations of yellow-eyed junco flocks, Pulliam and Caraco divide fighting into two categories: short term squabbles over access to pieces of food and attacks in which dominant birds attempt to evict subordinates from good feeding sites in order to ensure a supply of food for the rest of the winter. The three activities in the time budget are assumed

Fig. 6.8. Knot (*Calidris canutus*) (top) feed by touch and live in dense flocks while ringed plovers (*Charadrius hiaticula*) (bottom) feed by sight and live in loose flocks or as solitary individuals (Goss-Custard 1970). The difference can be interpreted in terms of the costs and benefits of flocking.

to be mutually exclusive, that is a bird cannot, for example, scan and feed at the same time. In order to scan it has to point its head upwards, while pecking involves facing towards the ground. Finally they assume that scanning for predators takes precedence over feeding, since failing to see an approaching predator is more dangerous than failing to eat a seed. Dominant birds are assumed to give higher priority to satisfying their daily energy requirements than to long term eviction of sub-ordinates, while for a subordinate bird, aggression must take priority over feeding since a bird cannot feed while it is being attacked. Figure 6.9a shows a simplified version of Pulliam and Caraco's model. The main features are as follows.

1 The proportion of time spent scanning by an individual is assumed to decrease with increasing group size. The basis for this assumption is that a given level of vigilance can be maintained with less scanning per individual as group size increases (pp. 112)

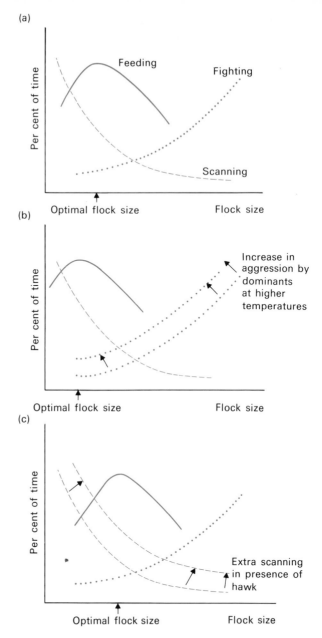

Fig. 6.9. A model of optimal flock size. (a) As flock size increases birds spend more time fighting and less time scanning. An intermediate flock size gives the maximum proportion of time feeding. (b) At higher temperatures (or when food is more plentiful) dominant birds can afford to spend more time attacking subordinates. The optimal flock size for the average bird therefore decreases. (c) When predation risk is increased by flying a hawk over the flock, the scanning level should go up and the optimal flock size is increased. Based on Pulliam (1976) and Caraco *et al.* (1980a).

2   As group size increases and encounters between birds become more frequent, the proportion of time spent in aggression increases.

3   The time spent feeding therefore is at a maximum in flocks of intermediate size.

Can this model of time budgets be used to predict the optimal group size? If the only benefit of flock feeding is to increase the time available for feeding while maintaining a certain level of vigilance, the optimal flock size is the one indicated in Fig. 6.9a. If there are other benefits of flocking such as dilution and increased vigilance (see pp. 111–16), the optimal flock size may be larger than the one shown in Fig. 6.9a. Therefore the model can be used to test whether or not maximizing food intake is the only benefit of flocking. However the picture is probably more complicated than Fig. 6.9a suggests, because the optimal flock size may be different for dominant and subordinate birds. Dominant birds obtain a long term benefit from evicting subordinates, so they should prefer to be in smaller groups.

Caraco (1979b) and Caraco *et al.* (1980a) have tested some of the assumptions of their model by recording the time budgets of yellow-eyed juncos in winter flocks in Arizona. They found that the proportion of time spent by individuals in scanning and fighting change with flock size in the directions assumed by the model. However the decrease in scanning time was much greater than the increase in fighting time over the range of flock sizes they studied, so that feeding time increased with flock size, as it does to the left of the peak in Fig. 6.9a.

In order to test whether time budgets influence flock size in the way suggested by the model, Caraco *et al.* predicted the effect of various environmental changes on flock size. The predictions were as follows.

1   As average daily temperature increases, dominant birds should have more time to evict subordinates because they can satisfy their energy requirements more rapidly. Flock size should therefore decrease (Fig. 6.9b). This prediction was supported by observations: at 2°C flocks contained an average of 7 birds, at 10°C they contained 2 birds. This decrease coincided with an increase in time spent fighting by dominant birds.

2   By a similar argument, an increase in food supply should produce a decrease in flock size and an increase in the proportion of time spent fighting by dominant birds. Again the results of field observations supported the prediction. When food was scattered in the canyon, the birds fed in smaller flocks.

3 An increase in the risk of predation should have exactly the opposite effect of the previous two changes. This is because a high risk of predator attack should cause the birds to spend more time scanning; they therefore have to feed in larger flocks to maintain a given rate of food intake (Fig. 6.9c). Caraco *et al.* (1980a) allowed a tame hawk to fly over the canyon, and as predicted the birds spent more time scanning and the mean flock size increased. It was 3.9 birds without the hawk and 7.3 with the hawk.

4 Finally, Caraco predicted that by adding more cover to the canyon in the form of a bush, the effective risk of predator attack would be reduced because the juncos would have easier access to a safe hiding spot. The birds should therefore spend less time scanning, which allows more time for feeding and fighting. This is exactly what happened when an experimental bush was placed near one of the favoured feeding sites, and as expected the flock size decreased.

What can we conclude from these results? First they show that flock size is influenced by time budgets as hypothesized in the model shown in Fig. 6.9a. Flocking allows more time for feeding because less time is spent scanning, and the maximum flock size depends on the time available for dominant birds to evict subordinates. Second, the results allow us to reject the simplest hypothesis about optimal flock size. The birds did not feed in flocks of the size that would have maximized feeding time: under normal conditions the average flock in the canyon contained 3.9 birds, but measurements showed that the time available for feeding would have been higher in a flock of 6 or 7. As we have mentioned already, the optimal flock size for dominant and subordinate birds probably differs since dominant birds benefit by evicting subordinates. The observed flocks may have been a compromise between the optimum for dominant and subordinate birds. A further complication is that birds in larger flocks benefit by the dilution effect and increased vigilance, as described earlier for ostriches (Caraco *et al.* 1980a).

The model in Fig. 6.9 is clearly too simple, but the study shows that time budgets can be used to analyse the effects of different costs and benefits on flock size. It also reminds us of the idea that flocking and resource defence may be two ends of a continuum. The model could be viewed as one which predicts the conditions under which it pays dominant birds to exclude subordinates and defend a territory. When food is plentiful or predation risk is low, dominants can afford the time to maintain a defended area, or in other words the territory is economically defendable.

Richard Sibly (1983) has pointed out that the groups of optimal size may rarely be found in the wild because if there was a group of this size, it would pay any solitary individuals around to join the group and therefore push the group above the optimal size. The idea is illustrated in Fig. 6.10. The curve shows a possible relationship between individual fitness (perhaps measured as feeding rate or probability of escaping predation) as a function of group size. The optimal group size, in which average fitness is maximal, is seven individuals. Imagine that individuals are free to join the group or forage alone as they arrive in the area. Clearly, if each individual chooses the option that maximizes its fitness, newcomers will join the group up to the point where group size is 14, twice the optimum! At this point the pay-off for foraging alone is the same as the pay-off for joining the group (note that this model is an example of the 'ideal free' distribution discussed in chapter 5). The same principle would apply if one imagined groups splitting and reforming as smaller units, but the argument is more complicated (Kramer 1985). A group of 12, for example would split into two units of 6, but then one individual would migrate to form a 7 and a 5 membered group, the seven would be joined by another because 8 is better than 5, and so on. The exact outcome depends on the shape of the curve in Fig. 6.10, and it is possible to draw fitness curves which will result in a stable optimal group size (Giraldeau & Gillis 1985). The general point, however, is that one should expect to find stable

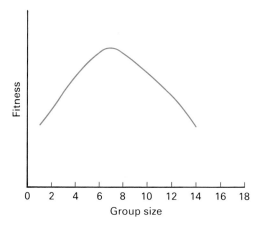

Fig. 6.10. Sibly's model of optimal and stable group size. Each individual joins the group that maximizes its fitness, so that the optimal size of seven is not necessarily stable — it will be joined by solitary individuals for example. After Sibly (1983).

groups rather than optimal groups in nature and very often, observed groups will be larger than the optimum (Pulliam & Caraco 1984).

INDIVIDUAL DIFFERENCES IN A GROUP

We have already emphasized the point that individuals in a group may benefit from group living to different extents. In foraging groups of jack, individuals at the front do better than those at the back. Similarly in starling flocks, birds at the edge spend more time scanning than those in the middle (Jennings & Evans 1980). How are such differences between individuals maintained? In many cases they are simply a reflection of dominance relationships. Within the group, older, more experienced, or bigger individuals are able to commandeer the best positions and force others to take what is left. Subordinate individuals put up with less pay-off as long as they could not do better by moving elsewhere (Vehrencamp 1983).

An alternative view of individual differences is that different individuals achieve similar pay-offs but in different ways

Fig. 6.11. Some mixed species flocks involve complex relationships. Barnard and Thompson (1985) studied golden plovers, *Pluvialis apricaria* (two in the foreground), lapwings, *Vanellus vanellus* and blackheaded gulls, *Larus ridibundus*, which often associate in winter feeding flocks in pastures in lowland Britain. Lapwings favour pastures where earthworm density is high and golden plovers select fields on the basis of the presence of lapwings, using them, in effect, as food finders. Blackheaded gulls are kleptoparasitic and steal worms from the lapwings and golden plovers, but also bring benefits because they give early warning of the approach of predators. After a painting by Donald Watson.

(chp. 10). In sparrow flocks, for example, some birds are good
at locating new sources of food while others are good at stealing
food once it is found. These two strategies, producer and
scrounger, may coexist in the flock with equal pay-offs for the
two (Barnard & Sibly 1981) (Fig. 6.11).

Summary

Two of the most important selective advantages of living in a
group are increased protection from predators and increased
likelihood of finding or capturing food. To be weighed against
these benefits are costs of group living such as increased competition for food and increased conspicuousness to predators.
The size of group in which animals live may reflect a compromise resulting from these various costs and benefits. One
way to analyse the trade-off between them is to use the common
currency of time to predict optimal group sizes.

Two limitations of the notion of optimal group size are (a)
that individuals in a group may attain different pay-offs and
may have different optima. If observed groups are a compromise between the optima for different individuals, they may
not be optimal for any one individual. (b) Optimal sized groups
may be unstable because they tend to be joined by individuals
from smaller groups.

Further reading

Bertram (1978) and Pulliam et al. (1982) are good reviews of
why animals live in groups. Hoogland (1979a,b) is a study of
prairie dogs in which both the costs and the benefits of group
living are investigated. Morse (1970) describes some of the
reasons why birds sometimes live in mixed species groups
and Rubenstein (1978) discusses in general terms how advantages of grouping with respect to food and predators may
interact. The book edited by Barnard (1984) discusses producers and scroungers in groups. Ian Patterson's (1965) paper
is a classic study of the benefits of synchronous and colonial
breeding.

Topics for discussion

1 How can different kinds of costs and benefits of group
living (e.g. feeding, predation, disease) be combined to predict
group size?
2 Why do subordinates remain in groups?
3 How would you test the hypothesis that groups are stable
rather than optimal size?

Chapter 7. Fighting and Assessment

In the last two chapters we saw that individuals often have to compete with others for scarce resources such as food, territories or mates. We now consider the evolution of contest behaviour. What factors determine which will be the most successful fighting strategies? It might at first be supposed that we could use simple optimality models to analyse this problem, for example assessing the costs and benefits and working out the strategy which results in greatest net benefit. We cannot do this, however, because the optimal way for one individual to behave turns out to depend on what other competitors in the population are doing. The appropriate technique to analyse situations like this is game theory, which was originally developed for the study of economics but which has become a very useful theoretical tool for the analysis of animal behaviour, particularly due to the work of Maynard Smith (1982). It is easiest to describe this method if we begin with an example.

The war of attrition, or 'the waiting game'

(a) The problem

Female dungflies, *Scatophaga stercoraria*, come to fresh cowpats in order to lay their eggs. Swarms of males are waiting for them on and around the cowpats (Fig. 7.1) and whenever a female arrives, the first male to encounter her copulates with her and then guards her while she lays (p. 55). Females prefer to lay in fresh dung and as the pat gets older, and a crust forms over it (thus making it less suitable for egg laying) fewer females arrive. The male's problem is, what is the optimum time to spend waiting for females at each cowpat? The answer is that the best waiting time for one male depends on what all the other males are doing. For example, if most males remain for short times then a male who stayed a little longer would have high mating success because he could claim all the late arriving females. If, on the other hand, most males were staying a long time then it would pay our male to move quickly to a new pat to claim the early arriving females there. We have to analyse, therefore, a competition among the males where the pay-offs for different waiting times are frequency dependent.

Fig. 7.1. A male dungfly guards the female while she lays her eggs in a fresh cowpat. Photo by G.A. Parker.

(b) The theory

This kind of game, where winners are decided simply by a contest involving waiting, or displaying, for different lengths of time is known as the war of attrition and the solution is as follows (Maynard Smith 1974). Imagine two individuals contesting for a resource of value V. The longer an individual waits, the greater the cost, m. Let animal A select a time x_A and animal B select a time x_B. The costs associated with these times are m_A and m_B. If $x_A > x_B$ then A wins the resource and so we have

$$\text{Pay-off to A} = V - m_B$$
$$\text{Pay-off to B} = \quad - m_B$$

Note that A only pays the cost of waiting for time x_B because B gave up first. We suppose now that waiting times are heritable and that individuals have offspring with identical waiting times to themselves. We suppose too that individuals reproduce in proportion to their pay-offs in the game.

How would evolution proceed in this game? Obviously B would have done better to wait for longer than x_A so it could have won. For example, a strategy 'wait for 1 min' would be beaten by a strategy 'wait for 1.1 min'. This might suggest that longer and longer waiting times would evolve (1.2 beats 1.1, 1.3 beats 1.2, and so on) but eventually these waiting times would become so long that the costs would exceed the value of the resource. It is clear that no 'pure' strategy can be an evolutionarily stable strategy, or ESS. (*An ESS is a strategy which when adopted by most members of the population cannot be beaten by any other strategy in the game.*) If an individual always played the same fixed waiting time, then other competitors could take advantage of this and play a time which would always beat it. In fact, it can be shown that the ESS for this game is to play a randomly chosen waiting time, in other words be unpredictable.

This stable solution to the waiting game could come about in two ways.

1 Every individual in the population could play the same variable strategy, or mixed ESS. Every individual would sometimes wait for a long time and sometimes for a short time, and the times would be selected from the appropriate random distribution.

2 There could be a polymorphism in the population with every individual playing a fixed waiting time, or pure strategy, but with the frequencies of the strategies following the random distribution. There would then be a mixture of pure strategies, with most individuals playing short waiting times and fewer playing longer times.

The first prediction from this model is therefore that *waiting times will be variable*, either within or between individuals. The second prediction is that *at the ESS, the pay-offs for the different waiting times will be equal.* Equality of success will come about through frequency dependent selection. Imagine, for example, that individuals with short waiting times are enjoying greatest reproductive success. Their progeny will inherit this successful strategy and the proportion of individuals adopting short stay times will increase in the population. As it does, however, the success of short stay times will decrease because of increased competition. As more and more male dungflies, for example, play short stay times, there will be more and more competition among these males for the early arriving females. The increase in proportion of the short stay times in the population will come to a stop when this strategy's reproductive success equals that of the other strategies. On the other hand, if a certain waiting time experienced lower reproductive success the frequency of this strategy would decrease in the population. As it did, its reproductive success would increase, because of reduced competition, and frequency dependent selection would again stabilize the proportion at that which gave equal reproductive success to the other waiting times. The solution to the game in the equation above is *stable* in the sense that once the population adopts this strategy then no individual who played a different strategy could gain higher reproductive success.

The model's assumption of asexual reproduction ('like gives rise to like') is not too restrictive. Provided the ESS phenotype can be produced by a genetic homozygote it will be uninvadeable (stable) in a sexually reproducing species with diploid inheritance just as in an asexual one (Maynard Smith 1982).

(c) Testing the theory

What do dungflies do? Parker (1970) measured the lengths of time that males spent waiting for females on cowpats and the results showed that, as predicted by the war of attrition model, the 'stay times' fitted a negative exponential (random) distribution (Fig. 7.2a). It is not yet known whether different individuals play different stay times or whether individuals play a mixed strategy. Parker also measured the reproductive success of males and showed that the distribution of male stay times was such that different waiting times resulted in similar mating success (Fig. 7.2b), again as predicted by the model.

The evolution of conventional fighting

The war of attrition model is likely to be most applicable to situations, like the dungflies, where individuals are engaged in exploitation competition (p. 93) or 'playing the field' (Maynard Smith 1982). Many animal conflicts, however, are examples of contest competition where two individuals meet and compete for a resource. Here we would not expect the contest to be solved simply by a waiting game! Indeed, in cases like this there tends to be direct confrontation in the form of fights. For example, after a male dungfly has paired with a female other males may approach and attempt to take her over. There is often a struggle before one eventually wins and the other gives way (Fig. 7.3).

Although fights like these can sometimes be vicious in animals (see below), in general disputes are often settled before there is serious injury. Why do animals often settle contests by display rather than by all out fighting? The long-accepted answer to this question was that escalated contests would result in many animals getting seriously injured and this would militate against the survival of the species (e.g. Lorenz 1966; Huxley 1966). This is a group selection argument and does not explain how natural selection acting on individuals can give rise to the evolution of conventional fighting. Since the mid 1960s many people realized that the answer must be framed in terms of costs and benefits of fights to individuals. Once again game theory models developed by Maynard Smith and others are useful for exploring the factors influencing the evolution of pair-wise contests.

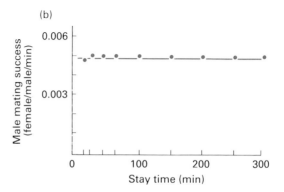

Fig. 7.2. (a) The number of male dungflies on a cowpat declines exponentially with time. This could come about through the population having a mixture of males with different fixed stay times or through all males being alike with flexible stay times; for example each may have the same constant probability of leaving per unit time. (b) Given this distribution of stay times, the result is that the mating success of males adopting different stay times is about equal, as predicted by an ESS model (see text). From Parker (1970).

HAWKS AND DOVES

Consider a game where there are just two strategies. 'Hawks' always fight to injure and kill their opponents, though in the process they will risk injury themselves. 'Doves' simply display and never engage in serious fights. Although simple, these two strategies are chosen to represent the two possible extremes we may see in nature.

In this evolutionary game, let the winner of a contest score +50, and the loser 0. The cost of a serious injury is −100 and the cost of wasting time in a display is −10. These pay-offs are

Fig. 7.3. A struggle between two male dungflies for the possession of a female. The attacking male (left) is attempting to push the paired male (right) off the female (hidden from view, underneath). Photo by G.A. Parker.

some measure of fitness and we will assume, for simplicity, that Hawk and Dove reproduce their own kind faithfully in proportion to their pay-offs. (The exact values do not matter and are chosen simply because this game is easier to explain with numbers rather than algebra.) The next step is to draw up a two by two matrix with the average pay-offs for the four possible types of encounter. These calculations are explained in Table 7.1.

How would evolution proceed in this particular game? Consider what would happen if all individuals in the population are Doves. Every contest is between a Dove and another Dove and the pay-off is on average +15. In this population, any mutant Hawk would do very well and the Hawk strategy would soon spread because when a Hawk meets a Dove it gets +50. It is clear that Dove is not an evolutionarily stable strategy, or ESS.

However, Hawk would not spread to take over the entire population. In a population of all Hawks the average pay-off is −25 and any mutant Dove would do better because when a Dove meets a Hawk it gets 0 (which is not very good, but still better than −25!). The Dove strategy would spread if the population consisted mainly of Hawks. Therefore Hawk is not an ESS either.

Table 7.1. The game between Hawk and Dove. After Maynard Smith (1976b)

(a) Pay-offs: Winner + 50   Injury   −100
               Loser   0   Display   −10

(b) Pay-off Matrix: average pay-offs in a fight to the attacker.

Attacker	Opponent	
	Hawk	Dove
Hawk	(a) $\frac{1}{2}(50) + \frac{1}{2}(-100)$ $= -25$	(b) +50
Dove	(c) 0	(d) $\frac{1}{2}(50 - 10) + \frac{1}{2}(-10)$ $= +15$

Notes:
(a) When a Hawk meets a Hawk we assume that on half of the occasions it wins and on half the occasions it suffers injury.
(b) Hawks always beat Doves.
(c) Doves always immediately retreat against Hawks.
(d) When a Dove meets a Dove we assume that there is always a display and it wins on half of the occasions.

    Nevertheless, a mixture of Hawks and Doves could be stable. The stable equilibrium will be when the average pay-offs for a Hawk are equal to the average pay-offs for a Dove. If the population moved away from this equilibrium then either Dove or Hawk would be doing better and so the population would not be stable. Each strategy does best when it is relatively rare and the tendency in this evolutionary game will be for frequency dependent selection to drive the frequencies of Hawk and Dove in the population so that they each enjoy the same success. For the values in Table 7.1, the stable mixture can be calculated as follows.

    Let h be the proportion of Hawks in the population. Therefore the proportion of Doves must be $(1-h)$. The average pay-off for a Hawk is the pay-off for each type of fight multiplied by the probability of meeting each type of contestant. Therefore,

$$\bar{H} = -25h + 50(1 - h).$$

Similarly, for Dove the average pay-off will be

$$\bar{D} = 0h + 15(1 - h).$$

At the stable equilibrium (the ESS), $\bar{H}$ is equal to $\bar{D}$. Solving the two equations above by setting $\bar{H} = \bar{D}$ gives $h = 7/12$, and therefore, by subtraction, the proportion of Doves $(1 - h)$ must be $5/12$.

    Just as in the war of attrition, the ESS could be achieved in two distinct ways:

1 The population could consist of individuals who played pure strategies. Each individual would either be Hawk or Dove, and the ESS would come about with 7/12 of the population being Hawks and 5/12 Doves.

2 The population could consists of individuals who all adopted a mixed strategy, playing Hawk with probability 7/12 and Dove with probability 5/12, choosing at random which strategy to play in each contest. Both these would produce stability in the game. If the population consisted of a different mixture of individuals from that in (1) or of individuals playing their strategies with different probabilities from those in (2), then there would be no equilibrium; either Hawk or Dove would enjoy a temporary increase in success until the population evolved to the ESS once more.

It is instructive to note that at the ESS, the average pay-off is 6.25 per contest. This is less than the average pay-off which individuals would enjoy if they all agreed to fight as Doves, namely 15! This makes clear the point we made in the opening paragraph to this chapter. The optimal strategy to maximize everyone's fitness would be an agreement among all to play Dove. However, this would not be stable because in a population of all Doves, Hawk mutants would invade. We expect evolution to lead to stable strategies because, in the words of Richard Dawkins, 'They are immune to treachery from within'.

HAWK, DOVE AND BOURGEOIS

Now imagine another strategy in this game, 'Bourgeois'. With this strategy the individual plays 'Hawk, if owner' and 'Dove, if intruder'. In other words, it fights hard if it's the owner but always retreats if it's the intruder. Let us keep the same pay-offs as before and, for simplicity, imagine that a Bourgeois individual finds itself owner half the time and intruder half the time. With three strategies in the game, the pay-offs are indicated in Table 7.2

In this game, Bourgeois is an ESS. If all the population are playing this strategy, no one ever engages in escalated fights because when two individuals contest for a resource, one is owner and the other is intruder; the result is that the intruder always gives way. With everyone playing the Bourgeois strategy the average pay-off for a contest is +25. This is stable against invasion by Hawks, who would only get +12.5, and also stable against Doves, who would only get +7.5. In fact Bourgeois is the only ESS in the game in Table 7.2. If all the population played Hawk, both Dove and Bourgeois could invade and do better. If all the population played Dove, then both Hawk and Bourgeois could invade.

Table 7.2. The Hawk-Dove-Bourgeois game. After Maynard Smith (1976b)

(a) Pay-offs (as in Table 7.1)

Winner + 50   Injury   −100
Loser   0   Display − 10

(b) Pay-off Matrix: average pay-offs in a fight to the attacker.

Attacker	Opponent Hawk	Opponent Dove	Bourgeois
Hawk	−25	+50	+12.5
Dove	0	+15	+7.5
Bourgeois	−12.5	+32.5	+25

Notes:
1 The top left four cells are exactly the same as in Table 7.1.
2 When Bourgeois meets either Hawk or Dove we assume it is owner half the time and therefore plays Hawk, and intruder half the time and therefore plays Dove. Its pay-offs are therefore the average of the two cells above it in the matrix.
3 When Bourgeois meets Bourgeois on half the occasions it is owner and wins while on half the occasions it is intruder and retreats. There is never any cost of display or injury.

SIMPLE MODELS AND REALITY

These models are so simple in comparison with what actually happens in nature that we may well ask what use they can possibly be in helping us to understand real animal contests. There are three main conclusions.

1 The most important point is that the best fighting strategy for any one individual must depend on what other competitors are doing, because the pay-offs for employing a strategy will be frequency dependent. Is Hawk a good strategy? The answer is yes if the population consists mainly of Doves and no if the population consists mainly of Hawks. Instead of asking whether a strategy is a *good* strategy, we should really instead be asking is it a *stable* strategy, or ESS?

2 The ESS will depend on the strategies in the game. In the first example, where there were just two strategies, the ESS was a mixture of Hawk and Dove. However, when we introduced another strategy, Bourgeois, into the game the ESS solution changed; Bourgeois turned out to be a pure ESS.

These three strategies are undoubtedly too simple to represent in detail the strategies that animals adopt in the wild. Nevertheless they are plausible alternatives which we could regard as simplified versions of strategies we might see in nature. It is interesting to find that neither Hawk nor Dove is an ESS in our simple games, but some mixture of Hawk and Dove behaviour can be stable. This is exactly what we see in real animals, a mixture of display and fighting. Because of its simplicity, we don't expect to actually see Hawk and Dove strategies in nature or to test the model in Table 7.1. The model should be used simply for gaining some insights into how contest behaviour might evolve.

More detailed models will have to consider a greater variety of more complicated strategies and we will have to rely on biological intuition to help us define the range of possibilities. ESS models cannot tell us what strategies will evolve, only what will be stable given a defined set of alternatives. There is no doubt that an animal with a machine gun would invade the Hawk-Dove game and soon spread, but until we see real animals behaving like this there is little point in putting the strategy into our models.

3   The ESS will also depend on the values of the pay-offs in the game. If we changed the pay-offs in Table 7.1 then the ESS mixture of Hawks and Doves would change. In fact, in this game it can be shown that as long as the cost of injury exceeds the value of winning then the ESS is a mixed one. The relative amount of Hawk-like behaviour we would expect will depend on the costs and benefits in the contest. If the value of the resource was greater than the cost of injury then pure Hawk is an ESS.

Therefore if game theory models are going to make precise predictions about contest behaviour, we will not only need to know the range of possible strategies but also the costs and benefits for the pay-off matrix. In the real world, ecological circumstances such as resource abundance and competitor density will determine the pay-offs in the evolutionary game. In practice it is not going to be easy for the field worker to go out and measure on a common scale (effect on fitness) the cost of displaying, the cost of serious injury and the value of winning. However we can see in an intuitive way that fighting strategies in nature vary depending on the value of the resource.

Examples of animal contests

(a) Serious fights

Conventional display does not always occur; sometimes fights are fierce and there is injury or death. Some weapons, such as horns and antlers, have evolved in relation to their efficiency in attack and defence (Geist 1966). In the musk ox, from 5 to 10 per cent of the adult bulls may die each year from fights over females (Wilkinson & Shank 1977) and figures for mule deer indicate that up to 10 per cent of males more than 1.5

years of age show signs of injury each year (Geist 1974). Nar-whals (*Monodon monoceros*) use their tusks for fighting and in one study over 60 per cent of the adult males had broken tusks; some had tusk tips embedded in their jaws and most adult males were covered in head scars (Silverman & Dunbar 1980). Smaller animals may also fight viciously. Some male fig-wasps have large mandibles which are able to chop another male in half. When several males occur inside the same fig fruit they may engage in lethal combat for the opportunity to mate with females in the fig (Hamilton 1979). One such fruit was found to contain 15 females, 12 uninjured males and 42 damaged males who were dead or dying from fighting in-juries. Damage included legs, antennae and heads completely bitten off, holes in the thorax, and eviscerated abdomens.

In all these examples the value of the resource probably exceeds the cost of injury. We would therefore expect Hawk-like strategies because failure in a contest could mean failure to pass on genes to future generations; individuals are really fighting for genetic life or death. Where, on the other hand, the resource is less valuable, or where the costs of fights are very high, then would we expect Bourgeois (owner wins — intru-der retreats) to evolve as a strategy for settling contests?

(b) Respect for ownership

One possible example comes from a study of lions by Packer and Pusey (1982). A coalition of several males (up to 7) de-fends the pride (Fig. 7.4a). Sometimes the males are close relatives, often brothers, but 42% of the coalitions contain unrelated males. Competition among males within the pride for access to oestrus females consists of competition for tem-porary ownership of a female. When a female comes into oestrus, one male forms a consortship with her, guarding her from the approaches of other males and copulating with her frequently for a period of about 4 days (see Fig. 7.4b and chapter 1).

What determines which males get consortship with a parti-cular female? Contests between unrelated males were no more intense than between related males. Two important factors in settling contests were age and size; small males and very old or very young (less vigorous) males were less likely to gain access to an oestrus female. Among prime males, however, the main factor influencing who won a contest was simply respect for ownership. Usually the first male to encounter an oestrus female guarded her and the other males then gave way. Most serious fights occurred when ownership was undecided, for example when two males simultaneously came into the vicin-ity of an unconsorted and potentially oestrus female, or when

(a)

(b)

Fig. 7.4. (a) A group of male lions with a pride of females, photo by Jonathan Scott, Planet Earth Pictures. (b) Lions mating, photo by © Craig Packer.

ownership was unclear, for example when the consorting male moved further from the female than a rival male. There was then usually a race between the males to arrive first at a female. On arrival the loser would defer to the winner, even though in subsequent consortships the roles might be reversed with the previous consorting male now a rival to the new consort. Lions may use this Bourgeois-like strategy to settle contests because the costs of serious fights are so high and the benefits relatively low (p. 8).

Another example where ownership settled contests is a study of competition for territories among male speckled wood butterflies (Davies 1978). Observations showed that contests were brief and always won by the owner, and experiments showed that in a contest between two individuals the outcome could be reversed simply depending on who was the owner at the time. Finally, when two contestants were tricked experimentally into simultaneous ownership of a territory, an escalated contest ensued. In this last experiment the asymmetry of owner versus intruder was absent and both contestants behaved like Hawks. In another study, however, (Wickman & Wiklund 1983) contests were longer and appeared to involve assessment, with intruders sometimes beating owners. Like the lions, therefore, contests may sometimes be settled by fighting ability and sometimes by ownership.

(c) The influence of resource value

The observation that owners tend to win contests versus intruders does not, of course, necessarily mean that the Bourgeois strategy is being adopted. Owners may win for at least three reasons.

Hypothesis 1. They are better fighters. This may be the reason why they got ownership in the first place!

Hypothesis 2. They have more to gain from a fight and so are prepared to fight harder. If you were attacked in the street you would probably fight harder if you had a lot of money in your pocket than if you had very little. Similarly, owners may value the resource more than intruders. For example in disputes over a territory the owner will be familiar with the territory boundaries and will know the location of the good feeding and nesting sites.

Hypothesis 3. Arbitrary asymmetry of ownership settles the contest. Ownership is simply a conventional settlement, as in the Hawk-Dove-Bourgeois game, similar to the conventions we use in deciding positions in a queue at a supermarket or occupancy of a seat on a bus.

Krebs (1982) tested these three hypotheses in a study of territorial defence by male great tits *Parus major*. He removed owners and kept them in a cage. Newcomers took over the vacancy and then, after various time intervals the original owner was rereleased onto the territory. The prediction from hypothesis 1 is that the original owner will win back his territory, through his superior fighting ability. Hypothesis 2 predicts a gradual reversal of dominance in favour of the newcomer with increased time on the territory, which allows the newcomer to learn the territory characteristics and so gain greater value from winning a contest. Hypothesis 3 predicts that the newcomer will win simply because it is now in the role of owner. The data supported the second hypothesis, with newcomers escalating and being more likely to win after they had had time to learn the territory characteristics and settle boundaries with the neighbours.

(d) Contests of strength

So far we have considered, in a general way, how resource value and the behaviour of other individuals will influence fighting behaviour. However, as we saw from the lion example another important factor is the fighting ability or strength of the various contestants. When an individual fights for a re-source its decision on how to fight must depend not only on the value of the prize but also on its own strength compared with that of its opponent. Animals in a population will differ in their strength and the costs of fighting a large individual will certainly be greater than those of fighting a small one. Therefore we might expect displays to signal fighting ability and so allow the contestants to settle their dispute quickly without recourse to a costly fight (Parker 1974). Further-more, displays used in assessment should be reliable signals of strength, otherwise weak individuals would be able to mimic the signal and gain an advantage.

In the autumn, red deer stags (*Cervus elaphus*) compete with each other for females. Reproductive success depends on fighting ability; the strongest stags are able to command the largest harems and enjoy the most copulations. Although fight-ing brings great potential benefits, it also entails serious costs. Almost all males suffer some slight injuries, and between 20 and 30 per cent of stags will become permanently injured sometime during their lives, through broken legs or being blinded by an antler point for example. Competing stags mini-ize fighting costs by assessment of each other's fighting poten-tial and so avoid contests with individuals they are unlikely to beat (Clutton-Brock *et al.* 1979; Clutton-Brock & Albon 1979).

In the first stage of the display, the harem holder and chal-lenger roar at each other (Fig. 7.5a). They start slowly at first and then escalate in rate. If the defender can roar at a faster rate then the intruder usually retreats. Roaring is a good signal of fighting ability because to roar well a stag has to be in good physical condition. At the peak of the rut, harem holders may be roaring at intruders throughout the day and night and,

(a)

(b)

(c)

Fig. 7.5. Stages of a fight between two red deer stags. The harem holder roars at the challenger (a). Then the pair engage in a parallel walk (b). Finally they interlock antlers and push against each other (c). Photos by Tim Clutton-Brock.

because they do not feed much during this time, they show a steady decline in body weight. Some stags literally become 'rutted out' and are unable to roar strongly. Their females may then be taken over by other stags. In the second stage of a contest if a challenger outroars the defender, or matches him, then he approaches and both stags engage in a parallel walk (Fig. 7.5b) This presumably enables them to assess each other more closely. Many fights end at this stage, but if the contestants are still equally matched then a serious fight ensues where they interlock antlers and push against each other. Body weight and skillful footwork are important determinants of victory, but there is a chance that even the winner may get injured. The important point is that these escalated fights are rare, and most contests are settled at an earlier stage by displays.

Many animal contests proceed like this and are direct or indirect trials of strength. Buffalos charge at each other and assess their fighting potential in head-on clashes (Sinclair 1977). Beetles engage in pushing contests in which the larger one emerges as victor (Eberhard 1979). Male frogs and toads have wrestling matches where the large ones win the best territories or the most females (chapter 8).

In many species of frogs and toads the pitch of a male's croak is closely related to his body size; the larger the male, the larger the vocal cords and so the deeper the croak. Common toads (*Bufo bufo*) assess the body size, and hence fighting potential, of their rivals by the pitch of their croaks. An attacker is much less likely to attempt a take-over of a female when deep croaks are broadcast from a loudspeaker next to the pair than when high pitched croaks are played (Fig. 7.6).

Often, both strength and ownership are cues used to settle contests. When crabs compete for burrows (Hyatt & Salmon 1978) or when spiders compete for webs (Riechert 1978, 1979) the larger individuals usually win the fights, but if the contestants are equal in size then it is often the owner who wins. In conclusion, the game theory models we discussed earlier are too simple because in the real world animals will not play fixed strategies but, rather, will adopt strategies conditional on the fighting potential of their opponents.

Contests with differences in both resource value and fighting ability

So far we have identified both resource value and fighting ability as important factors influencing the outcome of contests. In many contests the opponents will vary both in fighting ability and how much they value the resource. How will fighting strategies evolve in these cases?

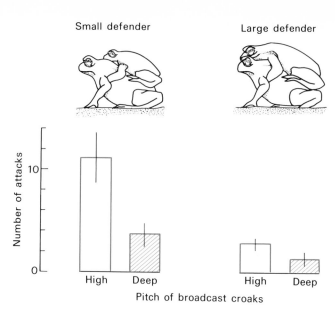

Fig. 7.6. An experiment on fighting assessment in toads, *Bufo bufo*. Medium-sized males attack either small or large paired males, which are silenced by means of a rubber band passing through their mouths. During an attack, tape recorded croaks are broadcast from a loudspeaker next to the pair. For both sizes of defender there are fewer attacks when the deep croaks of a large male are played than when the high pitched croaks of a small male are broadcast. Therefore croak pitch is used to assess a rival's body size, which is a good predictor of fighting ability because large males are more difficult to displace. Croaks cannot be the only assessment cue, however, because for either croak pitch there are fewer attacks at large defenders. The strength of a defender's kick may also be important (Davies & Halliday 1978).

THEORY: THE ASYMMETRIC WAR OF ATTRITION

The question of whether we expect to see contests in nature settled simply by the arbitrary convention of ownership, as opposed to real differences in fighting ability, depends on how individuals can vary their degree of risks of injury in a fight. In the Hawk-Dove-Bourgeois game, contestants can engage in conflicts at only two levels of risk, namely display (low risk) and escalated fighting (high risk). This discontinuity is crucial for the generation of the Bourgeois strategy as an 'uncorrelated' asymmetry to settle contests. (Uncorrelated means ownership need have no correlation with fighting ability or resource value, for example; it is simply an arbitrary convention to settle disputes without escalation.) If the contestants can control their degree of risk along a continuum, however, and therefore adjust their escalation in relation to the value of the resource, then the uncorrelated Bourgeois rule turns out to be no longer evolutionarily stable.

A contest in which there are differences in fighting ability, and in which individuals can vary their cost of escalation along a continuum is referred to as an 'asymmetric war of attrition'. The theory behind this game is complex and brave readers are referred to Parker & Rubenstein 1981 and Hammerstein & Parker 1982. The ESS for this kind of contest is for individual A to give up and retreat when

$$\frac{V_A}{K_A} < \frac{V_B}{K_B}$$

where V is the value of the resource to contestants A and B, and K is the rate at which the two individuals accrue costs during a contest. K will be related to fighting ability; good fighters will accrue costs at a slower rate than poor fighters.

Therefore if ownership is used to settle contests of the war of attrition kind, it is because there are asymmetries in V and/or K which favour residents, and residents are willing to escalate further. A verbal summary of this rule for settling contests is 'Withdraw if you will run out of fitness budget before your opponent'. This is the common sense solution to the game; the winner is always the one with the greatest benefit: cost ratio.

DATA: CONTESTS BETWEEN MALE SPIDERS

Steven Austad (1982, 1983) studied contests for females between male bowl and doily spiders, *Frontinella pyramitela*. Once mature, the males feed very little and spend most of their time going around female webs in search of mating opportunities. They only live for about 3 days in the final stage and so the way they compete for females is vital for their reproductive success. Laboratory experiments showed that there was first male priority in sperm competition. These experiments involved allowing two males to mate with a female, one normal male and one irradiated male. The eggs fertilized by the normal male developed normally while the eggs fertilized by the irradiated male failed to develop because of genetic abnormalities induced by the irradiation. If the sequence of matings was first the normal, then the irradiated male, 95% of the eggs developed (therefore 95% were fertilized by the first male). If the sequence was first the irradiated and then the normal male, only 5% of the eggs developed (therefore 95% were again fertilized by the first male). Note that this is the reverse of what happens in dungflies, where the *second* male has the advantage, (chapter 3). The second finding was that there was no sperm displacement as a result of a take-over by another male; if the first male had transferred enough sperm to fertilize 30% of the female's eggs he would fertilize these even if another male then took over and copulated. This means that if a male lost a fight half way through the mating, he would lose only future possible gains; the take-over would not affect his past success.

(a) Measuring resource value, V

Copulation proceeds as follows. First there is a pre-insemination phase during which the male assesses the female, for example to discover whether she is sexually mature. Then copulation itself occurs, the male passing sperm from his pedipalps to the female. The proportion of eggs fertilized increases at a rapid rate early on in copulation and then at a slower rate as time goes on (Fig. 7.7). The result is that the female is of different value to the male at different stages of the mating sequence. The average number of eggs fertilized, for all females encountered by a male, is 10 eggs, so when a male first meets a female her expected value is 10. Imagine this female is in fact a virgin, which is the most profitable female that a male encounters, being worth on average 40 eggs. During the pre-insemination phase the value increases to 40 as the male discovers that the female is a virgin. Copulation then begins and the value of the female declines as her eggs are fertilized. After 7 min of copulation, 90% of the eggs have been fertilized so the female is now worth only 4 eggs. After 21 min of copulation, the value is less than 1 egg because 99% of the eggs have now been fertilized (Fig. 7.7).

Imagine now that an intruding male comes along at different stages in this mating sequence by the resident male. Intruders are unaware of the state of the female unless they themselves are able to assess her. Therefore the intruder will always assume that the female is of average value, 10 (Fig. 7.7). This meant that Austad could vary V in a fight simply by introducing intruders at different stages. At the end of the pre-insemination phase, for example, V is higher for the resident, who has the knowledge that the female is a virgin and particularly valuable. After 7 min of copulation, V is higher for the intruder, because only the resident has the knowledge that most of the eggs are already fertilized. We may expect residents to fight harder for the female at the end of the pre-insemination phase than after 7 min of copulation.

(b) Measuring the costs of a fight, K

This is the expected reduction in lifetime reproductive success per minute of fighting. Austad found that costs, measured

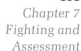

Fig. 7.7. (a) The proportion of eggs fertilized by a male bowl and doily spider during the process of copulation. The stages A to E were used for the introduction of intruders in staged fights for the female. (b) The value of the female to the owner and intruder varies depending on the stage at which the intruder is introduced. From Austad (1983).

in terms of the probability of serious injury (which is a measure of loss of future reproductive success), were linearly related to the duration of the contest. In fights, the males grab each other and grapple with their jaws and legs. Long fights led to body wounds, loss of legs and eventually certain death. A male's fighting ability depended on his body size. Therefore it was possible to vary K by staging encounters between males of different size.

(c) *Results of staged encounters*

Many contests were arranged, varying V and K as described above. The results were as follows:

1 When V was the same for both males (both introduced simultaneously at the start of the pre-insemination phase) fights were settled by differences in K. Larger males won 82% of the fights and grapple duration increased with closer matching of body size between the contestants presumably because it then took longer to assess differences in fighting ability (Fig. 7.8).

2 When K was the same (same size opponents), fights were settled by differences in V. Thus residents fought for longer and were more likely to win at the end of the pre-insemination phase, whereas they fought for less time and thus intruders were likely to win at the stage of 7 min of copulation (Fig. 7.9).

3 When residents were smaller than intruders, they persisted more in fights when V was greater. For example, at the end of the pre-insemination phase, when V was at its highest for the resident, small residents would persist in fights for so long that 90% of the grapples led to them sustaining serious injury. After 7 min of copulation, however, they gave up much sooner and only 30% of the fights led to injury.

4 The most serious escalations occurred where V/K was identical for both contestants. In these cases neither male was prepared to give way and most of the fights ended in serious injury and death for one of them. This occurred, for example, in fights between small residents versus large intruders at the

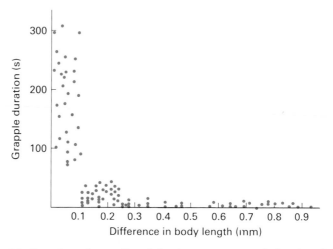

Fig. 7.8. Duration of grappling fights between two male bowl and doily spiders introduced simultaneously onto the webs of virgin females. After Austad (1983).

Fig. 7.9. Contests between two male bowl and doily spiders who have the same fighting ability (are the same size). Who wins tends to be determined by differences in resource value, V. For predicted curve, see values of V for residents and intruders in Fig. 7.7b. From Austad (1983).

end of the pre-insemination phase. Here V and K were both large for the resident, and were both small for the intruder.

Austad's experiments show clearly that both V and K influence duration of contests and who wins. Similar results have been obtained by Sigurjonsdottir and Parker (1981) who studied struggles between male dungflies for the possession of females on cowpats.

Fighting for dominance in flocks

Some animals live in groups and individual differences in fighting ability determine who will have priority of access to food or mates. Many displays within groups probably involve reliable assessment of strength; however the exact link between the display and fighting ability is not always obvious.

In the Harris sparrow (*Zonotrichia querula*) there is enormous plumage variability in winter when the birds go around in flocks searching for food (Fig. 7.10). Individuals with the blackest plumage are the dominants and they always displace pale birds from food supplies (Rohwer & Rohwer 1978). If blackness signals dominance then why don't the subordinates

Fig. 7.10. Plumage variability in the Harris sparrow. The darker males are dominant in the flocks and win most of the fights.

grow blacker feathers and enjoy a rise in status? In the spring all the birds come into breeding plumage and even the pale males develop black colouration, so the dark feathering is not an obviously uncheatable signal of strength like the roar of a rutting stag or the deep croak of a large male toad.

The Rohwers attempted to create 'cheats' in the flock by experimental treatment of subordinate birds (Table 7.3). In the first experiment they simply painted subordinates black. These treated birds were attacked by others and failed to rise in status. Secondly, some subordinates were injected with testosterone but their pale plumage was left unaltered. These birds behaved more aggressively and attempted to assert their dominance but they did not rise in status because their opponents did not retreat during disputes. Finally, subordinates were painted and injected so that they both looked and behaved like dominants. This time the cheating worked: the birds won more fights and were respected by others in the flock.

Table 7.3. Summary of the experiments on signals of dominance in flocks of the Harris sparrow. After Rohwer & Rohwer (1978).

Experimental treatment of subordinates	Look dominant	Behave as dominant	Rise in status?
1. Paint black	Yes	No	No
2. Inject testosterone	No	Yes	No
3. Paint black and inject with testosterone	Yes	Yes	Yes

Therefore the earlier attempt to create a cheat by painting alone did not fail simply because the other birds didn't like the paint! It failed because although the darkened birds looked dominant, they did not behave so. The conclusion is that plumage alone is not a passport to high status in the flock; the signal must be backed up by dominant behaviour as well. Nevertheless the dark plumage still acts as a badge to reduce the amount of fighting in the flock. In another experiment dominant birds were bleached so that they signalled low rank. They were attacked a lot by others who tried to displace them, they didn't yield however and eventually exerted their superiority but only after a lot of squabbling.

It is clear from the experiments above that cheating is prevented because a subordinate cannot become dominant through darker plumage alone. But why can't a low ranking bird increase its testosterone level as well? This same question arises from work on red grouse (Watson 1970) where males were injected with testosterone and immediately doubled their territory size and increased the number of females they attracted. Presumably the answer to this problem is that an increase in androgen levels would push the behaviour of a subordinate bird beyond its real capabilities. Although it may gain a short term increase in success, perhaps the long term effect would be that all its strength is sapped and it suffers a decrease in fitness (Silverin 1980, Røskaft *et al.* 1986).

Not all cases of plumage variability are concerned with status signalling. For example, a wading bird, the turnstone (*Arenaria interpres*) has very variable plumage. Whitfield (1986) removed male territory owners and replaced them with models painted to either resemble the removed owner or to be lighter or darker. Neighbours were much less likely to attack the models if they looked like the removed owner than if they looked different. However the attack intensity towards models that looked different was not related to whether the models were light or dark (Fig. 7.11). In this species, plumage variability may simply facilitate individual recognition.

Information transmission and animal contests

During a contest there are two sorts of information that a display could signal to an opponent.

(a) Information about strength

For example 'I am 2 m tall'. This kind of information is often present in animal displays. It is difficult to fake and we have already encountered examples, such as roaring and deep croaks, which are reliable signals of fighting potential.

(b) Information about intentions

For example, 'I am going to attack you' or 'I am going to display for 3 min'. This kind of information will be easier to fake and game theory predicts that it will not be transmitted in displays. Returning to the war of attrition model described

Fig. 7.11. Plumage variability among male turnstone. For this species experiments suggest that the variability is not concerned with status signalling but may simply facilitate individual recognition, see text. From Whitfield (1986).

at the beginning of this chapter, imagine that individual A chooses a cost m_A which is larger than the cost m_B chosen by B. Would it not pay both to signal their intentions at the start of the contest? B would then retreat at once and both contestants would be better off; A would get V instead of $V-m_B$, and B would get O instead of $-m_B$. The problem with signalling intentions is, of course, that of cheating. In a population of honest signallers, it would pay an individual to lie and signal a higher cost so that it could win every time. Eventually everyone would be lying and it would no longer pay to believe the message. Mutants which ignored the signals and just obeyed the war of attrition distribution would do better. The end result, the ESS, is for no one to give any information about intentions and for all to display according to the war of attrition model. As Maynard Smith points out, all this really means is that you should not believe what an opponent at poker tells you!

The prediction from this line of argument, that displays should not evolve to transmit information about intentions, is to a large extent borne out by the ethological literature. In blue tits (*Parus caeruleus*) a particular display can be followed by several different types of behaviour and there is no one posture that is followed with a high probability by attack (Caryl

1979). In the Siamese fighting fish (*Betta splendens*) there is no consistent difference between the behaviour of the eventual winner and eventual loser until near the end of the contest, so it is impossible to predict which individual will win from its displays (Simpson 1968). The evidence suggests, therefore, that information about intentions is not transmitted.

However there are two lines of research that need further investigation. Firstly, it is clear that animals use a great variety of displays in agonistic encounters and at present game theory has little to say about how the variety might have evolved. Secondly, although it is clear that no one display is a good predictor of future behaviour, sequences of displays may be important. Film analysis of Muhammed Ali in the boxing ring shows that about half of his left jabs are of shorter duration than the visual reaction time. Nevertheless, his opponents still managed to evade most of the punches. They could not have done this by responding to the preliminary stages of the jab. Somehow, therefore, they must have decoded Ali's behavioural sequences so as to predict the punch before it actually starts its movement (Stern 1977). We return to the discussion of information and variability in displays in chapter 14.

Summary

Game theory is a useful way to analyse animal contests because the best fighting strategy for one animal to adopt depends on what others in the population are doing. Evolution is expected to lead to an evolutionarily stable strategy, or ESS. The war of attrition model applies to examples of scramble competition (e.g. among male dungflies waiting for females on a cowpat) where the pay-offs for different waiting times are frequency dependent. In pair-wise contests, however, there is often direct confrontation between opponents. The Hawk-Dove-Bourgeois model shows how conventional fighting may be an ESS. Most real contests involve assessment and are settled by differences in fighting ability (red deer), resource value (great tits) or both (spiders). Plumage variability in birds may be concerned with status signalling or individual recognition. Although animals are expected to signal their fighting ability by displays, they are not expected to signal their intentions.

Further reading

Parker (1974) discusses assessment strategies, and Caryl (1980) and Maynard Smith (1982) review the application of game theory to contests. Yasukawa and Bick (1983) describe some neat experiments on dominance hierarchies in junco flocks. Caldwell (1985) shows that mantis shrimps can distinguish individuals on the basis of odour and are more likely to avoid odours of opponents that had defeated them previously than odours of opponents they had beaten.

Topics for discussion

1 How has game theory helped us to understand animal contests?

2 Why do territorial residents win?

3 Is assessment always based on reliable signals?

Chapter 8. Sexual Conflict and Sexual Selection

Ethologists used to view courtship rituals and mating as harmonious ventures in which male and female co-operated to propagate their respective genes. Admittedly some animals were obviously not co-operative, for example the praying mantis in which the female eats the male during mating, but on the whole courtship was seen as serving functions of common interest to male and female: to 'synchronize the sexual arousal of the sexes', to 'establish the pair bond' to 'allow species identification', and so on. However this view is no longer so widely held and more emphasis is placed on the idea that there are conflicts of interest between male and female in courtship and mating. The sexes are seen as forming an uneasy alliance in which each attempts to maximize its own success at propagating genes. They co-operate because both pass on their genes via the same progeny and therefore each has a 50 per cent stake in the survival of the offspring. But choice of mating partner, provisioning of the zygote with food, and caring for the eggs and young are all issues over which the sexes may disagree. The outcome of this sexual conflict is often more akin to exploitation by one sex of the other than to mutual co-operation.

In order to understand why sexual reproduction should be viewed in this way, we have to go right back to the beginning, to the fundamental difference between male and female.

Males and females

Sexual reproduction entails gamete formation by meiosis and the fusion of genetic material from two individuals. It almost always, but not invariably, involves two sexes called male and female. In higher animals the sexes are often most readily distinguished by external features such as genitalia, plumage, size, or colour, but these are not fundamental differences. In all plants and animals the basic difference between the sexes is the size of their gametes; females produce large, immobile, food-rich gametes called eggs, while male gametes or sperm are tiny, mobile, and consist of little more than a piece of self-propelled DNA. Sexual reproduction without males and females occurs in many protists such as *Paramecium* where the 'gametes' which fuse during sex are of the same size. This is

162

Chapter 8
Sexual
Conflict
and Sexual
Selection

referred to as *isogamous* sexual reproduction. The fusion of two gametes of unequal size, one large and one small is, however, much commoner and occurs in virtually all sexually reproducing multicellular plants and animals. It is called *an-isogamous* sex.

It is thought that anisogamy evolved from isogamy by an evolutionary process in which smaller than average gametes successfully parasitized those that were larger than average. This eventually led to the two distinct specializations we see today: small active sperm and large passive eggs (Parker *et al.* 1972).

As we shall show in the rest of this chapter, the fundamental asymmetry in gamete size and associated investment in offspring has far-reaching consequences for sexual behaviour. Because females put more resources than do males in each offspring, male courtship and mating behaviour is to a large extent directed towards competing for and exploiting female investment. Where the usual difference between the sexes in parental investment is reversed, males providing more care than females, so the roles in sexual competition are reversed and females compete for access to males.

FEMALES AS A SCARCE RESOURCE

Anisogamous sexual reproduction, then, involves parasitism of a large egg by a small sperm. Females produce relatively few large gametes and males produce many small ones. In addition, females often invest more than males in other forms of care. Because of this males can potentially fertilize eggs at a faster rate than they are produced (illustrated by the fact 5 ml of human semen contains enough sperm to fertilize in theory eggs amounting to twice the population of the USA), and females are therefore a *scarce resource* for which males compete. Even in species where males temporarily deplete their sperm supply when offered a surfeit of females, their potential for producing offspring is greater than that of females (Nakatsuru & Kramer 1982).

A male can increase its reproductive success by finding and fertilizing many different females, but a female can only increase her success by turning food into eggs or offspring at a faster rate (Fig. 8.1). This point is graphically demonstrated by mammals such as man, in which a female spends many months producing a single child, during which time a male could potentially fertilize hundreds of other mates. Only by speeding up her production of young can the female have more children in a lifetime. The same argument holds whenever females invest more than males in each offspring, whether the

163

*Chapter 8
Sexual
Conflict
and Sexual
Selection*

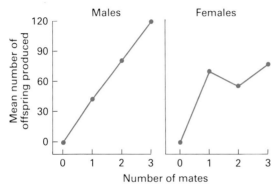

Fig. 8.1. A.J. Bateman (1948) put equal numbers of male and female fruit flies (*Drosophila melanogaster*) in bottles and scored the number of matings and offspring produced by each individual, using genetic markers to assign parentage. For males reproductive success goes up with number of matings, for females it does not, beyond the first mating. After Trivers (1985).

investment is in the form of food in the egg or care of the eggs and young later on.

This point was neatly summarized by Robert Trivers (1972), who was the first person to emphasize the relationship between the investment of resources in gametes and other forms of care, and sexual competition. He wrote 'Where one sex invests considerably more than the other, members of the latter will compete among themselves to mate with members of the former'. The term 'investment' was used by Trivers to refer to the effort put into rearing an individual offspring from the parent's limited pool of resources. The sum of parental investment in all offspring during a parent's lifetime is referred to as 'parental effort'. Females generally put most of their reproductive effort into 'parental effort' while males put most of theirs into 'mating effort' (Fig. 8.2).

The consequence is that males have a much greater reproductive potential than females (Table 8.1) and are therefore under strong selection to be good at seeking out and competing for females: the pay-off for a successful male in terms of offspring fathered is enormous. Much of male reproductive behaviour can be understood with these ideas in mind.

THE SEX RATIO

If one male can fertilize the eggs of dozens of females why not produce a sex ratio of, say, one male for every 20 females? With this ratio the reproductive success of the population would be higher than with a 1:1 ratio since there would be more eggs around to fertilize. Yet in nature the ratio is usually

164

Chapter 8
Sexual
Conflict
and Sexual
Selection

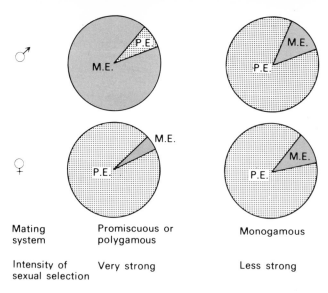

Fig. 8.2. The total resources of time and energy used by an animal in reproduction is referred to as reproductive effort. This is represented by a circle. Reproductive effort can be partitioned into parental effort (provisioning and rearing offspring) and mating effort (acquiring mates). These are represented by the stippled and coloured areas of the circles respectively. In general, males put relatively more into mating effort than do females, but this varies between species. The intensity of sexual selection (see p. 169 for a definition) therefore also varies. The differences in relative parental effort of the sexes is often related to the mating system. In monogamous species male and female effort is more similar than in polygamous and promiscuous species (see chp. 9). After Alexander and Borgia (1979).

very close to 1:1 even when males do nothing but fertilize the female. As we saw in chapter 1, the adaptive value of traits should not be viewed as being 'for the good of the population', but 'for the good of the individual', or more precisely 'for the good of the gene'. As R.A. Fisher (1930) first realized the 1:1 sex ratio can readily be explained in terms of selection acting on the individual; his argument is simple but subtle.

Suppose a population contained 20 females for every male. Every male has 20 times the expected reproductive success of a female (because there are on average 20 mates per male) and therefore a parent whose children are exclusively sons can expect to have almost 20 times the number of grandchildren produced by a parent with mainly female offspring. A female biassed sex ratio is therefore not evolutionarily stable (p. 136) because a gene which causes parents to bias the sex ratio of their offspring towards males would rapidly spread, and the sex ratio will gradually shift towards a greater proportion of males than the initial 1 in 20. But now imagine the converse. If males are 20 times as common as females a parent producing

Table 8.1. In polygamous or promiscuous species males have a much higher potential reproductive rate than females. The data for Man came from the Guinness Book of Records: the male was Moulay Ismail the Bloodthirsty, Emperor of Morocco, the woman had her children in 27 pregnancies. The data for elephant seals are from Le Bouef & Reiter (in press), for red deer from Clutton-Brock *et al.* (1982). In the monogamous kittiwake, where male and female invest similarly in each offspring, the difference in maximum reproductive output is negligible (Clutton-Brock 1983)

165

Chapter 8
Sexual
Conflict
and Sexual
Selection

Species	Maximum number of offspring produced during lifetime	
	Male	Female
Elephant seal	100	8
Red deer	24	14
Man	888	69
Kittiwake gull	26	28

only daughters will be at an advantage. Since one sperm fertilizes each egg, only one in every 20 males can contribute genes to any individual offspring and females therefore have twenty times the average reproductive success of a male. So a male biassed sex ratio is not stable either. The conclusion is that the rarer sex always has an advantage, and parents which concentrate on producing offspring of the rare sex will therefore be favoured by selection. Only when the sex ratio is exactly 1:1 will the expected success of a male and a female be equal and the population stable. Even a tiny bias favours the rarer sex: in a population of 51 females and 49 males where each female has one child, an average male has 51/49 children. This *average* value is the same whether one male does most of the fathering or whether fatherhood is spread equally among the males.

The argument that the sex ratio should be 1:1 can be refined by re-phrasing it in terms of investment. Suppose sons are twice as costly as daughters to produce because for example, they are twice as big and need twice as much food during development. When the sex ratio is 1:1 a son has the same average number of children as a daughter. But since sons are twice as costly to make they are a bad investment for a parent: each of its grandchildren produced by a son is twice as costly as one produced by a daughter. It would therefore pay the parents to concentrate on making daughters. As the sex ratio swings towards a female bias, the expected reproductive success of a son goes up until at a ratio of two females to every male an average son produces twice the number of children produced by an average daughter. At this point sons and daughters give exactly the same return per unit investment; a son costs twice as much to make but yields twice the return.

166

Chapter 8
Sexual
Conflict
and Sexual
Selection

This means that when sons and daughters cost different amounts to make, the stable strategy in evolution is for the parent to *invest* equally in the two sexes and not to produce equal numbers. An example to illustrate this point is Bob Metcalf's (1980) study of the sex ratio of two species of wasp; *Polistes metricus* and *P. variatus*. In the former females are smaller than males, while in the latter they are similar in size. As predicted, the population sex ratio is biassed in *P. metricus* and not in *P. variatus*. In both species the investment ratio is 1:1.

The prediction that parents should invest equally in sons and daughters does not always hold, and demonstrations of these deviations from 1:1 investment are among the most convincing pieces of evidence that the sex ratio has evolved in the way suggested by Fisher. We will pick out some examples in the following paragraphs.

(a) Local mate competition

Fisher's theory predicts a different outcome when brothers compete with each other for mates (so-called 'local mate competition'). Suppose, for example, that two sons have only one chance to mate and that they compete for the same female. Only one of them can be successful in mating, so from their mother's point of view one of them is 'wasted'. This is an extreme example, but it illustrates the general point that when sons compete for mates their value to their mother is reduced. The mother should therefore bias her ratio of investment towards daughters. The exact degree of bias predicted by Fisher's theory depends on the degree of local mate competition. Extreme competition is to be expected in species with limited powers of dispersal (because brothers will stay together in the same place) and therefore it tends to be associated with inbreeding. In the extreme case of inbreeding, a mother 'knows' that all her daughters will be fertilized by her sons. The best sex ratio in this instance is to produce just enough sons to fertilize the daughters, since any other males will be wasted. The crucial difference between this and the earlier argument for a 1:1 sex ratio is that here the ratio of males to females in the rest of the population does not matter. A female biassed ratio within a brood will not give other parents a chance to benefit by concentrating on sons. An example which supports this prediction is the viviparous mite *Acarophenox* which has a brood of one son and up to 20 daughters. The male mates with his sisters inside the mother and dies before he is born (Hamilton 1967).

Jack Werren (1980) has tested the prediction that the degree of bias depends on the extent of local mate competition. He

studied the parasitoid wasp *Nasonia vitripennis* which lays
its eggs inside the pupae of flies such as *Sarcophaga bullata*. If
one female parasitizes a pupa, her daughters are all fertilized
by her sons and as predicted the sex ratio of her clutch of eggs
is biassed towards females. Only 8.7 per cent of the brood is
male. If a second female lays her eggs in the same pupa, what
should her sex ratio be? If she lays few eggs she should produce mainly sons, since the first female has laid predomin-
antly female eggs. But as the proportion of the total number of
eggs in the pupa that come from the second female increases,
the chance that sons of the second female will compete
for mates also increases. Therefore her brood should have a
female biassed sex ratio. Werren found exactly this pattern:
when the second female's clutch was 1/10 the size of the first
female's it contained only males, but when it was twice as
large as the first female's it contained only 10 per cent males,
and the quantitative details of the change in sex ratio with
relative brood size were much as predicted.

167

Chapter 8
Sexual
Conflict
and Sexual
Selection

(b) Local resource competition or enhancement

Anne Clark (1978) found that the South African prosimian
Galago crassicaudatus has a male biassed investment ratio
among its offspring. She pointed out that this could be explained by the species' life history. As with most mammals,
female *Galago* disperse less far than males, and often end up
competing both with their mother and with each other for rich
sources of food such as gum and fruit trees in the mother's
home range. This local resource competition among females
reduces their value as offspring: in the extreme case only one
daughter might be able to survive on the food available near
home, and so investment in other daughters would be wasted.

Exactly the opposite effect could arise if the sex that stays at
home, rather than hindering one another or their parents,
actually helps. As we shall see in chapter 12, in some bird
species, males but not females stay at home and help. The
consequence of this is to make males slightly more valuable
than females as an investment (since they help the parent in
its future reproduction) and hence a male-biassed investment
ratio might be expected (Emlen *et al.* 1986).

(c) Maternal condition

We saw in chapter 7 that male red deer compete for females
by prolonged roaring and antler wrestling contests. In these
contests it is an advantage for a male to be big, and size
depends among other things on how well fed the male was as
a youngster, which in turn depends on his mother's ability to

168

Chapter 8
Sexual
Conflict
and Sexual
Selection

compete for good sources of food and hence produce a plentiful supply of milk. In other words there is a direct link between a mother's competitive ability while lactating and her son's expected reproductive success. Now if a mother 'knew' that her sons would be highly successful harem holders it would pay her to invest heavily in sons rather than daughters: the pay-off in terms of grandchildren would be much greater. Similarly a mother who 'knew' that her sons would not grow up to be big and strong would do better to have daughters, since a daughter's future reproductive success does not depend so much on her mother's milk. Exactly this pattern has been found in red deer: dominant females who are able to gain access to good feeding sites while lactating and hence produce strapping sons, tend to have sons, while subordinate females have daughters (Clutton-Brock et al. 1984). It is not known how the sex ratio is adjusted by the mother in the red deer or the galagos studied by Clark, but the fact that they are adjusted is striking because agricultural geneticists have failed to select for adjustment of the sex ratio or to separate male and female sperm of domestic mammals (imagine the value of a female biassed sex ratio to the milk farmer!), and it has often been concluded that mammalian sex ratios are very inflexible (Williams 1979; Maynard Smith 1980). Adjustment of the sex ratio in hymenoptera such as the wasps studied by Werren is not at all a problem because the mother can determine whether an egg becomes male or female by whether or not she fertilizes it (chp. 13).

(d) Population sex ratio

When the population ratio of investment deviates from 1:1, a compensatory bias in favour of the rarer sex should occur. In Metcalf's study of P. metricus he found that some nests produced only male offspring. As explained in chapter 13 these offspring are the product of unfertilized eggs and are produced by workers in nests where the queen has died. In the remainder of the nests in the population Metcalf found a female biassed sex ratio, so that the ratio of investment in the population as a whole is 1:1.

Finally it is worth pointing out that the theory of sex ratios discussed here is one example of a more general theory of sex allocation (Charnov 1982). Other examples of the problem of allocation of resources to male and female reproduction include the division of resources into eggs and sperms by simultaneous hermaphrodites and the timing of sex change in sequential hermaphrodites (see chp. 10).

Sexual selection

169

*Chapter 8
Sexual
Conflict
and Sexual
Selection*

The combination of females investing more than males and 1:1 population sex ratio means that males usually compete for females. The potential pay-off for male success is high, so selection for male ability to acquire matings is very strong. Selection for traits which are solely concerned with increasing mating success is usually referred to as *sexual selection*. It can work in two ways: by favouring the ability of one sex (usually males) to compete directly with one another for fertilizations, for example by fighting (*intra-sexual selection*), or by favouring traits in one sex which attract the other (*inter-sexual selection*). Often the two kinds of selection act at the same time.

The intensity of sexual selection depends on the degree of competition for mates. This in turn depends on two factors: the difference in parental effort between the sexes (Fig. 8.2) and the ratio of males to females available for mating at any one time (referred to as the operational sex ratio). When parental effort is more or less equal, as for example in monogamous birds where both male and female feed the young, sexual selection is less intense than in species with very different levels of parental effort. This follows from the point made earlier that the sex making little investment competes for the sex which makes a large investment (p. 163). If equal numbers of the two sexes come into breeding condition at the same time, the degree of sexual selection is reduced because there is less chance for a few males to control access to very large numbers of females. In contrast, when females come into breeding condition asynchronously there is a chance for a small number of males to control many females one after the other. With such high potential pay-offs, sexual competition is very intense (see chp. 9).

ARDENT MALES

The most dramatic and obvious way in which males compete for mates is by fighting and ritualized contests, and often males have evolved weapons for fighting. Males may dispute over direct access to females or over places where females are likely to go, as for example when male speckled wood butterflies defend patches of sunlight (chapter 7). Fighting is often a risky business, as illustrated by the injuries sustained by red deer stags referred to in chapter 7. The most intense fights in many species occur when females are ready to be fertilized and once a male finds a female he often guards her (Fig. 8.3).

Males often compete in ways which are less conspicuous than fights, but are no less effective and often more bizarre.

170

*Chapter 8
Sexual
Conflict
and Sexual
Selection*

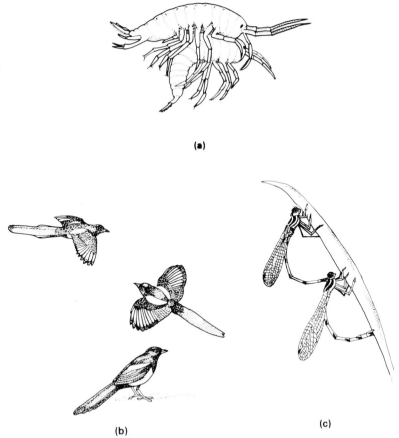

Fig. 8.3. Mate guarding as a form of sexual competition. (a) Precopulatory mate guarding in the freshwater amphipod *Gommarus*. The mature female in this species is ready to be fertilized immediately after she moults. Males guard females in the few days preceding this moult (Birkhead & Clarkson 1980). (b) Male European magpies (*Pica pica*) assiduously guard their mates against intruding males just before and during the period of egg laying (Birkhead 1979). (c) After copulation the male damselfly guards the female while she lays her eggs, by clasping her thorax with the tip of his abdomen in the 'tandem' position (Corbet 1962).

The invertebrates are a particularly rich seam of examples. Female dragonflies, as with many other insects, mate with a number of males and store the sperm in a special sac (the spermatheca) in the body for use at a later date. The males compete for fertilizations by trying to ensure that previous sperm is not used by the female. The penis of a male *Orthetrum cancellatum* is equipped with a barbed whip at the end which is used to scrape out of the female any sperm left by previous males before he injects new sperm into the sperm sac. *Crocethemis erythraea*, another dragonfly, uses an inflatable penis with a horn-like appendage to pack the sperm of

Fig. 8.4. Two sperm displacement mechanisms in Odonata. Photos by M. Siva-Jothy. *Crocothemis erythraea*: (a) Uninflated penis. (b) Inflated penis. The horn-like structure repositions sperm of previous males in the spermatheca. *Orthetrum cancellatum*: (c) The whip-like flagellum is everted during copula. (d) It carries barbs which remove sperm from the narrow ducts of the spermatheca (Siva-Jothy pers. comm.).

previous males into corners of the spermetheca (Fig. 8.4; see also Waage 1979).

In some invertebrates (especially insects) the male cements up the female's genital opening after copulation to prevent other males from fertilizing her. The males of *Moniliformes dubius*, a parasitic acanthocephalan worm in the intestine of rats, produces a chastity belt of this kind but in addition to sealing up the female after copulation, the male sometimes 'copulates' with rival males and applies cement to their genital region to prevent them from mating again (Abele & Gilchrist 1977). No less remarkable are the habits of the hemipteran insect *Xylocoris maculipennis*. In normal copulation of the species the male simply pierces the body wall of the female and injects sperm, which then swim around inside the female until they encounter and fertilize her eggs. As with the acanthocephalan worms, males sometimes engage in homosexual 'copulation'. A male *Xylocoris* may inject his sperm into a

171

172

Chapter 8
Sexual
Conflict
and Sexual
Selection

rival male. The sperms then swim inside the body to the victim's testes, where they wait to be passed on to a female next time the victim mates (Carayon 1974).

Competition between males to prevent each other's sperm from fertilizing eggs is sometimes referred to as 'sperm competition'. Another insect example was described in chapter 3: the sperm of a second male displaces that of the first male to mate with a female dungfly. Sperm competition also occurs in vertebrates. For example, during courtship male salamanders and newts deposit little sperm-capped rods of jelly (spermatophores) on the bottom of the pond and then try to manouever the female onto the spermatophore to achieve fertilization. In the salamander *Ambystoma maculatum* males compete by depositing their spermatophores on top of those of other males. The top spermatophore is the one that fertilizes the female's eggs (Arnold 1976).

A fourth example of the arcane methods of male-male competition found among invertebrates is the use of anti-aphrodisiac smells. Larry Gilbert (1976) noticed that female *Heliconius erato* butterflies always smell peculiar after they have mated. He was able to show experimentally that the scent does not come from the female herself, but is deposited by the male at the end of mating. Gilbert also found that the scent discourages other males from mating with the female, perhaps because it resembles a scent used by males to repel one another in other contexts.

RELUCTANT FEMALES

Since females in the great majority of species are the chief providers of resources for the zygote, they might be expected choose their mates carefully in order to get something in return. To put it another way, each egg represents a relatively large proportion of a female's lifetime production of gametes when compared with a sperm, so the female has more to lose if something goes wrong. Mating with the wrong species could cost a female frog her whole year's supply of eggs, but would cost the male very little apart from lost time — he could still go on to mate successfully with a member of the correct species the next day. Not surprisingly therefore, females are on the whole choosier than males during courtship. Choosiness extends not only to discriminating between species, but also to discriminating between males within a species. Females often select males on the basis of material resources they can offer and perhaps sometimes to obtain good genes for their offspring.

(a) Good resources and parental ability

173

Chapter 8
Sexual
Conflict
and Sexual
Selection

In many animal species males defend breeding territories containing resources which play a crucial role in the survival of a female's eggs or young (see also chp. 9). For example male North American bullfrogs (*Rana catesbiana*) defend territories in ponds and small lakes where females come to lay their eggs (Fig. 8.5). Some territories are much better for survival of eggs than others and these are the ones which females prefer. One factor which has an important influence on survival of eggs is predation by leeches (*Macrobdella decora*). Two environmental features of a territory influence leech predation: if the water is warm the eggs develop faster and are therefore exposed to predation for fewer days, and if the vegetation in the water is not too dense the eggs can form into a ball which the leeches find hard to attack. In territories with a dense mat of vegetation the eggs lie in a thin film on top of the plants and are more easily attacked. The bullfrogs also show that female choice and male-male competition may go hand in hand. The preferred territories are hotly contested by males and the largest, strongest frogs end up in the best sites.

Food is a resource which often limits a female's capacity to make eggs and during courtship females may choose whether or not to mate with a male on the basis of his ability to provide food. In some birds and insects for example, males may provide food for the female during courtship ('courtship

Fig. 8.5. Sexual selection in male bullfrogs. Males compete by wrestling and calling (left and middle) for good territories, in which the females prefer to lay their eggs (right). The good territories have high survival of eggs because they are warm and because the vegetation is not too dense (Howard 1978a,b).

174

Chapter 8
Sexual
Conflict
and Sexual
Selection

feeding') which makes a significant contribution to her eggs. Female hanging flies (*Hylobittacus apicalis*) will mate with a male only if he provides a large insect for her to eat during copulation. The larger the insect, the longer the male is allowed to copulate and the more eggs he fertilizes (Fig. 8.6). The female gains from a large insect by having more food to put into her eggs. In birds, the male usually helps to feed the young and courtship feeding may play the additional role of indicating to the female how good the male is at bringing food for the young. In the common tern (*Sterna fuscata*) there is a correlation between the ability of the male to bring food during courtship feeding and his ability to feed the chicks later in the season. Pairs often break up during the courtship feeding period and it is possible that females are assessing their mates and rejecting poor quality partners (Nisbet 1977).

(b) Good genes

If some males have 'better' genes than others, could a female improve the success of her progeny by choosing males with good genes? Good genes are ones which increase the ability of her offspring to survive, compete and reproduce. One of the very few studies in which this idea has been tested is an experiment by Linda Partridge (1980). She took groups of female fruit flies (*Drosophila*) and either allowed them to mate freely with a population of males or forced each female to mate with a randomly chosen partner. The offspring of the 'choice' and 'no choice' females were then tested for their competitive ability by rearing the larvae in bottles with a fixed number of standard competitors (these were distinguishable by a genetic marker). Partridge found that the offspring of the 'choice' group did slightly but consistently better than those of 'no choice' females in the larval competition experiments. This experiment suggests that females are able to increase the survival of their offspring by choosing good genes in their mates, but it must be born in mind that the results could also be in part explaind by intrasexual competition: in the 'choice' experiment, the males that mated may have been superior competitors against other males.

The theory of sexual selection is most famous as an attempt to explain the evolution of excessively elaborate adornments and displays of male peacocks, pheasants, birds of paradise and so on (Fig. 8.7) Some elaborate displays may have evolved for use in contests between males, but the most widely accepted (and still controversial) account of how such traits evolve depends on selection by females for good genes. In this case, however, the genes are not ones which increase the ability of a

175

*Chapter 8
Sexual
Conflict
and Sexual
Selection*

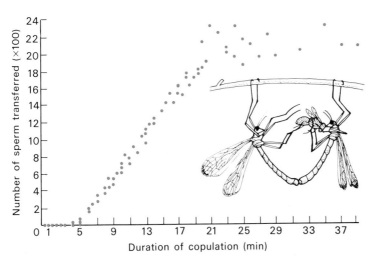

Fig. 8.6. Female choice for good resources. Female hanging flies (*Hylobit-tacus apicalis*) mate with males for longer if the male brings a larger prey item to eat during copulation. The male benefits from long copulation because he fertilizes more eggs (Thornhill 1976).

female's offspring to survive and compete (in fact the bizarre adornments usually seem to be a hindrance to survival) but rather for genes which increase the sexual attractiveness of her sons.

176

*Chapter 8
Sexual
Conflict
and Sexual
Selection*

Fig. 8.7. The male satin bower bird (*Ptilonorhynchus violaceus*) (left) builds a bower consisting of two parallel rows of vertical fine twigs on a court of cleared ground. The male displays and attracts females to the bower and mating takes place in or by the bower. The bower is decorated with flowers, feathers, leaves, snail shells, snake skins and where available with the debris of human society such as toothbrushes and Fosters' beer cans. Gerald Borgia (1985) filmed with automatic cameras all the matings at 22 bowers and found that males with more decorations get more females (snail shells, blue feathers and yellow leaves were especially attractive). In part a male's decoration reflects his competitive ability, since males steal decorations from each other's bowers.

R.A. Fisher (1930) was the first to clearly formulate the idea that elaborate male plumage may be sexually selected simply because it makes males attractive to females. This may sound circular, and indeed it is, but that is the elegance of Fisher's argument. At the beginning, he supposed, females preferred a particular male trait (let us take long tails as an example) because it indicated something about male quality. Perhaps males with longer tails were better at flying and therefore collecting food or avoiding predators. An alternative starting point is to suppose that longer tails were simply easier for females to detect and that they therefore attracted more mates. For example, male natterjack toads *Bufo calamita* with louder croaks attract more females simply because females locate males by moving passively down sound gradients (Arak 1983). If there is some genetic basis for differences between males in tail length the advantage will be passed on to the female's sons. At the same time, a gene which causes females to prefer longer than average tails will also be favoured since these females will have sons better able to fly or more readily detected by potential mates. Now once the female preference for longer tails starts to spread, longer tailed males will gain a double advantage: they will be better at flying and be more likely to get a mate. The female similarly gets a double advantage from choosing: she will have sons that are both good fliers and attractive to females. As the positive feedback between female preference and longer tails develops, gradually the benefit of attractive sons will become the more important reason for female choice, and the favoured trait might eventually decrease the survival ability of males. When the decrease in survival counter-balances sexual attractiveness, selection for increasing tail length will grind to a halt.

177

Chapter 8
Sexual
Conflict
and Sexual
Selection

Fisher's verbal argument has been formalized and extended in population genetics models by Russell Lande (1981) His analysis underlines the importance of the assumption, implicit in Fisher's argument, that the genes causing variation in preference and in tail length become associated with one another, or in other words show covariance (Box 8.1).

Although 'tail length' is the classic textbook trait in discussions of sexual selection it is in fact only recently that evidence for sexual selection of tail length in real life has been collected. Malte Andersson (1982) showed that females of the long-tailed widow bird *Euplectes progne* in Kenya prefer males with long tails. This highly polygynous species is an ideal candidate for sexual selection; the male is a sparrow-sized bird with a tail up to 50 cm long. The female's tail is about 7 cm long, presumably close to the optimum for flight purposes. Andersson studied 36 males which he divided into four groups. In one group he docked the tails to about 14 cm, while in another group he attached the severed bits of group 1 tails with Super glue. This increased the tail length of group 2 males by an average of 25 cm. The remaining two groups were controls: one lot were left untouched and the others had their tails cut and glued without altering the length. By counting the number of nests in each territory, Andersson showed that before his experimental manipulations there was no difference in mating success of the different groups, while afterwards the long tailed males did significantly better than the controls or the shorter tailed birds. (Fig. 8.8).

Another nice experimental study of a sexually selected elaborate display is that of Clive Catchpole (1980; Catchpole *et al.* 1984) on the song of the European sedge warbler. The song consists of a long stream of almost endlessly varying trills, whistles and buzzes and is sung by the male after arriving back on the breeding territory from the winter quarters: as soon as a male pairs, it stops singing. Catchpole's measurements showed that the males with the most elaborate songs

178

*Chapter 8
Sexual
Conflict
and Sexual
Selection*

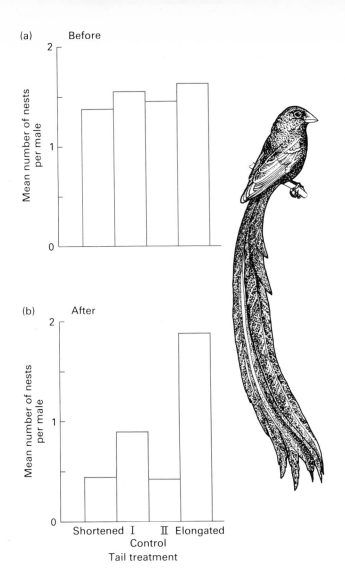

Fig. 8.8. Sexual selection for tail length in long-tailed widow birds. The top line shows that there was no difference between the four groups before the tails were altered. The bottom line shows that after the tails were cut and lengthened the mating success went down and up respectively. The two kinds of control birds were (I) unmanipulated, and (II) cut and glued back without altering length. Mating success is measured as the number of active nests in each male's territory (Andersson 1982).

are the first to acquire mates (Fig. 8.9), as expected from Fisher's hypothesis of sexual attractiveness. Further, when female warblers were brought into the laboratory and treated with oestradiol to make them sexually active, they were more responsive to large than to small repertoires. In contrast to the long tail of the widow bird it is not obvious what might be the counter-selection limiting the elaboration of song; one possi-

179

Chapter 8
Sexual
Conflict
and Sexual
Selection

bility is that elaborate songs are more readily detected by predators, as been found in the leopard frog (Ryan *et al.* 1982).

Perhaps the most controversial aspect of the idea that females choose males for good genes is the problem of genetic variance. This can be illustrated with an hypothetical example. Suppose a farmer wants to select for larger body size in a population of turkeys. He takes the heaviest males and females to start the next brood and repeats this procedure for several generations. What will happen? Assuming that there is some genetic variance for body weight, selection will at first be fairly effective, but soon the stock will become less variable with respect to genes for body weight, because only a few genotypes (the heaviest) have been allowed to breed. When the genetic variance is 'used up' selection will cease to be effective in changing body size. In the same way, females cannot improve the genetic quality of their offspring indefinitely by choosing males for good genes. So there is something of a paradox. Probably the genetic theory is too simple. Genetic variation for fitness-related traits might well be maintained in natural populations either because of new mutations or, more likely, because the optimal genotype is different in different places or at different times so that selection does not favour the same genes for long enough to exhaust all the genetic variation, or migration between populations introduces new variation.

Bill Hamilton and Marlene Zuk (1982) have suggested that one kind of selection in particular may have the desired property of favouring different genotypes at different times: selection for resistance to disease. The point can be brought home with a simple hypothetical example. Suppose that a disease-causing organism occurs in two genotypes X and x. Pathogens of type X can infect hosts of genotype F but not f, and similarly pathogens of type x can infect f but not F. In other words there is a gene-for-gene correspondence between pathogenicity and susceptibility to the disease. Now let us start with X as the common form of the disease and F as the common host genotype. Because of its resistance to X, f is at a strong selective advantage and will become commoner, but as f increases in frequency so disease type x gains an advantage relative to X. This in turn swings the pendulum back in favour of host F and thence pathogen X, and so on. It is easy to imagine (and imagination has been made respectable by computer simulations) that the host-pathogen interaction will produce endless cycles of genetic change with constant pressure on both sides to keep ahead or catch up with the other. What does this have to do with sexual selection? The idea is that females might select males on the basis of their resistance to disease, resis-

180

Chapter 8
Sexual
Conflict
and Sexual
Selection

Box 8.1. *Sexual selection for nose length: how Fisher's runaway process works. After Arnold (1983), explaining Lande (1981).*

1 Imagine that at the start there was a range of nose lengths and of female preferences in the population. Females with a preference for slightly longer than average noses would be mated to males with longer noses and vice versa. The crucial fact to note is that offspring of these matings would have *both* the nose and preference genes: either genes for long nose plus long preference or short nose and short preference. The preference is expressed only in females and the nose in males, but everyone carries both kinds of gene. In short, there will arise an association or *covariance* between nose and preference genes. You could look at a female's preference and predict what kind of nose genes she carries to give to her sons (Fig. a)

(a)

(a) Genes for long nose and long preference go together in the offspring. The slope of the line represents the degree of association or covariance

2 How will evolution proceed, given this covariance? If equal numbers of females have preference above and below the mean nose length (x), there will be no change. But if, by chance there was a slight predominance of females on one side of the mean (it could be long or short but let us take long), then positive feedback will start. This is shown by the arrows on Fig. a. Females select for long noses (long nosed males have a higher chance of mating) and thereby, *because of the covariance*, select for long preference. This in turn produces a further push to long noses and hence an increase in preference.

3 What will happen in the end? As nose length increases through evolutionary time it will exceed the 'utilitarian' optimum size (the optimum under natural selection) for doing its normal job of sniffing. How far it will go beyond this optimum will depend on the pull exerted by female choice: a very strong pull will drag the nose length far from the utilitarian optimum before counter-selection for smaller, more efficient sniffing noses is big enough to just balance the female pull. If on the other hand the female pull is weak, only a slight deviation above the optimum will produce enough counter-selection to balance the female choice. In short, female pull will drag nose length upwards until counter-balanced by utilitarian selection, but there is no single equilibrium point: there is an equilibrium *line* of combinations of female pull and counter-selection (Fig. b). Once nose length has reached the equilibrium line it could drift up and down by chance, since all points on the line are of equal net selective value.

181
Chapter 8
Sexual
Conflict
and Sexual
Selection

182

Chapter 8
Sexual
Conflict
and Sexual
Selection

4 In Fig. b nose length reaches an equilibrium but there may be no equilibrium if the covariance between nose length and preference is stronger. This is shown in Fig. c: as selection proceeds along the stepped trajectory, the equilibrium line is missed altogether.

5 Summary. The main points of Lande's model are (i) Covariance between nose length and preference is the key. (ii) When covariance is high, unrestrained runaway selection is more likely than when it is low. In the latter case there is no single outcome, but many equilibria. Therefore different outcomes of sexual selection between species need not necessarily be related to ecological differences. (iii) The process of sexual selection could be started by a chance shift in the distribution of female preference.

tance which would be passed on to progeny and help in the never-ending race to keep ahead. As Zuk (1984) points out, aviculturalists and poultry breeders can readily tell a sick bird by looking at its plumage (for example, a turkey suffering from erysipelas has a swollen purple snood), so why should females have not evolved a similar ability? Seen in this light, courtship displays and their associated body structures are a sort of medical examination by which one partner assesses the other.

As yet there is no evidence that females of any species are skilled diagnostic veterinarians, but Hamilton and Zuk have some evidence that is consistent with their idea. They argue that sexual selection for external features indicating vigour and disease resistance will have been strongest in those species that are exposed to the most diseases (simply because in these species it pays the female to scrutinize her partner closely). Therefore if traits like the showy plumage and elaborate courtship songs of male passerine birds have evolved as indicators of disease resistance, these traits should be most highly developed in the most diseased species. In a survey of the blood parasites, plumage and song of 109 North American passerine bird species this is generally what Hamilton and Zuk found.

THE HANDICAP PRINCIPLE

Amotz Zahavi (1975, 1977) has suggested an alternative view of elaborate male displays. He points out that the long tail of the peacock is a handicap in day to day survival, a fact which

183

Chapter 8
Sexual
Conflict
and Sexual
Selection

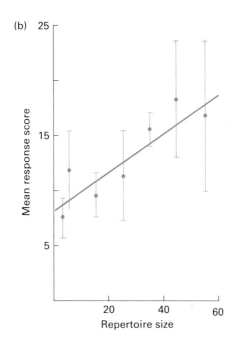

Fig. 8.9. (a) Male sedge warblers with the largest song repertoires are the first to acquire females in the spring. The size of song repertoire is estimated from sample tape recordings of each male. The results were collected in such a way as to control for the possibilities that older males, or males in better territories, both mate first and have larger repertoires (Catchpole 1980). (b) The mean ± s.e. response score of 5 females to repertoires of different sizes. The response score measures sexual behaviour. (Catchpole *et al.* 1984.)

184

*Chapter 8
Sexual
Conflict
and Sexual
Selection*

few would dispute. He then goes on to suggest that longer tails are preferred by females precisely *because* they are a handicap and therefore act as an advertisement of male quality. The tail demonstrates a male's ability to survive in spite of the handicap, which means that he must be extra good in other respects and is able to pass on any genetic component of this quality to his sons and daughters. The difficulty with this argument is that the offspring inherit not only the 'good genes' but also the handicap, and under most conditions they will be better off without either than with both. For this reason, Zahavi's idea is unlikely to work in its original form, but if we modify it to say that the handicap is not a fixed trait but is simply conditional upon male vigour its major problem disappears (Kodric-Brown & Brown 1984; Nur & Hasson 1984). In this modified version, the size of a male's handicap is a direct index of his current vigour, as in Hamilton and Zuk's parasite hypothesis where a male's plumage or other displays indicate his degree of disease resistance. In fact the idea that the intensity of a male's sexual displays are not genetically fixed (as in Fisher's hypothesis) but are simply a phenotypic expression of his current condition, can be extended beyond the handicap principle. Any display or trait (antler length, croaking rate, and so on) whether or not it is a handicap, could be a reliable index of a male's well-being and be used in assessment by females or rival males (Kodric-Brown & Brown 1984).

MALE INVESTMENT

We have so far assumed that females are investors and males are competitors. While this picture describes most animal species, there are exceptions. In many birds, some amphibians, and arthropods, both male and female invest about equally in the eggs or young by feeding, guarding or brooding. Sometimes the usual sex roles are completely reversed so that males do the investing and females the competing (chp. 9). The ideas about sexual conflict and sexual selection can still be applied in modified form to species with equal or primarily male investment. When both sexes care equally for the offspring, for example, courtship may involve assessment and choice by males as well as by females. Males of species with internal fertilization can never be absolutely sure that they have fathered the children of their mate, and one role of courtship may be as an insurance against cuckoldry. A prediction of this idea is that courtship allows males to assess whether or not females have previously mated with others. This was tested by Erickson and Zenone (1976). They found that male barbary doves (*Streptopelia risoria*) attack a female instead of courting

her if she performs the 'bow posture' (an advanced stage of
courtship) too quickly. Since the females which responded
in this way had been pre-treated by allowing them to court
with another male, the reaction of the test males in rejecting
eager females is adaptive if courtship plays a role in assessing
certainty of paternity, before investing in offspring. It would
not have been predicted by the older view that male courtship
serves to sexually arouse the female!

185
Chapter 8
Sexual
Conflict
and Sexual
Selection

In species with high male investment, females tend to be the
competitive sex and males may be choosy. In moorhens (*Gal-
linula chloropus*) males do almost three-quarters of the incubation and females play an active role in competing for the
chance to mate with good incubators. These ideal husbands
are small and fat: well equipped to survive on their reserves
during long incubation stints (Petrie 1983). In other species,
investing males may actually reject low quality females (Fig.
8.10).

SEXUAL CONFLICT

Let us now return to the starting point of this chapter, sexual
conflict. Recall the view of the origin of anisogamy as the
primaeval example of sexual conflict. The conflict was one
about mating decisions. Macrogametes might have done better
had they been able to discriminate against microgametes, but
in the evolutionary race microgametes won. Similar, but more
directly observable, conflicts of interest between the sexes are
still apparent today, not only with respect to mating decisions
but also in the contexts of parental investment, multiple
matings and infanticide.

(a) Mating decisions

As we have emphasized earlier in the chapter, females have
more to lose and therefore tend to be choosier than males.
Thus for a given encounter it will often be the case that males
are favoured if they do mate and females if they do not (Parker
1979). An extreme manifestation of this conflict is enforced
copulation as exemplified by scorpionflies (*Panorpa* spp.).
Male scorpionflies usually acquire a mate by presenting her
with a nuptial gift in the form of a special salivary secretion or
a dead insect (this is very similar to the *Hylobittacus* described
earlier). The female feeds on the gift during copulation and
turns the food into eggs. However, sometimes a male enforces
copulation: he grasps her with a special abdominal organ (the
notal organ) without offering a gift (Thornhill 1980). Enforced
copulation appears to be a case of sexual conflict. The female

Fig. 8.10. A female Katydid (bush cricket, *Conocephalus nigropleurum*). In some species of crickets the male produces a huge, protein-rich spermatophore (it may weigh up to 27% of the male's body weight) which is eaten by the female and used to make eggs. With this large investment the male can afford to be choosy. Females prefer large males because they make large spermatophores while males reject small females because they are capable of laying fewer eggs than are large females. (a) Mated female with spermatophore. (b) Female eating spermatophore. (c) Close up of spermatophore. From Gwynne (1982).

loses because she obtains no food for her eggs and has to search for food herself, while the male benefits because he avoids the risky business of finding a nuptial gift. Scorpionflies feed on insects in spiders' webs and quite often get caught up in the web themselves, so foraging is certainly risky (65 per cent of adults die this way). Why do not all males enforce copulations? The exact balance of costs and benefits is not known, but it appears that it results in a very low success rate in fertilizing females, so perhaps males adopt this strategy only when they cannot find prey or make enough saliva to attract a female.

(b) Parental investment

This is a topic to which we will return in the next chapter. Here it is sufficient to note that in species with investment beyond the gamete stage, each sex might be expected to exploit the other by reducing its own share in the investment. The outcome of this sexual conflict may depend on practical considerations such as which sex is the first to be in a position to desert the other. When fertilization is internal, for example, a male has the possibility of deserting the female immediately after fertilization and leaving her with the eggs or young to care for.

187
*Chapter 8
Sexual
Conflict
and Sexual
Selection*

(c) Infanticide

As we saw in chapter 1, male lions may slaughter the cubs in a pride shortly after they take over as group leaders. This behaviour (which is also seen in some primates) probably increases male reproductive success, as explained in chapter 1, and clearly decreases female success. This seems to be a case of sexual conflict in which the males have won, but it is perhaps surprising that females have not evolved counter-adaptations. They could, for example eat their own young once they have been killed in order to recoup as much as possible of their losses (chp. 12).

(d) Multiple matings

As Bateman's experiments with *Drosophila* showed (p. 163) females often gain little by mating with more than one male. However, because of sperm competition males may gain by mating with already fertilized females. Multiple matings are likely to be costly to the female at the same time as being advantageous to the male. This is dramatically illustrated by the dungflies described in chapter 7. When two males struggle for possession of a female, the female is sometimes drowned in cowdung by the fighting males on top of her!

Conflicts of interest between the sexes will lead to an evolutionary race of the sort envisaged by Parker *et al.* for sperms and eggs. There is no simple answer to the questions 'Which sex is more likely to win the chase?' As we discussed earlier, factors such as the strength of selection and the amount of genetic variation will determine how fast the two sexes can evolve adaptation and counter-adaptation, but it is not possible to make any more specific statements about the outcome of sexual conflict races.

The significance of courtship

As we have mentioned earlier in the chapter, some aspects of courtship behaviour can be interpreted in terms of sexual conflict and sexual selection. However this is not true of all courtship signals: many are designed for species identification, and here the interests of the two sexes are similar because both benefit by mating with a member of the same

188

Chapter 8
Sexual
Conflict
and Sexual
Selection

species. Some of the clearest examples of this role of courtship come from studies of frog calls. When several species of frogs live in the same pond, each has a characteristic and distinct mating call given by the male, and females are attracted only to calls of their own species. In some frogs (e.g. the cricket frog *Acris crepitans*) it has been shown that the female's selectivity of response results from the fact that the auditory system is tuned to the particular frequencies in the male call (Capranica *et al.* 1973).

Courtship displays may also play a role in competition between males within a species for mating opportunities. Often the same displays simultaneously serve to repel other males and attract females. An example for which this has been demonstrated experimentally is the mating call of the pacific tree frog (*Hyla regilla*) (Whitney & Krebs 1975a,b). Males are repelled and females attracted by loudspeakers broadcasting the mating call and females select out of a group of loudspeakers the one which calls for the longest bouts. Females may choose between displays purely on the basis of sexual attractiveness, as explained by Fisher's theory of sexual selection, but there is also the possibility that differences in courtship between males may indicate habitat quality, for example males with territories containing a lot of food might be able to afford to spend more time displaying.

A third role of courtship to which we have referred is assessment. In a species with male parental care females may assess the ability of the male to look after young and males may assess whether a female has previously been fertilized. Early work by ethologists on birds and fish showed that at the beginning of courtship males are often aggressive and females are coy or reluctant. Courtship was seen, therefore, as serving to synchronize sexual arousal of the partners. A possible explanation of *why* it should be necessary to overcome aggression and reluctance is that the early phases involve assessment by both partners before investing in offspring.

Throughout this chapter we have emphasized the role of females as investors in the zygote and offspring, but we have also mentioned that sometimes males invest as much as or more than females. Why should this happen in some species but not in others? In order to answer this question we will turn in the next chapter to the influence of ecological pressures.

Summary

Conflict lies at the heart of sexual reproduction. The fundamental difference between male and female is the size of gametes. Males produce tiny gametes and can be viewed as successful parasites of large female gametes. Because sperm is cheap, males can increase their reproductive success by mating with many females. Females can only increase their success by making eggs or young at a faster rate. Females are a scarce resource for which males compete and much of male courtship can be understood in terms of competition for matings. Females may be reluctant to mate unless they can choose partners with good resources or (perhaps) good genes. Sometimes the general rule of high female investment is reversed and males are the main investors: here females are competitive and males are choosy.

189

Chapter 8
Sexual
Conflict
and Sexual
Selection

According to the most widely accepted view, the extremely elaborate sexual displays of some species arose through Fisher's runaway process, which assumes that genes affecting variation in female preference and genes affecting variation in male displays become linked. As a result there is positive feedback in which selection by females for an increase in the male trait causes a concomittant increase in choosiness. The benefit to females of choosing is simply that they have sons that are sexually attractive. An alternative view is that variation in male displays is not genetic and reflects variation in general body condition. Females selecting for males with elaborate displays are, according to this idea, selecting for males in good condition.

Further reading

An important collection of papers on sexual selection and mate choice is the book edited by Bateson (1983). Halliday's introductory chapter and Arnold's review of models of the runaway process are especially notable. Maynard Smith (1984) and Partridge & Halliday (1984) are good reviews of evolutionary problems associated with sex and with mate choice respectively. Charnov's (1982) book is the definitive work on sex allocation, including sex ratio theory.

The papers by Burley et al. (1982), Burley (in press) and Burley & Coopersmith (in press) present a remarkable set of data on mate choice in a monogamous species of bird, the zebra finch. Burley discovered that the coloured leg rings placed on the birds for individual identification in her captive colony influenced mating success. Males with red rings and females with black ones had a higher reproductive success than those with some other colours, green and blue being especially unattractive rings on males and female respectively. It appears that individuals with attractive rings get partners who are willing to do more parental work in return for mating with a desirable member of the opposite sex.

190

Chapter 8
Sexual
Conflict
and Sexual
Selection

Sutherland (1985) discusses the classic paper of Bateman (1948) and shows that Bateman's finding that the variance in male mating success is higher than that of females could arise from random variation in mating success. The difference in variance between the sexes is therefore not good evidence for sexual selection.

A study of a species with male investment is that of Smith (1979) on a giant water bug, in which the male carries the eggs on his back. To ensure paternity of the eggs he carries, the male repeatedly inseminates the female during egg laying, the record being 100 copulations in 36 hours.

Endler's (1980) study of colouration in wild guppy (*Poecilia reticulata*) populations is an important example to illustrate the balance between sexual selection for bright colours and predator selection for crypsis. Other studies showing selection for conspicuous colour and tail length in the same species are those of Kodric-Brown (1985) and Bischoff *et al.* (1985).

McKinney *et al.* (1983) give an excellent review of forced copulation in waterfowl.

Topics for discussion

1 How might sexual selection have given rise to (a) antlers and (b) the peacock's tail?

2 What hypotheses might account for bright colouration in birds (see also Baker & Parker 1979; Lyon & Montgomery 1985 and references therein).

3 Why do males usually court females rather than vice versa?

Chapter 9. Parental Care and Mating Systems

In the last chapter we explored the fundamental differences between males and females. A male has the potential to father offspring at a faster rate than a female can produce them. Therefore females are a limiting resource for male reproductive success and males are expected to compete to maximize their number of matings. On the other hand, because females put a lot of investment into the zygote we expect them to resist male ardour and to be selective in their choice of mate.

However, care for the zygote does not end with gamete investment. In many animals there is some form of parental care, for example guarding the eggs or feeding the young. Parental care may be done by both parents (e.g. starling, *Sturnus vulgaris*), by the female alone (e.g. red deer) or by the male alone (e.g. sea horse, *Hippocampus*, where the male carries the fertilized eggs around in his brood pouch).

Often, differences between species in parental care are associated with differences in mating system. Mating systems can be categorized under the following broad headings.

1 *Monogamy*. A male and a female form a pair bond, either short or long term (part of or a whole breeding season or even a lifetime). Often both parents care for the eggs and young.

2 *Polygyny*. A male mates with several females, while females each mate with only one male. A male may associate with several females at once (simultaneous polygyny) or in succession (successive polygyny). With polygyny it is usually the female that provides the parental care.

3 *Polyandry*. This is exactly the reverse of polygyny. A female either associates with several males at once (simultaneous polyandry) or in succession (successive polyandry). In this case, it is often the male who does most parental care.

4 *Promiscuity*. Both male and female mate several times with different individuals, so there is a mixture of polygyny and polyandry. Either sex may care for the eggs or young. *Polygamy* is often used as a general term for when an individual of either sex has more than one mate.

These divisions are not hard and fast but they will help as a general guideline.

The ideal way for a male to maximize his reproductive success may be to go round copulating with lots of females, each of which stays at home to care for his offspring. For a female, the ideal may be to mate and then leave the male to care for the offspring while she forms reserves for more eggs. In practise, as we shall now see, there are two factors that will influence how the sexes resolve this conflict. First, different groups of animals have different physiological and life history constraints which may predispose one sex to do more parental care than the other. Second, ecological factors will influence the costs and benefits associated with parental care and mating behaviour.

Proximate constraints on parental care

These can be illustrated by contrasting three classes of vertebrates, fish, birds and mammals (Table 9.1). Are there any basic differences in the physiology and life histories of these three groups which could account for the differences in parental care and mating system? While considering this question we must bear in mind that both male and female will be selected to maximize their reproductive success and may behave at the expense of the other sex (Trivers 1972).

Table 9.1. Frequent types of parental care and mating system in three classes of vertebrates*

	Parental care	Mating System
Birds	Both male and female	Monogamous
Mammals	Female only	Polygynous
Fish	Male only	Polygamy/ Promiscuity

*These are very broad generalizations and many exceptions are discussed later in the chapter.

BIRDS

As we saw in the first chapter, reproductive success in birds can be limited by the rate at which food is delivered to the nest. At least in those species where parental feeding of the young is important, we could imagine that two parents will be able to feed twice as many young as a single parent. Therefore both male and female increase their reproductive success by staying together. If either sex deserted, they would approximately halve the output of the last brood and would also have to spend time searching for a new mate and nest site before being able to start again. So monogamy and parental care by both male and female are not hard to understand. In some seabirds like the kittiwake, *Rissa tridactyla* (Coulson 1966) and the manx shearwater, *Puffinus puffinus* (Brooke 1978), long term mate fidelity seems to pay because pairs that have bred together previously have higher reproductive success than new pairs.

When the constraint of having to have two parents feeding the young is lifted, it is usually the male who deserts and the female who is left caring for the brood. Comparative evidence shows that polygyny often occurs in fruit and seed eaters, probably because these food supplies become so seasonally abundant that one parent can feed the young almost as efficiently as two (e.g. weaver birds, chp. 2). Why is it the male who deserts? There are two factors which may be important. First, the male has the opportunity to desert before the female. With internal fertilization, she is left literally holding the babies inside her. Second, the male can gain more by desertion than the female because his lifetime reproductive success depends more on his number of matings (see chp. 8).

MAMMALS

In mammals, females are even more predisposed to care for the young. The offspring often have a prolonged period of gestation inside the female, during which the male can do little direct care (though he can protect and feed the female). Once the young are born they are fed on milk and only the female lactates. Because of these constraints on the opportunity to care for offspring and also because, with internal fertilization, the male can desert first, it is not surprising that most mammals have a polygynous mating system and parental care by the female alone.

Monogamy and biparental care occurs in a few species where the male contributes to feeding (carnivores) or to carrying the young (e.g. marmosets). Perhaps it is surprising that male lactation has not evolved in these cases (see Daly 1979).

FISH

In the bony fish (teleosts), most families (79%) have no parental care (Gross & Sargent 1985). In those families which do care for the eggs or young, it is usually done by one parent; biparental care occurs in less than 25% of the families which show care. Compared with the elaborate care of offspring by birds, parental care in fish is a simple affair often consisting of just guarding or fanning eggs. These tasks can usually be done effectively by one parent alone. Which parent will provide care? Table 9.2 shows that female care is commonest with internal fertilization (86 per cent female care) and male care with external fertilization (70 per cent male care). The overall predominance of male parental care in fish is related to the prevalence of external fertilization.

Three hypotheses can be proposed to explain this result (see Gross & Shine 1981).

Hypothesis 1. Paternity certainty

Trivers (1972) suggested that reliability of paternity will be affected by mode of fertilization. Since external fertilization occurs at the time of oviposition, reliability of paternity may be greater than with internal fertilization where sperm competition may take place inside the female's reproductive tract (chp. 8). According to this hypothesis, with internal fertilization, therefore, a male should be less prepared to provide parental care because he is less certain than the female of genetic representation in the offspring. However, paternity certainty alone should not influence a male's decision whether to care or desert because he will presumably be equally uncertain of his paternity if he deserts and mates again. A male will only gain by desertion if this increases his reproductive output above that which he would achieve by staying. Furthermore, it may not necessarily be true that external fertilization leads to greater paternity certainty. In some cases, for example sunfishes *Lepomis*, cuckoldry takes place during oviposition (see chp. 10).

Hypothesis 2. Order of gamete release

Dawkins and Carlisle (1976) suggested that, as with birds and mammals, internal fertilization gives the male the chance to desert first and thus leave the female to care. With external fertilization they suggested that the roles may be reversed. Because sperm are lighter than eggs the male must wait until the eggs are laid before he can fertilize them, or else his gametes will float away. Therefore, the female has the opportunity to desert first and swim off while the male is still

Table 9.2. Distribution of male and female parental care with respect to mode of fertilization in teleost fishes. The table shows number of families; a single family may appear in more than one category, but is not listed under 'no parental care' unless care is completely unknown in the family. From Gross and Shine (1981)

	Internal fertilization	External fertilization
Male parental care	2	61
Female parental care	14	24
No parental care	5	100

fertilizing the eggs! This is an ingenious idea but it must be rejected on empirical grounds. In fact, the most common pattern of gamete release is simultaneous release by male and female. In these cases both sexes have an equal chance to desert, but 36 out of 46 species which have simultaneous gamete release and monoparental care, have care provided by the male. Secondly, in some families of fish (Callichthyidae, Belontiidae) the male builds a foam nest and releases sperm before the female lays eggs. In these cases, the 'opportunity for desertion' hypothesis predicts that males can desert first, but nevertheless parental care is provided by the male. Therefore, male parental care remains correlated with external fertilization independently of the order of gamete release or opportunity to desert.

Hypothesis 3. Association

Williams (1975) suggested simply that association with the embryos preadapts a sex for parental care. For example, with internal fertilization the female is most closely associated with the embryo and this may set the stage for the evolution of embryo retention and live birth, followed by care of the young fry. With external fertilization, on the other hand, the eggs are often laid in a male's territory and it is the male who is most closely associated with the embryos. Defence of the territory in order to attract further females becomes, incidently, defence of the eggs and young, and therefore provides a preadaptation for more elaborate parental care by males. This hypothesis is the best predictor of the data in Table 9.2. Male parental care is more common in territorial species and the prevalence of male parental care with external fertilization results from the fact that male territoriality is particularly common with external fertilization.

ANCESTRAL ORIGIN OF UNIPARENTAL CARE

Another factor that may influence which sex cares for the zygote is the ancestral origin of uniparental care. In birds, the ancestral state is probably biparental care because in most species both male and female look after the eggs and young. This means that the female cannot use up all her resources in egg laying since she has to care for the young later on. It is well known that female birds can often readily lay replacement eggs if their first clutches fail. Therefore, if the male should desert, the female will still have some resources left to care for the brood.

In fish, the ancestral state is probably no parental care. The

female may therefore use up all her resources in egg laying
and so cannot afford to guard. This may mean that it is the
male who often has to look after the young if care is necessary.
Furthermore, Gross and Sargent (1985) suggest that in fish,
female fecundity usually increases more with size than does
male fecundity. The predominance of male parental care may
arise because the loss of growth that results from parental
effort has less cost (in terms of decreased future reproductive
success) for males than females. However, this may not always
be the case (see Fig. 10.10).

An ESS model of parental investment

So far, we have seen that each sex is faced with the decision of
whether to stay and care for the young or desert. Ecological
factors will determine the costs and benefits of these two
options and proximate constraints will also be important (e.g.
female mammals lactate, males do not). However, it is also
important to realize that the best decision for one sex will
depend on the strategy adopted by the other sex. For example,
if the female stays, it may pay the male to desert but if the
female deserts it may pay the male to stay. Maynard Smith
(1977a) has developed a useful model for considering what
will happen in such cases of conflict.

   The model looks for a pair of strategies, say I_m for males and
I_f for females, such that it would not pay a male to diverge
from strategy I_m so long as females adopt I_f, and it would not
pay females to diverge from I_f as long as males adopt I_m. In
other words, we seek the evolutionarily stable strategies for
male and female (see p. 136). Suppose that the reproductive
success of a pair during the breeding season depends on the
amount of parental care and on the number of eggs laid by the
female. It is assumed that the more the female invests in eggs,
the less she is able to invest in care and vice versa.

   Let P_0, P_1 and P_2 be the probabilities of survival of eggs
which are not cared for, cared for by one parent and cared for
by two parents, respectively; $P_2 > P_1 > P_0$. A male who
deserts has a chance p of mating again. A female who deserts
lays W eggs and one who cares lays w eggs; $W > w$.

The pay-off matrix for this game is illustrated in Table 9.3.
There are four possible ESSs.

ESS 1   *Female deserts and male deserts*

   This requires $WP_0 > wP_1$, or the female will care and P_0
$(1 + p) > P_1$, or the male will care.

ESS 2   *Female deserts and male cares*

   This requires $WP_1 > wP_2$, or the female will care, and P_1
$> P_0 (1 + p)$, or the male will desert.

This requires wP$_1$ > WP$_0$, or the female will desert, and P$_1$

ESS 3 *Female cares and male deserts*

This requires $wP_1 > WP_0$, or the female will desert, and P_1 $(1 + p) > P_2$, or the male will care.

ESS 4 *Female cares and male cares*

This requires $wP_2 > WP_1$, or the female will desert, and $P_2 > P_1 (1 + p)$ or the male will desert.

For given values of the parameters in the model, ESS 1 and ESS 4 can be alternative possibilities, as can ESS 2 and ESS 3. For example, ESS 2 is favoured if the female can lay many more eggs if she does not invest in caring ($W \gg w$) and if one parent is much better than none ($P_1 \gg P_0$) but two parents are not much better than one ($P_2 \simeq P_1$). This situation probably applies to many fish, as discussed above, where the female tends to desert and the male often cares. However, ESS 3 is an alternative possibility, especially if a male who deserts has a much better chance of mating again; this may apply to some species of birds and mammals (see below). If two parents can raise twice as many young as one ($P_2 \gg P_1$), or if the chance that a deserting parent will remate is small, then ESS 4 is the likely outcome, as in many species of passerine birds.

Maynard Smith's model is useful in identifying the factors which will lead to different patterns of parental investment and mating systems. In the rest of this chapter we will discuss how ecological factors, such as the distribution of resources and predators, and the availability of mates influence the values of the various parameters in the model and hence the kind of mating system which results.

Ecological correlates of mating systems

We will now examine in detail how particular environmental factors influence male and female reproductive success and so determine the form of parental care and mating system. The ideas can best be illustrated with reference to different kinds of polygamy in birds and mammals. The male can usually benefit more than the female can by maximizing his number of

Table 9.3. An ESS model of parental investment (Maynard Smith 1977a). Each sex has the possibility of caring or deserting. The matrix gives the reproductive success for males and females. See text for details

Male		Female	
		Care	Desert
Care	Female gets	wP_2	WP_1
	Male gets	wP_2	WP_1
Desert	Female gets	wP_1	WP_0
	Male gets	$wP_1 (1 + p)$	$WP_0 (1 + p)$

matings, and so the major ecological influence on mating systems will be how environmental factors determine the ability of a male to gain access to females.

A male could increase his reproductive success either by defending females themselves or by controlling some critical resource that the females need and then mating with them when they visit the resource (e.g. spawning sites in frogs, chp. 8). As we saw in chapter 5, the ability of an individual to control access to limiting resources will depend on the economics of defence. If resources or females are easily defendable then there will be high potential for polygyny. The economics of female or resource defence will be determined by their distribution in space and time (Emlen & Oring 1977; Wittenberger 1980).

(a) Spatial distribution

When mates or resources are evenly distributed there is little opportunity for one individual to command more than others. However, when they are patchily distributed some individuals may be able to defend more mates or better quality resources (e.g. nest sites, food) than others. A patchy environment can therefore be said to have a high polygamy potential (Fig. 9.1).

(b) Temporal distribution

The key factor here is the 'operational sex ratio' (see also chapter 8), which is the ratio of the number of receptive females to the number of sexually active males at any one time. If all the females were receptive at exactly the same time, then with a real sex ratio of 1:1 in the population, the

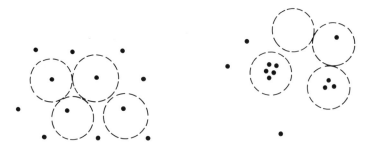

Even distribution.
Little polygamy potential

Patchy distribution.
High polygamy potential

Fig. 9.1. The influence of the spatial distribution of resources (food, nest sites) or mates on the ability of individuals to monopolize more than others. Dots are resources and circles are defended areas. With a patchy distribution of resources or mates there is greater potential for some individuals to 'grab more than their fair share'.

operational sex ratio at the time of breeding would also be 1:1.
There would be little opportunity for a male to mate with
more than one female because by the time he had mated once,
all the other females would have finished breeding. With female synchrony, therefore, there is little potential for polygyny.
Nancy Knowlton (1979) has suggested that female synchrony
may evolve as a mechanism to enforce monogamy on males.
If, on the other hand, the females became receptive in sequence there would be a great opportunity for a male to mate
more than once. It is clear that the greater the operational sex
ratio deviates from unity, the greater the potential for polygamy.

The influence of female synchrony on male mating success
can be illustrated by comparing two species of anuran amphibians. The common toad, *Bufo bufo*, is an explosive breeder;
all the females lay their eggs in a period of about a week. This
means that a male only has time to mate with one female, or at
best two, before the breeding season has ended. Bullfrogs,
Rana catesbeiana, in contrast, have a prolonged breeding season with females arriving at the pond over several weeks.
Males which can defend the most attractive spawning sites
may mate with as many as six females in a season.

We shall now examine how the evolution of polygamy is
determined by the spatial and temporal distribution of resources
and females. To make the discussion clearer we will divide
polygamy into various ecological categories.

Resource defence polygyny in birds

There are many examples where the males that defend the
best quality food patches or nest sites gain access to several
females while males with poor resources never mate. Male
orange-rumped honeyguides, *Indicator xanthonotus*, defend
bees' nests and when a female comes to feed on the wax the
male copulates with her. In effect the male is trading food for
sex; the more bees' nests he can defend, the more females he
will attract. One territorial male was seen to mate with 18
females while non-territorial males failed to mate at all
(Cronin & Sherman 1977).

The problem for a female who mates with an already paired
male is that she may suffer a decrease in reproductive success
because the resources or care by the male have to be shared
with other females. In the yellow-bellied marmot (*Marmota
flaviventris*), males who defend the best burrow sites attract
the most females. However, the more females that breed in the
same burrow system, the lower the success per female (Downhower & Armitage 1971). Female birds who mate polygynously have to share the male's parental care. If they mate
monogamously they can claim the full-time help of the male.

Jared Verner and Mary Willson (1966) suggested that polygyny could arise by female choice for good resources (e.g. nest sites, food) if the difference in quality between male territories was great enough to mean that a female could rear more young by sharing a good territory with another female rather than by mating monogamously with a male on a poor territory. Gordon Orians (1969) summarized this idea in a graphical model which is explained in Fig. 9.2. The model makes the following predictions.

Prediction 1. First, if it is differences in the quality of male territories that influences the evolution of polygyny then we would predict that polygyny would occur most often in 'patchy' habitats where differences in territory quality will be most marked (Fig. 9.2). Verner and Willson (1966) tested this prediction by comparing the habitats and mating systems of all the North American passerine birds. Of the 291 species, 14 were regularly polygynous or promiscuous. Thirteen of these bred in marshes, prairies or savannah-like habitats where there are often big differences in productivity even between small adjacent areas. There are two problems in the interpretation of these data. One, that we have encountered before, is

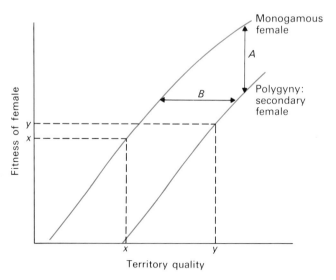

Fig. 9.2. The polygyny threshold model (from Orians 1969). It is assumed that reproductive success of the female is correlated with environmental factors, such as the quality of the male's territory in which she breeds, and that females choose mates from the available males. In the model, a female suffers a decrease in fitness A by going as a secondary female to an already mated male compared with the fitness she could expect if she had a male all to herself. Despite this drop in fitness, provided the difference in quality between the territories is sufficient (B = the polygyny threshold), a female may expect greater reproductive success by breeding with an already mated male. For example, a female who shared a male on territory y would do better than a female who had a male all to herself on territory x.

the danger of invoking 'patchy resources' as an explanation without rigorous measurements of resource distribution and abundance. The other is the problem of which is the appropriate taxonomic level for analysis of this comparative data. Nine of the 14 polygynous species are in one family, the Icteridae, and it is difficult to know whether we should treat these as nine independent cases or one (see chp. 2 for a discussion of this problem). Nevertheless, Verner and Willson's analysis does suggest that differences in male territory quality may be an important ecological correlate of polygyny. *Prediction 2.* The second prediction from the model is that a male's territory quality will influence his reproductive success, in particular the number of females he attracts. The two most important resources that will influence success are probably food availability and nest sites. In the long-billed marsh wren (*Telmatodytes palustris*) males with the largest territories, which contain the most food, attract two females while those with small territories attract only one mate or remain as bachelors (Verner 1964). In red-winged blackbirds (*Agelaius phoeniceus*) males whose territories contain the type of vegetation that provides the safest nest sites are the ones that attract the most females (Holm 1973).

One of the clearest examples of how a male's territory quality influences female reproductive success is Wanda Pleszczynska's (1978) study of the lark bunting (*Calamospiza melanocorys*). This bird is a summer visitor to the grasslands of North America. The males arrive first in the spring, compete for territories and then display with song flights to attract females which arrive soon afterwards. Once a female settles on a territory she builds a nest and the male helps her care for the offspring. Pleszczynska found that some males attracted a second female even when there were unmated males available in the population. These 'secondary females' got no help with breeding from the male, who was already fully occupied with his first mate, and their breeding success was lower. Nevertheless they still chose to mate polygynously. The resource on a territory which was most important for reproductive success was a shady nest site. The main cause of nestling mortality was over-heating in the hot sun and observations showed that nests built in the shade had much greater fledging success. This was also demonstrated experimentally by putting up small strips of plastic above some exposed nests to act as umbrellas and so protect the young from the sun. The success of these experimental nests was significantly greater than that of unshaded controls. As we might predict, the males who attracted two females were those which defended territories with the most cover.

There have been surprisingly few experiments to test whether a male's reproductive success can be increased by manipulating the aspects of territory quality which are suspected to influence his ability to attract mates. An experimental approach provides the most direct test of prediction 2. Pleszczynska and Hansell (1980) increased the number of females in some male lark bunting territories by providing extra shade, while Ewald and Rohwer (1982) did likewise for male red-winged blackbirds by the provision of extra food.

Prediction 3. The third prediction from the model is that, for females settling at the same time, the reproductive success of those who choose to mate with an already-paired male will be at least as great as those who choose an unpaired male (as explained in Fig. 9.3). For example, in the lark bunting a female who mates polygynously is expected to make this choice because access to a better nest site offsets the disadvantage of having to rear the young without any male help. As predicted, Pleszczynska found that secondary females had fledging success at least equal to that of monogamous pairs breeding at the same time. It is worth pointing out the difficulties of measuring reproductive success to test this prediction. As we saw in the great tit study, described in chapter 1, the number of young fledging from a nest may not always give a good indication of success because fledging weight will influence survival. Furthermore, a female is presumably selected to maximize her lifetime reproductive success which may not be the same as maximizing the number of young produced each year because her chances of surviving to breed again decreases with the brood size she produces.

THE 'SEXY SON' HYPOTHESIS, DECEPTION AND SEXUAL CONFLICT

The polygyny threshold model has been very influential for our understanding of the evolution of polygamy. However, so far the only study which has provided good support for prediction 3 above is the lark bunting. In most cases secondary females have lower reproductive success than monogamous females (Table 9.4). In both of the species in Table 9.4, the polygynous males are those with the best quality nest sites in their territories (see Askenmo 1984). Nevertheless, the choice of a good nest site by the secondary female does not compensate her for the loss of male help with chick feeding. She would, therefore, have done better to mate with an unpaired male rather than choosing to mate polygynously.

One possible objection to this interpretation is that the average success of the two classes of female (monogamous and

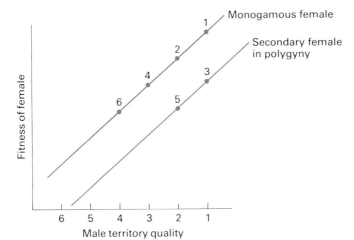

Fig. 9.3. An example of how females are expected to settle on male territories according to the polygyny threshold model (from Garson *et al.* 1981). There are six male territories, with 1 being the highest quality and 6 the lowest. Females settle on the territory which gives them the highest fitness. The numbers on the fitness curves give the predicted settlement order of six females. Females who settle at about the same time are expected to have similar fitness (e.g. females 3 and 4; 5 and 6). In this example, two males will be polygynous, two will be monogamous and two will remain unmated.

polygynous) does not represent the reproductive pay-offs which are available to a particular female when she has to decide whether to mate monogamously or polygynously. For example, settlement order may influence reproductive success, and young females who often have lower reproductive success through inexperience may settle later and thus be more likely to be secondary females. The key experiment to perform would be to remove the primary female so that the male was now free to help the secondary female with chick rearing. If it is lack of male help which causes the lower reproductive success of secondary females, rather than poor female quality, then their success should now increase to the same value as primary females. While recognizing these difficulties of measuring reproductive pay-offs, we now consider two other hypotheses for the data in Table 9.4.

Hypothesis 1. The 'sexy son' hypothesis

Weatherhead and Robertson (1979) have suggested that although the secondary female may produce fewer offspring, this may be offset if she has sons who inherit their father's ability to be polygynous. The female therefore loses out in the first generation, but then makes up for this in the second

Table 9.4. Mean number of young fledged per nest by monogamous and polygynous females in two species of birds. In both cases the first-mated (primary) females in polygynous mating systems get the male's help with chick rearing and produce as many young as a monogamous female. The second-mated (secondary) females, however, get little or no help from the male and have lower reproductive success

	Monogamous female	Polygynous females	
		Primary	secondary
Pied flycatcher, *Ficedula hypoleuca* (Alatalo *et al.* 1981)	5.4	5.4	3.3
Great reed warbler, *Acrocephalus arundinaceus* (Catchpole *et al.* 1985)	3.3	3.2	1.1

generation when her 'sexy sons' sire lots of grandchildren compared to the sons of monogamous females. According to this hypothesis (which is a derivative of Fisher's sexual selection argument, chp. 8), secondary females are still making the correct choice when they decide to mate polygynously, but male quality is another factor which must be taken into account as well as territory quality in the x axis of the polygyny threshold model (Fig. 9.2).

Heisler (1981) has developed a model to explore the conditions under which this hypothesis may hold. It is assumed that males tend to be of the same mating status as their father and that females mate at random with respect to their parentage. Finally, it is assumed that the frequency of polygyny in the population remains constant between generations. This is equivalent to assuming that environmental factors limit the number of breeding situations in which a polygynously mated female can produce the threshold number of offspring to make polygyny pay her. If the frequency of polygyny, the sex ratio, and the average number of females per polygynous male are known, then it is possible to estimate the degree of father-son resemblance necessary to compensate for an observed disadvantage associated with polygyny in terms of number of offspring produced (see Fig. 9.4).

Can the sexy son hypothesis explain why the secondary females in Table 9.4 choose to mate polygynously? The answer is no. For the pied flycatcher, the heritability of male mating status would have to be 0.85 for 'sexiness' of sons to offset the loss in offspring numbers. Heritability of male mating status is not known because young birds disperse and breed away from their natal area. However the repeatability of a male's status between different years is only 0.29, which must give an upper limit to the heritability value. For the great reed warb-

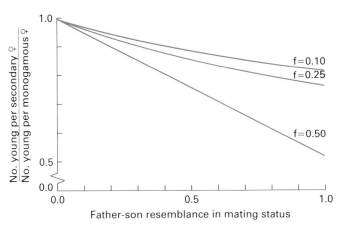

Fig. 9.4. The 'sexy son' hypothesis. A secondary female can produce fewer offspring compared to a monogamous female and still have greater fitness provided her sons inherit their father's tendency to be polygynous. The lines show the threshold values for the sexy son effect to compensate the secondary female for her loss of reproductive success. The model assumes that the sex ratio is 1:1 and that polygynous males have two females. The three curves are for different values of frequency of polygynous males in the population (f). From Heisler (1981).

ler, even if heritability of male mating status was 1, the secondary female would have to produce at least 0.76 times the number of offspring as monogamous females and in fact they produce only 0.33 times as many. For both these species, therefore, the costs of being a secondary female are too high to make polygyny of benefit to the female (Alatalo *et al.* 1981; Catchpole *et al.* 1985). Furthermore, recent theoretical work has questioned the feasibility of the sexy son hypothesis (Kirkpatrick 1985).

Hypothesis 2. Sexual conflict

Although the secondary females suffer in Table 9.4, the polygynous males clearly do better than monogamous males because their total reproductive success is the sum of that of the primary and secondary females. The polygyny threshold model may have limited value, therefore, because it ignores conflicts of interest. Polygyny may occur not only where females choose to mate polygynously but where males are able to gain an advantage at the expense of females. In both the pied flycatcher and great reed warbler it appears that females have difficulty in assessing male mating status and males may actively deceive females into mating polygynously. Great reed warblers breed in dense reed beds and it is possible that a secondary female is unaware that the male she is courting has another female (Catchpole *et al.* 1985). In the pied flycatcher, once a male has attracted one female and she has laid her eggs, he then goes to another nest hole and tries to attract another female (Fig. 9.5). Males do not, however, simply advertise from the nearest nest site but go, on average, 200 m away and even up to 3.5 km! They may do this to make it impossible for a new female to tell that they are already mated (Alatalo *et al.* 1981). By the time the secondary female has laid her clutch and the male has deserted her to go back to his first female, it is too late in the season for it to be profitable to start another clutch and so she has to make the best of a bad job and rear her offspring alone. Polygyny in these species, and possibly others too, is perhaps best interpreted as an example of successful deception of females by males.

SEXUAL CONFLICT AMONG DUNNOCKS

Conflicts of interest within a breeding population have led to a very variable mating system in another passerine bird, the dunnock *Prunella modularis* (Davies & Houston 1986). Here, mating combinations include simple pairs (monogamy), a male with two females (polygyny), a female with two males, both of

Fig. 9.5. Once a male pied flycatcher has attracted one female, he flies off to another nest site some distance away and tries to attract another. Secondary females suffer because they get little or no help from the male in chick rearing. However females probably are unable to assess whether the male they pair with has another female because of the large distance between a male's two nest sites.

whom mate with her and feed her young (co-operative poly-andry), and combinations where two males share two or more females (polygynandry). Males have greatest reproductive suc-cess with polygyny (Table 9.5). However, a female has to share male parental care with polygyny and has lower reproductive success than with monogamy. Polygynous females fight with each other and the dominant one attempts to drive the other female away and thus claim the male to herself. The male on the other hand tries to keep between his squabbling females so that both remain with him!

Although a female has greater reproductive success with monogamy than with polygyny, she does best of all with co-operative polyandry, where she has the assistance of two males to rear her young. Monogamous females, therefore, solicit co-pulations from other males who approach them in the hope that they will remain and provide parental care. The cost of polyandry from a male's point of view, however, is shared paternity and it works out that this is the least profitable mating system for a male. As expected, the males in poly-androus combinations are constantly fighting during the mat-ing period, with the dominant male attempting to drive the subordinate away or at least to prevent him from mating with the female.

The different mating systems of the dunnock can be viewed as a continuum reflecting different abilities of individuals to monopolize mates. Polygyny occurs where males can gain an advantage at the expense of females, and co-operative poly-andry where females can gain at the expense of males. Mono-gamy and polygynandry represent mating systems where nei-ther sex is able to gain an advantage at the expense of the other. In polygynandry involving two males and two females, for example, the dominant male is unable to drive the other male off to claim both females (polygyny) and the dominant female is unable to drive the other female off and so claim both males (polyandry). Who is able to win the sexual conflict in particular cases depends on differences in individual com-petitive ability and food distribution, which determines range size and thus ability to monopolize mates (Davies & Lundberg 1984).

Female defence polygyny in mammals

In mammals, where males usually provide no parental care (97% of mammal species), differences between species in mating systems are related to differences in the economic defendability of females by males. Female dispersion is usually

Table 9.5. Reproductive success of male and female dunnocks in different mating combinations. The cost of polygyny to females is shared male parental care. The cost of co-operative polyandry to males is shared paternity (Davies & Houston 1986)

Mating system	No. adults who care for young	Reproductive success No. young fledged per breeding season	
		Per female	Per male
Polygyny, ♂2♀♀	♀ + part time help of 1♂	3.8	7.6
Monogamy, ♂♀	♀ + full time help of 1♂	5.0	5.0
Polyandry. 2♂♂♀	♀ + full time help of 2♂♂	6.7	α4.0* β2.7

*It is assumed that the two males share paternity in proportion to the copulations each achieves, dominant (alpha) male 0.6; subordinate (beta) male 0.4.

determined by predation pressure and the spatial and temporal distribution of resources, particularly food (see chp. 2), and female dispersion in turn influences the ease with which males can monopolize females. The three main factors which will influence the economics of female monopolization are female range size, female group size and the seasonality of reproduction, which will determine the degree of competition for mates at any one time. The following comparative survey is based on the reviews by Clutton-Brock and Harvey (1978) and Clutton-Brock (in prepn.).

1   *Females solitary: range defensible by male.* In over 60 per cent of mammalian species females are solitary and a male defends a territory which overlaps one or more female ranges. If female ranges are small relative to the area which a male can defend then the male can be polygynous. If female ranges are larger, then the male may only be able to defend one female, hence monogamy (e.g. most rodents and nocturnal prosimians; Kleiman 1977). Usually the male simply mates with the female then leaves her to care for the offspring alone. More rarely (3% of mammalian species) the male may help defend the young against predators (e.g. Klipspringer, *Oreotragus oreotragus*) or carry them (e.g. siamangs, marmosets, tamarins) or help to feed them (e.g. jackals, wild dogs). Such obligate monogamy occurs where female ranges are small enough for a male to defend but where a male is unable to defend a large enough area to have more than one female (Rutberg 1983). In

these cases a male may then maximize his reproductive success by providing parental care. Species with obligate monogamy tend to have large litter sizes. For example this mating system is common in canids, which have large litters, and rare in felids, which have smaller litters. Marmosets, which have male parental care, produce twins whereas most monkeys produce only single offspring. If a male marmoset dies then the female often deserts the young so male assistance with parental care seems to be important.

2 *Females solitary: range not defensible by male.* Where females wander more widely then males may rove over wide ranges, associating with females temporarily while they are in oestrus. This occurs in moose *Alces alces* and orang utans *Pongo pygmaeus*; in the latter species the females move over large ranges following the fruiting seasons of different species of plants (Mackinnon 1974).

3 *Females social: range defensible by male.* Where females occur in small groups in a small range, then a single male may be able to defend them as a permanent harem within his territory (e.g. black and white colobus *Colobus guereza*, Hanuman langurs *Presbytis entellus*). When a new male takes over the territory, he often kills the young offspring fathered by the previous male, thus bringing the female into oestrus sooner and hastening the day he has a chance to sire his own young (Hrdy 1977). Where females occur in larger groups several males (often relatives) may defend the territory together (e.g. red colobus *Colobus badius*, chimpanzees, lions). Joint defence by several males may increase the length of tenure of a harem and may also be necessary for economic defence of large groups of females wandering over a large range (Bygott *et al.* 1979).

4 *Females social: range not defensible by male.* Sometimes groups of females wander over ranges which are uneconomic for one or more males to defend. The ways in which males compete for females then depends on how predictable female group movements are in time and space.

(a) Daily female movements predictable

Sometimes the group of females wanders over a large range but they use regular routes to particular water holes or rich sources of food. In these places the males may defend small territories, much smaller than the females' range, and attempt to mate with them as they pass through (e.g. topi *Damaliscus lunatus korrigum*, Grevy's zebra, *Equus grevii*). Such defence of mating territories may occur where more direct competition between males, such as fights for harems, would be costly because males are unable to build up the food stores necessary for them to engage in intense male-male interactions (Owen-Smith 1977).

More intense competition for territories may lead to the defence of tiny patches of ground in areas where females pass through, so the result is something like the lek systems described for birds (see later). Male waterbuck (*Kobus elypsiprimnus*) defend territories of 0.25 to 0.5 km² and mate with the females as they pass through the territory in small groups. Puku (*K. vardini*) have smaller territories of 0.1 km² and finally Lechwe (*K. lechwe*) and Uganda kob (*K. kob thomasi*) have tiny territories which may be no more than 15 to 30 m across. All four of these species, in the genus *Kobus*, inhabit marshes where females may come to particular areas to feed on rich aquatic vegetation. In the Lechwe and Uganda kob, females move in large herds and males may only be able to economically defend very small territories (Leuthold 1966). These examples of ungulate leks may simply be extensions, therefore, of resource defence polygyny, with territories being small because of the high costs of defence.

(b) Daily female movements not predictable

Here males tend to follow the females, rather than waiting for the females to come to them. Where females live in small groups males may rove and associate with individuals in oestrus (e.g. mountain sheep, *Ovis canadensis*, elephants, *Loxodonta africana*). Where female groups are larger, the males may attempt to defend harems.

1 *Seasonal harems.* If all the females come into oestrus at a particular season then it may pay a male to put on energy reserves to enable him to have a burst of energy expenditure on harem defence. For example, male red deer (*Cervus elaphus*) stags compete to defend harems during the one month in which all females come into oestrus. A male's reproductive success depends on his harem size and the length of time for which he can defend the harem and this in turn depends on his body size and fighting ability. After the mating season the males are reduced to very poor body condition and are literally 'rutted out'! (Clutton-Brock *et al.* 1979).

As another example, female Northern elephant seals (*Mirounga angustirostris*) haul up on beaches to drop their pups and mate again for the production of next year's offspring. Because the females are grouped, due to the localized nature of the breeding grounds, they are a defendable resource and the males fight with each other to monopolize them. The largest and strongest males win the biggest harems and in any one year all the matings are performed by just a few males (Le Boeuf 1972, 1974; Cox & Le Boeuf 1977). The Southern elephant seal (M. leonina) has a similar mating system (McCann 1981, Fig. 9.6).

To be a harem master is so exhausting that a male usually only manages to be top ranking for a year or two before he dies. In the process of defending his harem against other males he sometimes tramples on his females' new-born pups. Although this is obviously not in the females' interests, these pups were probably not sired by the male himself because he is unlikely to have been a harem master the previous year. From the male's point of view, therefore, there is little cost in damaging or even killing the pups; his main concern is to protect his paternity.

2 *Permanent harems.* Where females do not all come into oestrus at a particular time, males may defend permanent harems for the whole duration of their reproductive life (e.g. hamadryas baboons, *Papio hamadryas* and gelada baboons, *Theropithecus gelada*; Dunbar 1984; Burchell's zebra, *Equus burchelli*). Often several groups (male plus harem) go around together forming a large 'super-group'. Where females go around in still larger groups, several males may associate with large groups of females and compete with each other for matings. Such 'multi-male' groups occur in buffalo, *Syncerus caffer* and olive baboons, *Papio anubis* (Altmann 1974).

Leks and choruses

Finally, polygyny can occur even without resource defence or female defence. In some species of birds and mammals the males congregate on display grounds where they defend tiny territories. In manakins, *Manacus m. trinitatis*, territories are just bare patches of ground a few metres across (Lill 1974). Females visit the male assemblies, or leks, solely to copulate. Males of lek species provide neither resources nor parental care.

In leks, the females appear to be free to pick a male of their choice. Most females choose the same few males with the result that one or two males on each lek perform most of the copulations. For example, male hammer-headed bats, *Hypsignathus monstrosus*, congregate in groups high up in the tree canopy and call to attract females. In one lek of 85 males studied by Jack Bradbury (1977), 79% of the copulations were accounted for by just 5 males (see also Fig. 9.7). In most cases it is not at all clear whether females are choosing a particular site on the lek or showing a preference for particular male

(a)

(b)

(c)

Fig. 9.6. (a) Female Southern elephant seals haul up on beaches to give birth. (b) The large males fight to defend harems. (c) Successful males copulate with the females. Photos by T.S. McCann.

characteristics, such as vigour of display. Why do the males aggregate into leks? Five main hypotheses have been proposed (Bradbury & Gibson 1983).

1 Males may aggregate to decrease predation.

2 They may aggregate to increase the efficiency of attracting females.

3 They may aggregate if suitable display sites are limited.

4 Males may clump near 'hotspots' through which the largest number of females are likely to pass.

5 Males may aggregate because females prefer clumped males, either because this facilitates mate choice or because it reduces predation on females.

At the moment there are not enough detailed studies to enable us to distinguish between these hypotheses. Furthermore, it is not at all clear what benefits females obtain from their choice of males on the lek. There seem to be two main possibilities; either they get 'good genes' (see chp. 8) or simply gain a safe copulation.

The aggregations of males into choruses in some species of frogs is similar to the aggregations of male birds on leks. In the neotropical frog, *Physalaemus pustulosus*, calling males suffer heavy predation by bats, *Trachops* spp. Calling males are safer in large choruses because of the dilution effect (chp. 6). Individual males also attract more females in large choruses (Fig. 9.8). This may be simply because large choruses only

213

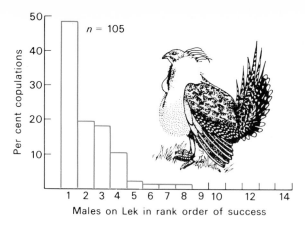

Fig. 9.7. In sage grouse (*Centrocercus urophasianus*) leks almost all of the copulations are performed by just a few of the males. Gibson & Bradbury (1985) have shown that male display characteristics are the most important predictors of mating success. From Wiley (1973).

form in areas where females are particularly abundant but another possibility, which needs testing experimentally, is that larger choruses present a more attractive stimulus to females. In some frog choruses, therefore, it may pay males to aggregate because this both decreases predation and increases mating success (Ryan *et al.* 1981). It would be nice to have similar data for bird leks but this will be difficult to obtain because females visit several leks and, with internal fertilization, it is difficult to assess a male's reproductive success because of sperm competition.

In summary, polygyny may evolve when males defend good resources, or when they are able to defend groups of females or when they compete directly for access to females in a lek.

Sex role reversal and polyandry in birds

Polyandry is much rarer than polygyny because it is usually the males who will gain the most by maximizing their number of matings. Nevertheless there are a few species of birds where the behaviour of the sexes is reversed and females compete with each other for access to males (Jenni 1974).

The evolutionary pathway to polyandry is most easy to understand in the Arctic waders (Oring 1981; Maxson & Oring 1980). In some species, like the Temminck's stint, *Calidris temminckii*, and the sanderling, *Calidris alba*, the male defends a territory. The female lays one clutch of eggs and the male incubates them while she lays another clutch and then incubates that herself. Therefore the pair produces two broods, one cared for by the male and one by the female. This is an efficient strategy for breeding in the arctic tundra where the

(a)

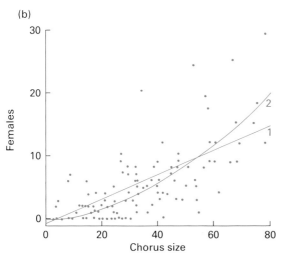

(b)

Fig. 9.8. Male frogs, *Physalaemus pustulosus* aggregate into choruses. In larger choruses individuals are safer from predatory bats (a). The number of females attracted also increases with chorus size (b). The curve (2) gives a better fit to the observed points than a staight line (1) which suggests that the number of females per male increases with chorus size (Ryan *et al.* 1981).

season is very short but very productive. The female is able to lay two clutches in rapid succession because the insect food supply is so rich, and the neat division of labour between the male and female enables them to raise two broods during the short period of high productivity. Two clutches are also a better insurance against predation, which can be heavy.

This system of 'double clutching' could give rise to the evolution of polyandry if the female is not faithful to the first male but, instead, goes off to lay her second clutch for another male. As food abundance increases so that a female can produce many clutches then she will do best by laying for a number of males in turn.

This is exactly what has happened in the spotted sandpiper, *Actitis macularia*, where productivity on the breeding grounds can be so high that the female becomes rather like an egg factory (Fig. 9.9). A female can lay 5 clutches in 40 days, a total of 20 eggs, which represents 4 times her own body weight! Her reproductive success is no longer limited by her ability to form reserves for the eggs but rather by the number of males she can find to incubate them. At some times most of the males are tied up with incubation duties. These ecological conditions have given rise to the evolution of sex role reversal. The females are 25 per cent larger than the males and they compete with each other to get males to care for their clutches. Females defend territories and some may have three males sitting on clutches at the same time. Sometimes the female may incubate the last clutch of the year herself.

Fig. 9.9. A polyandrous wader, the spotted sandpiper. This photograph by Lew Oring shows two brood males in an agonistic encounter at a territory boundary. Female aggression tends to be more intense and peaks at the start of the season when females are competing to acquire males.

Ecology and dispersal

217

Chapter 9
Parental Care
and Mating
Systems

As well as influencing the evolution of parental care and mating systems, ecological factors are also an important determinant of another aspect of reproduction, namely dispersal. The movement of a young animal from its place of birth to that of its first breeding attempt is referred to as *natal dispersal*. When birds and mammals are compared, some striking trends emerge (Greenwood 1980; Table 9.6).

Table 9.6. The number of species of birds and mammals where natal dispersal is more extensive in males or females. From Greenwood (1980)

| | No. species with predominant dispersal | | |
	By males	By females	No sex difference
Birds	3	21	6
Mammals	45	5	15

1 *In both birds and mammals, one sex usually disperses more than the other.*
The result is that close inbreeding is avoided. When an animal mates with a close relative there is a greater chance that any harmful recessive alleles will become homozygous in the offspring and cause low reproductive success.

For example, in the great tit population in Wytham Woods, Oxford, juvenile females disperse more than males so that from a total of 885 pairings only 13 (1.5 per cent) have been inbred pairs. These rare cases were where a male had dispersed more than average or where a female had dispersed less, with the result that a son mated with his mother or a brother with a sister. Such inbred pairs had lower reproductive success than outbred pairs (Greenwood *et al.* 1978). It is known that inbreeding also results in decreased reproductive success in mammals and it is one of the problems associated with trying to breed from small populations in zoos (Ralls *et al.* 1979).

Another way of avoiding inbreeding would be to recognize close relatives. This could come about through early learning because the individuals most closely encountered when young are likely to be kin (see chp. 11). This mechanism may operate in man; in a study of Israeli kibbutzim, where children are brought up communally, it was found that individuals never married anyone with whom they were associated when young, even when they were not close relatives (Shepher 1971).

2 *In birds, females disperse more than males.*
As we saw above, in many species of birds the male helps care for the young. Often he defends a territory and females may

choose a male on the basis of his territory quality. It may pay a male to remain near his birth site because it might be easier to set up a territory in the vicinity of relatives, for example by inheriting part of the father's territory (see chapter 12). Once this happens, it may pay the females to disperse to avoid inbreeding. They may also benefit by moving so that they are able to sample many male territories and choose the best.

3 *In mammals, males disperse more than females.*

Male mammals are more often polygynous than birds, and their mating system is usually based on mate defence rather than resource defence, with the male contributing little to the care of the offspring. In mammals, a male will benefit most from gaining access to a large number of females and so male dispersal may have been favoured.

It is clear that much more data are needed on the costs and benefits of dispersing to breed versus staying at home. For any one individual the pay-offs of the two options must depend on what others in the population are doing and so we really need a theory which will analyse the problem in terms of what will be the evolutionarily stable strategy for a male and a female. We should also remember that in many cases, particularly mammals where the male contributes little to parental care, the cost of inbreeding for a male will be less than that for a female. For example in the olive baboon (*Papio anubis*) inbreeding is avoided because males transfer from their natal troops whereas females do not. However, before they disperse, young males may still surreptitiously try to mate with females, even close relatives. The females combat this risk of incestuous mating by preferring to associate with transferred (strange) males rather than natal (familiar) males who are likely to be relatives (Packer 1979).

CONSEQUENCES OF SEX DIFFERENCES IN DISPERSAL

A consequence of differential dispersal of the sexes is that members of the sedentary sex will tend to be closely related to their near neighbours, and so we may expect more altruism between them than between members of the dispersing sex where individuals living near by will tend to be unrelated. (see chp. 11).

In mammals, where females are usually more sedentary, the females in an area are often relatives. Altruism between females is common; for example female ground squirrels give alarm calls to warn other females of the approach of a predator and in lionesses, females will suckle other females' cubs (chp. 11).

In birds it is the other way round. Males disperse less and

so in a particular area it is the males who are closely related and who show most altruism to each other. In chapter 12 we will encounter several species of birds where the males help look after another individual's offspring.

Conclusion

It is clear from our comparative approach to different groups of animals and different habitats that ecological factors are important in shaping parental care and mating system. The distribution of resources, such as suitable egg laying sites and food, and the distribution of the females in space and time all influence the way individuals can behave to maximize their reproductive success. Proximate constraints, like the mode of fertilization, may also determine the mating system by the way they predispose one parent to care for the offspring. We have also seen that the evolution of the behaviour we now see will depend on what came beforehand. Which sex guards the offspring in uniparental care in fish and birds, and the evolution of polyandry in waders are both easier to understand if we take account of the ancestral state of the behaviour.

Categorizing species in terms of their characteristic 'mating system' is fine when we are making broad comparisons across taxa or across habitat types. However we should not expect all individuals within a species to behave in exactly the same way. While some individuals are displaying or defending a resource to attract mates, others may be achieving reproductive success by more devious and sneaky methods. In the next chapter we will take a closer look at the evolution of these alternative strategies.

Summary

Differences between species in parental care and mating systems can be correlated with differences in physiological constraints and ecology. Internal fertilization and specializations such as lactation predispose females to perform parental care, while territory defence, external fertilization and the need for two parents to look after the young may predispose males to show parental care. In birds and mammals polygyny occurs when males defend good resources (e.g. lark bunting) or groups of females (e.g. elephant seals). There may often be conflicts of interest between males and females over the mating system which maximizes an individual's reproductive success (e.g. pied flycatcher and dunnock). There are sex differences in dispersal in birds and mammals which might also be related to ecological factors.

Further reading

Ridley (1978) reviews paternal care. Good reviews of mating systems in particular animal groups are those by Bradbury and Vehrencamp (1977) on bats, Wells (1977) on frogs and toads, Pitelka, Holmes & MacLean (1974) and Erckmann (1983) on arctic wading birds, and Christy & Salmon (1984) on fiddler crabs. Smith *et al.* (1982) showed experimentally how changes in the sex ratio can lead to increased polygyny in song sparrows. Alexander (1975) discusses chorusing behaviour and leks in insects. The paper by May (1979) examines the costs and benefits of staying at home and being incestuous versus dispersal and outbreeding.

Topics for discussion

1 What is the link between parental investment and mating systems?

2 What data would you need to collect to test the polygyny threshold model?

3 Could Maynard Smith's model of parental investment be tested experimentally?

Chapter 10. Alternative Strategies

In the last chapter we used mainly a comparative approach to try and understand differences between species and we characterized each by a 'typical' mating system. During recent years, however, it has become clear that there are marked differences between individuals within a species in the way they compete for scarce resources. A decade ago, if an animal was seen behaving in a different way from the majority of the population it was often thought to be abnormal. Male ducks that engaged in forced copulations instead of courting females by displays were said to be behaving abnormally due to overcrowding. If we observed a male bullfrog sitting silently in the middle of a chorus, while other males were croaking loudly to attract females, we would perhaps have thought that it was ill or having a rest.

Nowadays, whenever we see an individual doing something different we are tempted to label it as a 'strategy'. Silent male frogs may not be tired after all, they may be employing sneaky strategies. There are three main reasons for this change of emphasis. First, the realization that evolutionary arguments must be framed in terms of benefit to the individual, rather than for the good of the species, has led us to expect individuals to compete selfishly with each other. If some males are attracting females by calling then we now expect to find others parasitizing their efforts and behaving as sneaks. Once we have seen courtship as a conflict between individuals (chp. 8), rather than as a co-operative venture, we are not surprised to see some males attempting to force mating with females.

Second, the application of game theory to the study of behaviour (chp. 7) has shown that it is possible, in theory, to have stable equilibria with individuals in a population behaving in different ways. Finally, an increase in the number of field studies with individually marked animals has shown that indeed there are often several different strategies used within the same species to compete for a mate, a nest site or some other scarce resource.

In this chapter we will examine some examples of individual differences in competitive behaviour and discuss how they might have evolved.

Hypotheses for the occurrence of alternative strategies within a species

Some authors prefer to use the word 'strategy' to give a complete specification of what an animal will do when competing for a scarce resource, and the word 'tactic' for the behavioural components of a strategy. For example, if young male dragonflies searched for females by wandering over large areas while older males defended territories and waited for females to come to them, then the term strategy would describe the whole behaviour pattern 'if young wander, if older defend a territory', and the term tactic would be applied to 'wandering' and 'territorial defence'. Although this distinction is fine in theory, as we shall see it is often difficult to distinguish tactics from strategies and we prefer to use the word strategy somewhat loosely to describe any behaviour pattern or structure used by an individual to compete for a scarce resource.

We consider three main hypotheses for the occurrence of alternative strategies within a species, giving examples under each heading to illustrate the arguments.

CHANGING ENVIRONMENT

The best strategy may depend on the habitat, and if this is patchy in space or changes frequently in time then several strategies may persist. In some cases, individuals may change their behaviour depending on the habitat. When a male speckled wood butterfly searches for females in patches of sunlight on a woodland floor, it defends a territory and sallies out from favourite perches to inspect passing objects. When the same male flies up to the tree canopy, where females are scarcer, it adopts a different searching behaviour and patrols over a large area.

In other cases, different individuals play different strategies. An example is the different colour morphs of the three-spined stickleback (McPhail 1969; Semler 1971; Moodie 1972). Some males have bright red throats while others have dull throats. The red males are more attractive to females; in a choice test with a red male at one end of a tank and a dull male at the other, most females selected the red male. The attractiveness of a dull male can be increased simply by dying his throat red, so it is the colour itself that is attractive to females rather than something special about the behaviour of red males. Why, then, are not all males red?

The answer is that although a red throat brings a benefit in terms of increased attractiveness to females, it also imposes a cost because red males are more susceptible to predation by trout. Experiments showed that red males were particularly preyed upon in bright light, when their colour enabled the predator to spot them more easily. Samples from North American lakes reveal that in deep water most males are of the red type. Presumably in these dark waters the mating advantage of red males outweighs the disadvantage of increased predation. In shallow waters, however, nearly all the males are dull in colour. Here, in the brightest light, dull males do best because the red males are very conspicuous and quickly eaten by predators.

The colour of sticklebacks is therefore selected for by a trade-off between mating and predation. Both red and dull males persist because each does better in a different environment.

MAKING THE BEST OF A BAD JOB

Sometimes, perhaps because of its small size, an individual is unable to compete successfully by fighting or displaying. Instead, it must make the best of its poor circumstances by employing some alternative strategy. Even though it ends up with fewer rewards than others, it will still be doing the best it can given its own competitive ability.

Body size is often age dependent. Unable to compete with larger rivals, young animals may attempt to steal resources by sneaking. We have already seen in chapter 8 that the largest bullfrogs win the best territories and attract females by croaking. Small, young males are not strong enough to defend a territory and so they behave as satellites, sitting silently near a calling male and attempting to intercept and mate with any females he attracts (Fig. 10.1). They are not very successful; only 2 out of 73 matings were by satellite males (Howard 1978b). Nevertheless this is probably their best chance of getting a female when they are small, and the satellites make the best of a bad job by parasitizing the largest calling males on the best territories, where most females will be attracted.

Young male elephant seals have no hope of defending a harem themselves and instead they attempt to sneak copulations by pretending to be a female and joining a large bull's harem! (Le Boeuf 1974). Both young bullfrogs and young elephant seals will change their strategies from sneaking to fighting as they get older and grow larger and stronger.

In other cases, small body size may be fixed throughout an individual's lifetime and be a consequence of poor feeding conditions when young. An example is the bee, *Centris pallida*, where the largest males are three times the weight of the smallest males (Alcock *et al.* 1977). Large males search for

Fig. 10.1. A male bullfrog (left) calls from his territory and attracts a female who is entering the pond on the right. In the middle of the picture is a small male who sits silently in the larger male's territory as a satellite and attempts to intercept the female on her way to the caller.

females by patrolling over the ground, searching for buried virgin females about to emerge. When they discover an emerging female they dig her up and copulate. It takes several minutes to dig up a female, during which time other males are attracted to the site by the activity. There are often violent fights and only large males are able to defend their discoveries successfully.

It is not surprising, therefore, that only large males adopt the strategy of patrolling and digging. Small males search for mates by hovering above the emergence areas and pursuing airborn females who have escaped the diggers. Intermediate sized males may adopt both digging and hovering strategies. Observations showed that large males clearly had the greatest mating success and so it is probable that the smaller males are forced to adopt hovering throughout their lives to make the best of a bad job.

These are all examples of strategies that are conditional on an individual's phenotype, for example 'if big, fight; if small, sneak'. The largest and strongest individuals have the greatest success and others are forced by circumstance to adopt less successful, alternative strategies.

ALTERNATIVE STRATEGIES IN EVOLUTIONARY
EQUILIBRIUM

Even when there are no constraints from the environment, or phenotypic constraints such as body size, individuals may still differ because the pay-off for one strategy depends on what others in the population are doing. We have seen examples of this in chapter 7 where different fighting strategies, such as Hawk and Dove, may coexist as a mixed ESS.

If two strategies we observe in nature are an example of a mixed ESS, then at equilibrium we would expect them to enjoy, on average, equal success, just like Hawk and Dove in our hypothetical game in chapter 7. This is a very different prediction from the previous section where individuals employing a strategy that makes the best of a bad job are expected to have lower success than others. We consider three examples in detail.

(a) Salmon: hooknoses and jacks

Coho salmon (*Oncorhynchus kisutch*) spawn in freshwater streams from November to January on the west coast of North America. After spawning all the adults die. The young fry remain in the streams for a year and then migrate to the ocean as smolts. Females return to breed in the streams at the age of 3 years. For males, however, there are two life history pathways (Fig. 10.2). Small males, called 'jacks', mature at the age of 2 while others, called 'hooknoses', do not mature until 3 years of age. Hooknose males may be two or three times the size of the jacks.

On the breeding grounds females excavate nests in the gravel beds of the streams. The large hooknose males have exaggerated snouts and enlarged teeth and they fight for access to the spawning females. The winners fertilize the eggs externally as they are laid in the nest. By contrast, the small jacks lack secondary sexual characteristics and are relatively cryptic on the breeding grounds. Instead of fighting, they attempt to gain

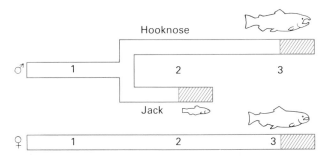

Fig. 10.2. Two life histories exist in male coho salmon: precocious maturity at age 2 as a small jack, or delayed maturity to age 3 as a large hooknose. The term 'hooknose' is derived from the exaggerated snout and enlarged teeth which develop at maturity and are used in fighting. By contrast, 'jacks' lack secondary sexual characters and are relatively cryptic on the breeding grounds. Both jack and hooknose males die after breeding, as do the females. From Gross (1985).

access to females by sneaking. They hide behind rocks or debris to avoid aggression from the hooknoses and surreptitiously gain close proximity to the female during egg laying. Proximity to the female influences the chances of fertilizing the eggs. Since jack density in the streams is too high for all to sneak successfully, some are forced to fight. Likewise sometimes the hooknoses attempt to gain access to females by sneaking. Observation has shown that the smaller males were more successful at sneaking and the larger males at fighting (Fig. 10.3).

The difference between the two male strategies appears to be genetic. Genes coding for an inferior life history in terms of reproductive success should be eliminated from the gene pool by natural selection. How then can we explain the coexistence of the two strategies? The most likely hypothesis is that the two strategies are maintained by frequency dependent selection such that the average reproductive success of each is the same. For example, an increase in the proportion of jacks in the population would increase competition among them for opportunities to sneak. More and more jacks would be forced to fight and the average reproductive success of this strategy would decrease. On the other hand, any decrease in the proportion of jacks would increase an individuals chances of sneaking and so the average reproductive success would increase. The same kind of argument holds for hooknoses; the pay-offs for this strategy too would decrease as the proportion of hooknoses increased and the degree of fighting increased.

This may, in theory, give rise to stable alternative life history strategies (Fig. 10.4). If the alternative male strategies represent a stable polymorphism in salmon then the lifetime

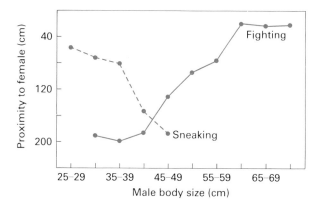

Fig. 10.3. Male proximity to spawning females by either fighting or sneaking in coho salmon. Fighting is most effectively done by large males and sneaking by small males. From Gross (1985).

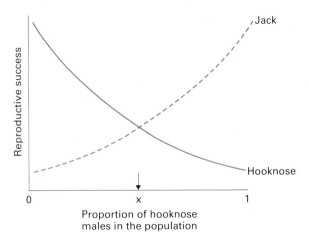

Fig. 10.4. Hypothetical relationship between the success of a strategy (e.g. hooknose versus jack in salmon) and its frequency in the population. The success of each strategy is highest when it is rare. Frequency dependent selection will result in the proportions of the two strategies being maintained at x, where their reproductive success is equal.

fitness of jacks and hooknoses should be equal at the frequencies of these strategies observed.

Gross (1985) calculated reproductive success as follows. Measurements of survivorship, from tagged individuals, showed that 13 per cent of jacks survived from leaving the streams as smolts to returning to breed as mature adults. For hooknoses, which spent longer in the ocean before returning, survival was only 6 per cent. On the breeding grounds jacks were reproductively active for on average 8.4 days and hooknoses 12.7 days. It was assumed that the number of opportunities for spawning was proportional to these times spent on the breeding grounds. The average proximity of jacks to spawning females (averaged over the mix of sneaking and fighting observed) was 124.6 cm, and for hooknoses it was 93.0 cm. It is assumed that relative proximity provides a measure of relative fertilization success for males. Jack success was, therefore, 66% (1 − [124.6 − 93.0] / 93.0) that of hooknose males.

The relative lifetime reproductive success of jack and hooknose is:

$$\frac{\text{Success jack}}{\text{Success hooknose}} = \text{(Survivorship to} \times \text{(breeding} \times \text{(mating}$$
$$\text{maturity)} \qquad \text{lifespan)} \qquad \text{success)}$$

$$= \frac{0.13}{0.06} \times \frac{8.4}{12.7} \times \frac{0.66}{1}$$

$$= 0.95$$

This calculation suggests that the fitnesses are indeed about equal.

Similar alternative strategies exist in the bluegill sunfish (*Lepomis macrochirus*) which breed in colonies in freshwater lakes in North America (Dominey 1980; Gross 1982). Some males mature at the age of 7 years, defend territories and build nests in which the females spawn. These 'parental' males then look after the eggs and young. Other males mature at 2 years of age and are 'cuckolders'; they do not build nests but instead attempt to sneak fertilizations as the female lays her eggs. Small cuckolders swim along the substrate and make rapid rushes at nests where females are laying eggs, releasing sperm as they swim past. Larger cuckolders, however, mimic female colouration and behaviour; they enter nests slowly and try to insert themselves between the parental male and spawning female and release sperm over the eggs. The different male strategies, parental and cuckolder, and probably (like the salmon) a genetic polymorphism, and calculations again suggest that the reproductive success of the two is equal (Gross & Charnov 1980).

(b) Figwasps: fighters and dispersers

Another dramatic example of male dimorphism with the same species occurs in some figwasps *Idarnes* spp. (Hamilton 1979). Some males are wingless and put their resources into fighting; they have large heads and mandibles that can chop another figwasp in half. These males remain inside the fig where they were born and fight to mate with newly hatched females which develop from larvae in the fig. Other males are winged and put their resources into dispersal; they have tiny heads and mandibles, are not aggressive and fly off to mate with females that have emerged from the fruits (Fig. 10.5).

It is not known whether the difference between male strategies is genetic but it is easy to see that the pay-offs for each strategy must be frequency dependent. If most males were dispersers, then individuals who remained in their natal figs would have high reproductive success because they would mate with the females before they emerged. If most males were wingless, however, competition for females in the figs would be high and any male which dispersed would have access to all the females which emerged unmated. The pay-offs for the two male strategies will also depend on the dispersion of females among figs. For example, in species where many females are likely to occur in a male's natal fig the advantages will be high for fighters. On the other hand if very few eggs were laid in each fig fruit so that a male may find no females to mate with in his natal fig, it would be more profitable to be a disperser. The relative proportion of fighters and

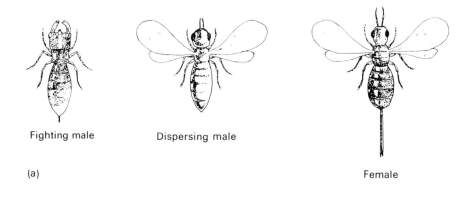

Fighting male   Dispersing male

(a)

Female

(b)

Fraction of females not exposed to wingless males

Fraction of males winged

Fig. 10.5. (a) Male dimorphism in figwasps. Some males have small heads and can fly; others are flightless and have enormous mandibles that can chop another male in half. (b) For ten species of figwasps, there is a good relationship between the fraction of males in the population that are winged and the fraction of females leaving their natal fig before mating. Since these females will be mated by winged males, the equality of the two fractions implies equal mating success for the two morphs. From Hamilton (1979).

dispersers, therefore, is expected to vary between species depending on how females lay their eggs.

Hamilton (1979) found that the proportions of fighters and dispersers in different species were such that the average reproductive success of the two strategies tended to be equal

229

(Fig. 10.5). This then is another example where alternative male strategies exist as a mixed ESS.

(c) Nesting strategies in female digger wasps: diggers and enterers

Female golden digger wasps, *Sphex ichneumoneus*, lay their eggs in burrows underground which they provision with katy-dids as food for the offspring (Fig. 10.6). Jane Brockmann discovered that females have two ways of obtaining a burrow: they either dig one for themselves or they enter an already dug burrow. Digging a burrow is hard work and takes on average 100 min so entering may seem a good strategy because the female gets a burrow without having to spend time and energy in digging. However the wasps do not seem to recognize whether the burrow they enter is empty (abandoned by its builder) or occupied. If it is empty, then they are able to lay and provision in peace, but if there is another female using the same burrow the two wasps eventually meet and there is always a fight with the result that only one female is success-ful in breeding in the burrow. Even if a female digs a burrow herself she may be joined by another female who is playing the entering strategy (Brockmann *et al.* 1979).

How can we explain the evolution of these alternative strate-gies, 'digging' and 'entering'? It can be readily seen how their success could be frequency dependent. Imagine that all the females in the population are digging all the time. There would

Fig. 10.6. A female digger wasp, *Sphex ichneumoneus* at a burrow en-trance. Photo by Jane Brockmann.

be plenty of empty burrows from previous and failed breeding
attempts and so it would pay any female that started entering
because she would save time not having to dig. Digging is
therefore unlikely to be an ESS. However entering would not
spread to take over the whole population because if all the
wasps were entering most would end up sharing, there would
be lots of fights, and it would certaionly pay a female to go off
and dig her own burrow, where the chances of sharing would
be decreased. Entering is therefore also unlikely to be an ESS.
However, because each strategy does best when rare (just like
Hawk and Dove), there could be a stable mixture of entering
and digging (a mixed ESS) brought about by frequency dependent selection, where the success of the two strategies was
equal.

Brockmann *et al.* (1979) measured the success of the two
strategies in terms of the number of eggs laid per unit time.
Because individual females employed both strategies, they
could not measure success by comparing *individuals* who
were enterers with those who were diggers. Instead they measured the success of 'entering' and 'digging' *decisions*. Going
back to the Hawk-Dove game in chapter 7, it will be remem-
bered that a mixed ESS can come about either by a polymorphism in the population, some individuals always playing
Hawk and others always playing Dove (as was the case for the
figwasps and salmon, above), or by individuals themselves
playing Hawk and Dove in the proportion that satisfies the
mixed ESS. Therefore, if the two strategies are an example of a
mixed ESS we would predict that digging and entering decisions should have equal success.

To test this hypothesis data on the nesting behaviour of 68
female wasps were analysed. This field work was based on
over 1500 h of observation and provided a nearly complete
record of the histories of 410 burrows. Calculating the eggs
laid per unit time from digging versus entering was a complicated exercise. Brockmann *et al.* (1979) had to work out the
possible outcomes of the two decisions in order to calculate
the overall success (Fig. 10.7). For example, if a female who
was entering joined an occupied burrow and was then later
evicted before she had the chance to lay any eggs, this had to
be included as time spent 'entering' for no rewards.

The results showed that a female's decision to enter or dig
was not conditional on any obvious phenotypic character,
such as body size, nor on the environment, for example time
of breeding season. Furthermore, there was no significant difference between the success rates of the two strategies. Fifty-
nine per cent of the decisions were 'dig' and the average
reproductive success of this strategy was 0.96 eggs laid per

Decisions   Outcomes

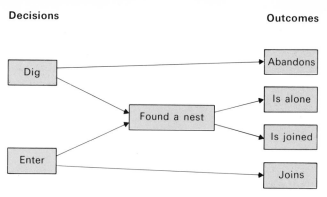

Fig. 10.7. A female digger wasp can either decide to dig a nest burrow herself or enter an already dug burrow. A digging decision may result in one of three outcomes: the wasp may abandon the burrow, she may start a nest and remain alone or she may be joined by another female. An entering decision may also result in one of three outcomes: a wasp may join another female in her nest, or she may have found a nest of her own in which case she may end up nesting alone or in company with another wasp. In order to calculate the overall benefits per unit time for the two decisions, the benefits and times spent in each outcome must be measured. From Brockmann *et al.* (1979).

100 hours; 41 per cent of the decisions were 'enter' and the sucess here was 0.84 eggs per 100 hours.

Therefore the hypothesis that best explains the data is that digging and entering are a mixed ESS. Individual females seem to be programmed with a simple rule such as 'dig with probability p, enter with probability $(1 - p)'$. The value of p has evidently been fixed by frequency dependent selection at 0.59, so that the two decisions have the same reproductive success.

Problems of measuring costs and benefits of alternative strategies

Whenever we find individuals within a population employing different strategies it is useful to ask two questions (Caro & Bateson 1986).

1 *What are the causes of the difference in behaviour?* Differences in behaviour may be due to genotypic differences, as is suspected for the salmon and sunfish, or it may simply be due to differences in environment. For example, all male *Centris* bees are likely to be potential 'hoverers' or 'diggers'. Which behaviour they adopt simply depends on their body size, which depends on larval feeding conditions. The interaction between genotype and environment in determining the strategy adopted may be complex. For example, some male field crickets, *Gryllus integer*, call to attract females by rubbing their forewings together. Other males sit silently near the callers and attempt to intercept females as they arrive! Breeding experiments have shown that 'caller' and 'satellite' behaviours are heritable, but although genotype differences influence calling differences, the amount of calling also varies with environment. An individual's calling frequency varies, for example, with the time of day and with degree of isolation from other males (Cade 1981; Cade & Wyatt 1984). Individuals can switch from 'caller' to 'satellite' but their genotype may influence the propensity with which they respond to environmental cues.

2 *Are the pay-offs of the different strategies equal or unequal?* There are two problems here. First, the definition of the alternatives may need care. Continuous variation in behaviour patterns may be the most common situation in nature (e.g. variation in stay times of male dungflies at a cowpat, chp. 7) and these may be divided up arbitrarily by the observer into discrete categories. Second, there is a difficulty in testing a theory, such as mixed ESS theory, which predicts equality in success. It is impossible to demonstrate statistically that two strategies have exactly equal pay-offs; all we can do is to infer equality if we fail to find a significant difference. The problem is that the smaller the sample size, the less likely we are to find a significant difference, so some supposed examples of mixed ESSs may simply reflect small sample sizes.

Another problem faced by the field worker in measuring pay-offs is that although it is often easy to measure the benefits, such as number of females attracted, it is difficult to measure costs. For example, even though satellite males attract fewer females, their net benefit may be just as great as that of callers because they may incur lower costs. Cade (1979) showed that louder cricket calls attracted not only more females but also more satellites and a parasitic tachinid fly, *Euphasiopteryx ochracea*, which laid larvae into the body of the male cricket or onto a loudspeaker broadcasting cricket calls (Table 10.1)! When Cade sampled crickets he found that 11 out of 14 calling males were parasitized by larvae of this fly while only 4 out of 29 satellites were infected. Parasitism must inflict a serious cost on callers because the cricket always dies when the adult flies emerge from its body (Fig. 10.8). It will therefore be difficult to distinguish among our three hypotheses for the occurrence of 'calling' and 'satellite' behaviour in the cricket. Both strategies may persist because each does better in a different environment, calling at times and places where parasitic flies are scarce and satellite behaviour

Table 10.1. When the song of a male cricket, *Gryllus integer*, is broadcast from a loudspeaker, it attracts not only females but also satellite males and a parasitic fly which will kill the cricket (Cade 1979)

| Broadcast | No. females | Attracted to speaker | |
		No. satellite males	No. parasitic flies
Silent	0	0	0
80 dB song	7	7	3
90 dB song	21	16	18

Fig. 10.8. Calling male crickets not only attract females but also a parasitic fly which lays live larvae into the cricket which will eventually kill it.

where parasites are abundant. Alternatively, calling may be the best way of maximizing reproductive success and satellites may be individuals in poor body condition making the best of a bad job. Finally, the two strategies may coexist as a frequency dependent equilibrium each enjoying, on average, equal success with 'calling' bringing high benefits and high costs, and 'satellite' low benefits and low costs.

Where it can be shown that two strategies really do have equal success, it is useful to recognize that this can come about in three different ways.

1 *Polymorphism.* Different individuals may play different strategies and the frequencies of the morphs will be fixed by frequency dependent selection. The polymorphism may arise because of genetic differences between morphs (e.g. salmon and sunfish). However it may arise because of environmental differences. For example, in some species sex is environmentally determined, males (one strategy) being born when the eggs are incubated above a certain temperature and females (the other strategy) when the eggs are incubated below this temperature (e.g. alligators, Ferguson & Joanen 1982; fish, Conover 1984; see Bull 1980 for a review). At equilibrium the reproductive success of males and females must be equal and the temperature threshold for sex determination will presumably have evolved so that males and females are born in the correct proportion to give them equal success.

2 *Individuals play the different strategies in the proportions that satisfy the mixed ESS.* This occurs, for example, in digger wasps where individuals both dig and enter. Selection could, in theory, have resulted in every individual playing the two strategies with a fixed proportion which results in equal success for 'dig' and 'enter'.

3 *Behavioural assessment.* The third possibility is that instead of having a fixed rule, every individual may have flexible behaviour and base its decision on assessment of what others in the population are doing. For example, imagine that every male frog is free to choose between 'caller' and 'satellite' behaviour. If everyone else was calling it would perhaps pay to be a satellite whereas if all the others were silent it may pay to call. In theory equal success could arise if all individuals simply varied their strategy in relation to the strategies others adopted.

The male green tree frog, *Hyla cinerea*, has both callers and satellites. Unlike the bullfrog discussed earlier, there is no difference in body size between callers and satellites. Indeed individuals often switch between the two. Field experiments showed that when a calling male was removed, the satellite often began to call. Alternatively, when calls were broadcast from a loudspeaker near a calling male, the caller would sometimes become a satellite to the speaker (Perrill *et al.* 1982). Individuals therefore clearly varied their strategy depending on what others were doing. Furthermore, and again unlike the bullfrogs, the success of satellites was high. Perrill *et al.* (1978) released females near callers and satellites and found that the calling male amplexed with the female on 17 occasions while the satellite did so on 13 occasions. It is the possible, therefore, that caller and satellite have equal success which comes about through behaviour varying by assessment of strategies adopted by others.

Of course assessment may also be involved in cases where satellites make the best of a bad job. If the caller is removed then a satellite may be able to take over the territory vacancy and enjoy the greater success of calling, as occurs in green frogs, *Rana clamitans* (Wells 1978).

Our final example illustrates the problems of interpretation when we discover that the two strategies we are studying appear to have unequal success.

Scorpionflies: hunters and stealers

In chapter 8 we described how a male scorpionfly, *Hylobittacus apicalis*, (also called the hanging fly) must first offer a female a nuptial meal before she will allow him to copulate. Randy Thornhill (1979) discovered that males employ two methods for obtaining a prey item to give to the female. They either hunt for the prey and capture it themselves or they steal from another male. Stealing often takes the form of flying straight at another male and knocking the prey forceably from his grasp. There is also the rather more devious method of flying up to a male and adopting a female-like posture to try and dupe him into giving up his prize. Such 'transvestite' behaviour sometimes succeeds but sometimes the male grabs his prey item back again when he discovers that the other individual is refusing to copulate!

Individual males used both strategies to get prey; sometimes they hunted while other times they stole. Therefore success has to be measured in terms of decisions. Thornhill's data showed that stealing was more profitable than hunting; for stealing the inter-copulation time was only 17 min while for hunting it was 26 min. How do we interpret these data? There are three possibilities.

1 The pay-offs for hunting and stealing could be frequency dependent. It is obvious that all males cannot steal all the time. However, if most males were hunters then stealing would be very profitable. If there were no phenotypic constraints (e.g. body size) on whether a male could hunt or steal then we might expect frequency dependent selection to stabilize the success of the two strategies as equal. If this mixed ESS idea is correct then the data suggest that the system is not very stable. Our first possible interpretation is, therefore, that hunting and stealing represent a mixed ESS and that the data do not correctly represent the costs and benefits of the two strategies. Perhaps there is some other unmeasured cost to stealing?

2 The second possibility is that the data are accurate reflections of the pay-offs, in which case the idea that hunting and stealing are a mixed ESS must be wrong. Hunting and stealing may be part of a conditional strategy such as 'if large, steal; if small, hunt'. Maybe just as only the strongest bullfrogs defend territories, so only the best male scorpionflies go in for stealing.

3 The third possibility is that the mixed ESS idea and the data are both correct but that the population has not yet reached a stable equilibrium. Perhaps in time the frequency of stealing will increase and the pay-offs stabilize as equal. This interpretation must be considered especially as many data are now collected in habitats recently modified by man, such as polluted streams and suburban gardens. This means that some animals may be using rules adapted to different environments from the ones in which we study them.

This list of possibilities is very similar to the one given in chapter 3 in relation to optimality models; failure to measure some of the costs or benefits, incorrect assumptions about constraints and the possibility that animals are not well adapted.

In conclusion, although it is clear that there are several evolutionary pathways to differences between individuals within a population, there will often be problems in sorting out which hypothesis best applies to a particular example. Measurements of costs as well as benefits, together with data on individual constraints such as age and body size, are needed before we can distinguish 'best of a bad job strategies' from examples of a mixed ESS. Table 10.2 summarizes the examples of alternative strategies which we have discussed.

Sex change as an alternative strategy

CHANGING FROM FEMALE TO MALE

Whenever there is intense competition between males for females, it is usually the largest strongest males who will enjoy the greatest reproductive success. We have seen that one way in which young and small males can avoid direct competition with stronger rivals is to adopt sneaky mating strategies. However there is another more startling way of overcoming the disadvantage of small size when young. This is to start reproductive life as a female and then change to a male when large enough to be a successful competitor (Fig. 10.9a). This system of sex change is common in fish and is known as protogynous hermaphroditism. It will be favoured whenever an individual reproduces best as a female when small and as a male when large, and when lifetime reproductive success is greater if an individual changes sex than if it remains as either a male or female throughout its life (Ghiselin 1969; Warner 1975). As mentioned in chapter 8, the problem of sex change

Table 10.2. Examples of alternative strategies discussed in this chapter

Species	Strategy 1	Strategy 2
Stickleback	Red	Dull
Butterfly	Perch	Patrol
Bullfrog, green treefrog, field cricket	Call	Satellite
Bee	Hover	Dig
Salmon	Hooknose	Jack
Sunfish	Parental	Cuckolder
Figwasp	Fight	Disperse
Digger wasp	Dig	Enter
Scorpionfly	Hunt	Steal

is closely related to that of the sex ratio. Both are part of the general problem of sex allocation.

The blue headed wrasse, *Thalassoma bifasciatum*, lives on coral reefs in the Western Atlantic. The males are brightly coloured and they defend territories on the reef. The females, which are dull in colour, choose the largest and brightest males for mating. The largest male on the reef may spawn 40 times a day at the peak of the breeding season. As we might expect from the fact that only the largest individuals are successful males, this species is a protogynous hermaphrodite; fish start reproductive life as females when small and only change to being males when large (Warner *et al.* 1975). Sex change is socially controlled. If the largest males on a reef are removed, the next largest individuals (females) will change sex and become bright-coloured males.

The story is a little more complicated than this because there are differences in behaviour on different sized reefs (Warner & Hoffman 1980). On the largest reefs, where populations of wrasse may reach 16 000 individuals, the few largest males have potentially an enormous reproductive success. However, with large numbers of females being attracted to a few males, there are also good opportunities for sneaking by small males. On large reefs some individuals are born as males and remain male throughout their lives (primary males). When small they either go in for sneaky matings, stealing into a large male's territory and attempting to spawn with the females he has attracted, or they may go round in gangs with other small males, chasing females and stimulating them to spawn with the group.

On small reefs, the largest males again attract the most females but the maximum number they can attract is less than on a large reef because the total population is small, perhaps only 20 wrasse. Therefore there is less opportunity for sneaking and small males can be excluded from breeding altogether by the large males. On small reefs there appears to be an advantage for sex-changing individuals who start life as a female and then become a male when they have grown larger. Warner and Hoffman (1980) suggest that both sex-changers and primary males may be maintained in the blue headed wrasse because of different advantages on different sized reefs.

CHANGING FROM MALE TO FEMALE

More rarely, an individual may be a male when small and then change to a female when large (protandrous hermaphroditism). This type of sex change may be favoured if male—male competition is not so intense and male size has little

Fig. 10.9. (a) When male—male competition is intense, only the largest individuals will be successful at mating. Although female fecundity also increases with size (larger females are able to lay more eggs) the influence of male size on mating success is much stronger. Under these conditions it may pay an individual to be a female when small, because all females will breed, and a male only when large enough to be a successful competitor (after Warner 1975). An example of this case is the blue headed wrasse (see text). (b) When male—male competition is less intense, female fecundity may be more dependent on body size than male fecundity. It may then pay an individual to start life as a male when it is small and change to a female when it is older and larger (after Warner 1975). An example of this case is the anemonefish, *Amphiprion* (see text).

effect on breeding success (Fig. 10.9b). An individual may then reproduce best as a male when small because it is able to spawn with some of the large, most fecund females.

An example of a fish that changes from male to female is the anemonefish, or clownfish, *Amphiprion akallopisos*, which lives on coral reefs in the Indian Ocean. It lives in close symbiosis with sea anemones and because there is usually only enough space for two fish to inhabit the same anemone this species lives in pairs (Fig. 10.10). In effect the habitat forces them to be monogamous. The reproductive success of a pair is limited more by the female's ability to produce eggs than by the male's ability to produce sperm and so each individual does better if the larger one is female (Fricke 1979). Like the wrasses, sex change is socially controlled. If the female is removed, the male is then joined by a smaller individual, so he changes sex and lays the eggs while the newcomer functions as a male (Fricke & Fricke 1977).

SEX CHANGE VERSUS SNEAKING

Where there is a size advantage for being one sex, the smaller individuals may sneak (as in elephant seals) or may be sequential hermaphrodites, changing sex as they grow larger (as in some fish). Why do not young elephant seals start life as females and then change to males when they are large and strong, like the blue headed wrasse? The most likely explanation is that in mammals the sexes are more differentiated than in many fish; for example there is internal fertilization and elaborate care of the young by the female in pregnancy and during lactation. It may therefore be too costly to change sex. It is striking that all species of fish known to change sex have relatively simple sex organs and external fertilization. Furthermore for a young mammal, like an elephant seal, experience

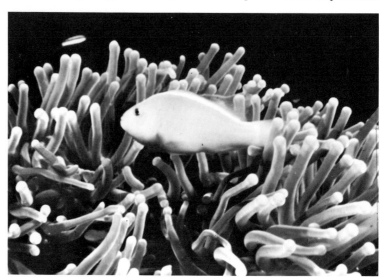

Fig. 10.10. An anemonefish, *Amphiprion akallopisos*, with its anemone. Photo by Hans Fricke.

may be needed to be a successful male. It may be better to forgo reproduction when young and instead put resources into growth and learning the techniques of successful harem defence (Warner 1978).

Summary

Individuals within a species often differ in the way they compete for scarce resources such as food, mates or nest sites. Different strategies may be favoured in different environments (e.g. red and dull male sticklebacks). Sometimes behaviour is conditional on size and strength so that while the largest individuals display and fight to attract mates, smaller individuals employ sneaky strategies which make the 'best of a bad job' (e.g. bullfrog satellites, hoverers in the bee *Centris pallida*). Finally, alternative strategies may exist as a mixed ESS, an evolutionary equilibrium where different strategies enjoy equal success. In these cases there may be a polymorphism in the population, with frequency dependent selection fixing the frequencies of the different types (e.g. hooknoses and jack salmon), or alternatively each individual may play the various strategies in the proportions that satisfy the ESS (e.g. digging and entering in digger wasps). In many cases there are not enough data on constraints and the success of the different strategies to say whether they are 'best of a bad job', or an example of a mixed ESS.

Some fish change sex as a way of increasing reproductive success when small in size. When male−male competition is intense so that only the largest individuals are successful as males, individuals may be sequential hermaphrodites, changing from female to male as they grow larger (e.g. blue headed wrasse). More unusually, the reproductive success of a female is more size dependent than that of a male, in which case individuals may change from male to female as they grow larger (e.g. anemone fish, *Amphiprion*).

Further reading

Dawkins (1980) and Dunbar (1983) give good reviews of alternative strategies. Rubenstein (1980) discusses alternative mating strategies giving both a theoretical treatment and a review of examples. Thornhill (1981) gives an excellent account of alternative mate searching strategies in scorpionflies. The two papers on nesting strategies in digger wasps (Brockmann & Dawkins 1979; Brockmann, Grafen & Dawkins 1979) give a clear account of the methods and problems of applying ESS ideas to alternative strategies in one species. Charnov (1982)

models the optimal timing of sex change and tests his theory
with data on shrimps.

Topics for discussion

1   In a chorus of male frogs, some are calling and others are
silent. What are the hypotheses for the occurrence of these
two behaviours? How would you test between them?

2   What explanations, other than the ones in this chapter, can
you suggest for variation in reproductive behaviour within a
species?

Chapter 11. On Selfishness and Altruism

So far throughout this book we have championed the view that natural selection designs individuals to behave in their own selfish interests and not for the good of their species or for the good of the group in which they live. For example, observed clutch size, foraging behaviour and mating patterns are what would be expected if selection optimized behaviour and life history strategies so as to maximize an individual's reproductive success.

However, it will be obvious to any naturalist that animals do not behave selfishly all the time. Often individuals apparently co-operate with others. Several lions may co-operate to hunt prey; in many species of birds and mammals individuals give alarm calls to warn others of the approach of a predator; sometimes an individual may help others to produce offspring rather than have young itself (see chps. 12 and 13). Before 1960 such co-operation did not demand special attention. It was seen to be 'good for the species' and therefore adaptive. Since the 1960s, however, people have appreciated again what Darwin said all along, namely that in evolution there is a struggle between individuals to outcompete others in the population. How, then, can we account for the evolution of co-operative behaviour in terms of advantage to individuals? If natural selection favours individuals who do the best and have the most surviving young, how can behaviour evolve which involves helping others to survive and have young at a cost to the helper's own chance of doing so? As E.O. Wilson (1975) put it, the central problem of sociobiology is how can altruism evolve? *Altruism* is defined as acting to increase another individual's lifetime number of offspring at a cost to one's own survival and reproduction.

In this chapter we will distinguish four hypotheses for the evolution of co-operation. Behaviour which appears to be altruistic at the phenotypic level may turn out to be genetically selfish (hypotheses 1 and 4 below). Some forms of co-operation do not involve any altruism (hypothesis 2) and others involve self-sacrifice that is both phenotypically and genotypically altruistic (hypothesis 3).

Kin selection

Theory. The most familiar example of an individual giving aid to another is, of course, parental care. We are not surprised to see a parent bird hard at work feeding its offspring because natural selection favours individuals who maximize their gene contribution to future generations. The young will have copies of their parent's genes and so parental care is genotypically selfish.

We can quantify the probability that a copy of a particular gene in a parent is present in one of its offspring. In diploid species, when an egg and a sperm fuse to form a zygote, each parent contributes exactly 50 per cent of its genes to the offspring. Therefore the probability that a parent and an offspring will share a copy of a particular gene identical by descent (in an outbreeding species) is 0.5. This quantity is called the *coefficient of relatedness*, often denoted by r.

Now offspring are not the only relatives to share copies of the same genes identical by descent. Again we can calculate the probability that a copy of a gene in one individual will be present, by virtue of descent from a common ancestor, in a brother, sister, cousin and so on. For brothers and sisters r is 0.5, for grandchildren it is 0.25 and for cousins it is 0.125 (Box 11.1). It was W.D. Hamilton (1964) who realized the important implication of this for the evolution of altruism, although the idea was anticipated by Fisher (1930) and Haldane (1953). Just as gene proliferation can occur through parental care so it can through care for siblings, cousins or other relatives. Table 11.1 shows various values of r for descendant kin and non-descendant kin. The term *individual selection* is often used to describe the process by which a behavioural act is favoured due to its beneficial effects on the production of offspring. The term *kin selection* (Maynard Smith 1964) is often used to describe the process by which a behavioural act is favoured owing to its beneficial effects on other relatives, such as siblings or cousins. As far as evolution is concerned, however, there will be no distinction between gene copies produced by helping offspring and siblings. Whichever route the genes take, the consequence of selection is changes in relative gene frequencies in the gene pool.

The conditions under which an altruistic act will spread by

Table 11.1. Coefficients of relatedness (r) for descendant and non-descendant kin

r	Descendant kin	Non-descendant kin
0.5	Offspring	Full siblings
0.25	Grandchildren	Half-siblings
		Nephews and nieces
0.125	Great-grandchildren	Cousins

Box 11.1. *Calculation of r, the coefficient of relatedness.* r is the probability that a gene in one individual is an identical copy, by descent, of a gene in another individual.

General method

Draw a diagram with the individuals concerned and their common ancestors, indicating the generation links by arrows. At each generation link there is a meiosis and so a 0.5 probability that a copy of a particular gene will get passed on. For L generation links the probability is $(0.5)^L$. To calculate r, sum this value for all possible pathways between the two individuals.

$r = \Sigma (0.5)^L$.

Specific examples

These diagrams show calculations of r between two individuals represented by solid circles; other relatives are indicated by open circles. The solid lines are the generation links used in the calculations; the dotted lines are the other links in the pedigrees.

(a) Parent and offspring

$r = 1 \ (0.5)^1$
  $= 0.5$

(b) Grandparent and grandchild

$r = 1 \ (0.5)^2$
  $= 0.25$

(c) Full sibs (brother, sister)

$r = 2 \ (0.5)^2$
  $= 0.5$

(Identical genes by descent can be inherited by two pathways, either mother or father)

(d) Half-sibs

$r = 1 \ (0.5)^2$
  $= 0.25$

(Identical genes by descent can only be inherited from one parent)

(e) Cousins

$r = 2 \ (0.5)^4$
  $= 0.125$

kin selection are as follows (Hamilton 1964). Imagine an inter-action between an altruist (or donor) and a recipient in which the costs and benefits of the interaction can be assessed in terms of survival chances of the donor and recipient. If the donor suffers cost C (through, for example, giving an alarm call) and the recipient gains a benefit B as a result of the altruistic act, then the gene causing the donor to act will increase in frequency if

$$\frac{B}{C} > \frac{1}{r}, \text{ or alternatively, if } rB - C > 0$$

where r is the coefficient of relatedness of the donor to the recipient. This result is known as 'Hamilton's rule'.

An intuitive understanding of this formula can be got as follows. As an extreme example of altruism imagine a gene that programs an individual to die in order to save the lives of relatives. One copy of the gene will be lost from the population in the death of the altruist, but the gene will still increase in frequency in the gene pool if, on average, the altruistic act saves the lives of more than 2 brothers or sisters (r = 0.5), more than 4 nieces or nephews (r = 0.25) or more than 8 cousins (r = 0.125). Having made these calculations on the back of an envelope in a pub one evening, J.B.S. Haldane announced that he would be prepared to lay down his life for the sake of 2 brothers or 8 cousins!

It is often useful to measure the costs and benefits in terms of offspring lost and gained, in which case we use the follow-ing form of Hamilton's rule,

$$\frac{B}{C} > \frac{r_{\text{donor to own offspring}}}{r_{\text{donor to recipient's offspring}}}$$

Two examples will help to make this clear. Imagine an indivi-dual has a choice between rearing its own offspring and help-ing its mother to produce offspring. The individual's own offspring and its mother's offspring, assuming they are full siblings, both have r = 0.5, so the expression above becomes B/C > 1. Therefore helping will be favoured by kin selection if by your help your mother produces more extra offspring than you have 'sacrificed' through providing help (i.e. through for-going the chance to produce your own offspring). If the in-dividual was faced with the alternative of rearing its own offspring or helping its sister to produce offspring, then the expression becomes B/C > 2 (0.5/0.25). In this case helping would evolve only if it resulted in two or more extra offspring produced by the sister for every one offspring lost by the donor. We will discuss some examples of this kind of altru-ism in chapter 12.

It is worth pausing for a moment to appreciate the importance of Hamilton's idea which is really an extension of Darwin's theory of natural selection. The central role of Hamilton's rule for our understanding of the evolution of social behaviour has been summarized beautifully by Alan Grafen (1985) as follows.

'In an uncomplicated world, the morphology and behaviour of an individual would affect the number of offspring it had in its lifetime but not the number of offspring other individual's had. Given that individuals do interact, though, the question is how will selection cause animals to value each other's reproduction? Will they place no value on it — that is, will selection favour forms according only to their own number of offspring? Or will they place equal value on their own reproduction and that of others — that is, will selection favour those forms that produce more offspring in total for the species? Hamilton's answer was intermediate between these two extremes. He showed mathematically that natural selection would act so that one individual would value the reproduction of another according to how closely the two individuals were related. Identical twins would value each other's offspring equally with their own, brothers and sisters would value each other's offspring half as much as offspring of their own, and so on. The evolution of a social trait then depends on an 'inclusive fitness' calculation, in which all the effects of the trait on reproduction are added together and weighted by the appropriate relatedness. Relatedness matters because relatives share more genes than non-relatives, and natural selection is a matter of the non-random survival and proliferation of some genes at the expense of others.'

EXAMPLES OF ALTRUISM BETWEEN RELATIVES

The social insects provide good examples of extreme altruism. Worker bees have barbed stings and attack predators which approach their nests. In the act of stinging the predator the barbs of the sting become embedded in the victim and the worker bee dies as a result. The evolution of such suicidal behaviour poses a problem until we discover that the beneficiaries of the altruistic act are in fact close relatives of the worker. Workers are altruistic in another way because they rarely reproduce themselves, but instead help others in the nest to produce offspring. Darwin regarded this observation as potentially fatal to his theory of natural selection. How can such altruism evolve if the altruists never reproduce? Hamilton's theory of kin selection immediately suggests a possible answer to this problem because the sterile workers usually help their mother (the queen) to produce offspring (chp. 13).

Not all acts of altruism are as extreme as suicide or sterility. We now consider two examples where the costs to the altruist are smaller, and where it is likely that kin selection has been a major force in the evolution of the altruism.

(a) Co-operation and alarm calls in ground squirrels and prairie dogs

Paul Sherman has made an extensive study of Belding's ground squirrel, *Spermophilus beldingi*, a diurnal social rodent which inhabits the subalpine meadows of the far western United States (Fig. 11.1). This species hibernates during the winter, emerging above ground in May. Soon after emergence the females become sexually receptive and mate. After mating the males wander off and leave the females to rear their young alone. A female establishes a territory surrounding its nest burrow and produces a single litter of 3 to 6 young per year. The pups first emerge above ground at the time of weaning, when they are 3 to 4 weeks old and then soon after this the juvenile males disperse while the juvenile females tend to remain near their natal area. This means that males seldom, if ever, interact with close relatives whereas females spend their whole lives surrounded by close female kin.

Sherman found that closely related females (mother and daughters, sisters) seldom fought for nest burrows and seldom chased each other from their territories. Indeed they co-operated to defend each other's young against infanticidal conspecifics. Eight per cent of all young born were dragged from their burrows and killed by other ground squirrels. The killers were not close relatives of the victims and were either young males wandering in search of an easy meal or immigrant adult females who were searching for new nesting burrows. These females attempted to take over occupied burrows, killing any young they found as a way of clearing the territory of potential competitors (Sherman 1981a,b). Such co-operation among close relatives, in contrast to the conflict among unrelated individuals, is exactly what would be predicted from the theory of kin selection.

Individuals also gave alarm calls whenever a predator, such as a coyote or weasel, approached. Callers probably suffered a cost from giving the alarm because they were more likely to be attacked by the predator, perhaps because the calls made them more conspicuous. Others, however, benefited from the early warning and were more likely to escape. Sherman (1977) found that females were much more likely to give alarm calls than

Fig. 11.1. Belding's ground squirrels. (a) Female giving an alarm call. (b) A group of pups at the mouth of a burrow. Photo by © George D. Lepp, Bio-Tech Images.

males, and furthermore, females with close relatives nearby were more likely to give calls than females without. Although in most cases the beneficiaries of the alarm calling were likely to be offspring, individuals also gave alarm calls when only parents or non-descendant relatives were nearby. For example, young females who had yet to produce their own young gave alarm calls to warn their mother and sisters of the approach of a predator. In another species of ground squirrel, *Spermophilus tereticaudus*, males were much more likely to give alarm calls before leaving their mother's home area, when they had relatives nearby, but when they dispersed and left their close relatives they were more likely to remain silent when a predator came along (Dunford 1977).

These data on alarm calls show clearly that individuals are more likely to incur the cost of calling when relatives are nearby to gain a benefit. However it could be argued that the major force responsible for the evolution of the calling is individual selection because the relatives which usually got the benefit were offspring. Non-descendant kin may simply be common secondary beneficiaries (Shields 1980). Warning offspring (i.e. parental care) and warning sisters are, of course, just different ways of increasing gene propagation to future generations. However it is still interesting to ask whether alarm calling could evolve mainly because of benefit to non-descendant relatives. The best evidence for this comes from a study by John Hoogland (1983) of another colonial rodent, the black-tailed prairie dog *Cynomys ludovicianus*.

Black-tailed prairie dogs live in social groups called coteries,

249

typically one adult male with 3 to 4 adult females and their offspring. Young females remain in their natal coterie all their lives while young males disperse in their second year. All the females and yearling males within a coterie are therefore usually close genetic relatives. Hoogland studied alarm calling responses by presenting a stuffed specimen of a natural pre-dator, the badger *Taxidea taxus*. This enabled him to get more data than could be obtained by waiting for natural predator attacks and also allowed him to control for proximity of the predator to the prairie dogs. Fig. 11.2 summarizes the results of over 700 experiments. The data show that individuals gave alarm calls just as frequently when there were only non-descendant kin in their home coterie as when there were offspring present. Factors other than the warning of relatives must also be involved, however, because immigrants who had no relatives nearby nevertheless sometimes called (Fig. 11.2). There may often be direct benefits to the caller itself in giving the alarm, for example signalling to the predator 'I've seen you', which may reduce the likelihood of attack because the predator then does not have the advantage of surprise. Another possibility is that it pays to warn others of the approach of a predator, even if they are not close relatives, because if a neighbour was caught successfully then the predator may be more likely to return to hunt in the same area again. Alarm calling may therefore reduce the likelihood of future attacks by the same predator.

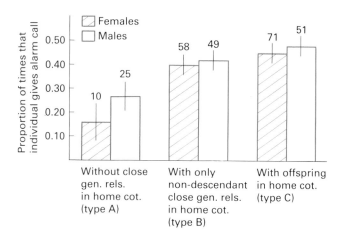

Fig. 11.2. Alarm calling by black-tailed prairie dogs to a stuffed badger. For both males (white histograms) and females (hatched), there are signifi-cant differences between type A and type B individuals and also between type A and type C. There was, however, no significant difference for either sex between type B and type C. Data are means ± 1 S.E., with number of different individuals observed indicated. From Hoogland (1983).

These studies of co-operation among relatives in ground squirrels and prairie dogs are consistent with the kin selection model but they do not constitute a quantitative test of Hamilton's rule. Indeed it would be difficult to assess the costs and benefits of alarm calls in terms of offspring lost and gained. We now turn to an example where these values can be estimated.

(b) Wife sharing in the Tasmanian native hen

The Tasmanian native hen *Tribonyx mortierii* is a flightless rail. There are more males than females in the breeding population and mating combinations include simple pairs (one male with one female) and trios (two males and one female). In trios, the two males are usually, but not always, brothers and both of them copulate with the female and help to raise young. One of the males is usually dominant to the other, yet he apparently allows the other male to mate with the female. Under what conditions could such co-operation evolve? Box 11.2 applies Hamilton's rule to derive the conditions under which co-operation would pay the dominant male if he was either unrelated to the other male or was a brother (adapted from Maynard Smith & Ridpath 1972). The conditions for co-operation with a brother are less severe than with an unrelated male, of course, because if a male helps its brother to produce two extra offspring then this is equivalent to producing one offspring itself as far as gene contribution to the next generation is concerned (r to own offspring is 0.5; to brother's offspring is 0.25).

Table 11.2 summarizes the data on reproductive success of pairs and trios. Lifetime success is not well known but it is likely that each bird has at most 5 breeding attempts and at least 2. It is assumed that survival is equal in pairs and trios, and so success for 5 and 2 attempts provide maximum and minimum estimates of lifetime success. If males are unrelated, then co-operation by the dominant males does not make sense because $1/2 N_2$ is always less than N_1 (Box 11.2). If they are brothers, however, then for the maximum estimate of lifetime success $3/4 N_2 = 21.8$ which is only a little less than the value of N_1, 23.1. For the minimum estimate of lifetime success, $3/4 N_2 = 7.2$ which is greater than N_1, 6.6, so the conditions for co-operation derived in Box 11.2 hold. These calculations suggest that co-operation between brothers may satisfy the conditions necessary for kin selection to favour the behaviour. The fact that unrelated males may co-operate is still a problem. Provided $N_2 > N_1$, however, it will pay a female to breed in a trio rather than in a pair so there may be conflicts of interest between the sexes (see also chp. 9).

Box 11.2. *Wife sharing in the Tasmanian native hen.* The number of offspring raised by a pair is N_1 and by a trio N_2. If the dominant male was selfish and drove the subordinate away we assume that the dominant male then produces N_1 young and the subordinate produces no young (females are scarce and it is unlikely that the evicted male would find another mate). If the two males shared the female in a trio, we assume they share paternity equally and so each produce $\frac{1}{2} N_2$ young. The pay-offs are therefore as follows:

| | Dominant male behaviour | |
Pay-off to	Co-operate	Selfish
Dominant male	$\frac{1}{2} N_2$	N_1
Subordinate male	$\frac{1}{2} N_2$	0

The benefit to the subordinate of remaining in the trio is the number of young he gains through the dominant male's co-operation.

$B = \frac{1}{2} N_2 - 0$

The cost of co-operation to the dominant male is the number of young he loses through co-operation with the subordinate compared to what he would get by being selfish

$C = N_1 - \frac{1}{2} N_2$

Applying Hamilton's rule, the dominant male should co-operate if $B \times r$ to subordinate's offspring $> C \times r$ to own offspring.

Consider two cases.

(a) *The males are unrelated* (r to subordinate's offspring $= 0$)

The dominant male should co-operate if

$0 > \frac{1}{2} (N_1 - \frac{1}{2} N_2)$

i.e. if $\boxed{\frac{1}{2} N_2 > N_1}$

(b) *The males are brothers* (r to subordinate's offspring $= \frac{1}{4}$)

The dominant male should co-operate if

$\frac{1}{4} (\frac{1}{2} N_2) > \frac{1}{2} (N_1 - \frac{1}{2} N_2)$

i.e. if $\boxed{\frac{3}{4} N_2 > N_1}$

Table 11.2. Reproductive success of Tasmanian native hens in pairs and trios. From Maynard Smith and Ridpath (1972)

1 Numbers of surviving young produced per season

	1st year breeders	Experienced breeders
Pair, $\male\female$, N_1	1.1	5.5
Trio, $2\male\female$, N_2	3.1	6.5

2 Number of surviving young produced per lifetime

	If 5 breeding seasons	If 2 breeding seasons
Pair, N_1	$1.1 + (4 \times 5.5) = 23.1$	$1.1 + 5.5 = 6.6$
Trio, N_2	$3.1 + (4 \times 6.5) = 29.1$	$3.1 + 6.5 = 9.6$

HOW DO INDIVIDUALS RECOGNIZE KIN?

Hamilton's theory of kin selection requires an individual to behave differently towards individuals of different degrees of relatedness, and this could involve assessing relatedness to other individuals. There is now a rapidly growing body of evidence that individuals can indeed recognize kin and even distinguish close kin from distant kin. How do they achieve this? An entertaining but theoretically unlikely possibility was proposed by Hamilton (1964) and coined the 'green beard effect' by Richard Dawkins. The idea is that there may be 'recognition alleles' which express their effects phenotypically so enabling bearers to recognize these alleles in others, and also cause bearers to behave altruistically towards others with the recognizable phenotypic effect. For example, if a gene conferred on its owner both a green beard and a tendency to be nice to others with the same feature, it would be favoured by evolution. This would provide a mechanism of kin recognition without learning.

A more likely mechanism is that individuals use a simple rule, for example 'treat anyone in my home as kin'. Parent birds, for example, may ignore their own young if they are placed just outside the rim of their nests, yet will readily accept a strange chick placed inside their nest. This can lead to odd results; for example a reed warbler, *Acrocephalus scirpaceus*, may mob an adult cuckoo, *Cuculus canorus*, which approaches its nest and then, a minute later, return to the hard work of feeding a baby cuckoo inside its nest! Usually, however, this simple rule will lead individuals to care for their own offspring.

Another mechanism for recognizing kin is to learn that those you grow up with are kin. Konrad Lorenz gave the name 'imprinting' to the phenomenon observed in young geese, of following the first conspicuous moving object they see after hatching. Usually this will be their mother and so result in them following someone who will keep them warm and protect them. In an experimental situation, however, young birds have been imprinted on humans and even flashing lights. Experiments by Holmes and Sherman (1982) show that sibling recognition in ground squirrels is also in part based on association in the natal nest. They captured pregnant females and used their pups to create four kinds of experimental rearing groups; siblings reared by one mother (their own or a foster mother), siblings reared apart by different mothers, non-siblings reared as a single litter and non-siblings reared apart. When they were older, animals from the four groups were placed in pairs in arenas and their interactions were observed. Holmes and Sherman found that regardless of true genetic relatedness animals that were reared together rarely fought. Fig. 11.3 shows that unrelated individuals reared together were no more aggressive to one another than true sibs reared together. This suggests, therefore, that individuals learn who are their kin from association in early life.

However, it was also found that among animals reared apart, genetic siblings were less aggressive to one another in an arena test than unrelated individuals (Fig. 11.3a). Interestingly, this effect only occurred among females; thus true sisters reared apart were less aggressive to each other than unrelated females reared apart but genetic relatedness did not affect aggression between male−male or male−female pairs. Only females, the sex that behaves altruistically in the field, show evidence of being able to recognize unfamiliar, but genetically related kin.

Of course, sisters reared apart may still learn to 'recognize' each other because of prenatal experience in their mother's uterus. Field observations by Paul Sherman, however, suggest that this may not be the whole story. Female Belding's ground squirrels mate with up to 8 different males (mean 3.3 males) on the one afternoon they are sexually receptive in spring. Analysis of polymorphic blood proteins collected from mothers, potential fathers and their offspring, showed that 78 per cent of the litters were sired by more than one male (Hanken & Sherman 1981; this method of assessing probable paternity is rather like the use of blood groups in man to settle paternity disputes in court cases). The exciting discovery was that littermates who were full sisters (same mother, same father) were less agonistic to one another and more co-operative than half-sisters (same mother, different father; i.e. half-sisters arise because of multiple matings). For example, when establishing nest burrows and defending territories full sisters fought and chased less often when they encountered each other than did half-sisters (Fig. 11.3b,c).

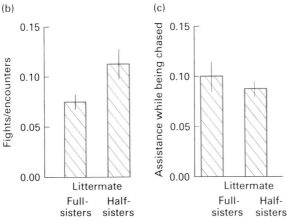

Fig. 11.3. Kin recognition in Belding's ground squirrels (from Holmes & Sherman 1982). (a) Laboratory experiments: mean number (± 1 S.E.) of agonistic encounters between pairs of yearling Belding's ground squirrels in arena tests. Non-siblings reared together (NS.RT) are no more aggressive than siblings reared together (S.RT). However non-siblings reared apart (NS.RA) are more aggressive than siblings reared apart (S.RA). (b) and (c) Field observations: aggression and co-operation among yearling females which were full or half-sisters (genetic relatedness determined by blood proteins). Full sisters are less aggressive to one another (b), and assist each other more (c).

Littermates all share the same nest burrow and the same uterus, whether they are full sisters or half-sisters, so some mechanism other than this common experience must be involved. One possible mechanism is known as 'phenotype matching'. A female may be most altruistic to nestmates who are like her phenotypically (e.g. match her own odour). In conclusion, a female Belding's ground squirrel seems to categorize others in two ways. First she recognizes and co-operates with individuals she shared a burrow with as opposed to those she did not; the former will be full sibs or half-sibs. Second, she may be particularly co-operative with nest mates who are like herself phenotypically, and hence more likely to be full sibs rather than half-sibs.

CONCLUSIONS ABOUT KIN SELECTION

We have dealt at length with kin selection because in many ways it is the most theoretically interesting and possibly the most important mechanism for the evolution of altruism. As predicted by Hamilton's model, much of the altruism seen in nature does occur between closely related individuals and recent work shows that individuals can distinguish close kin from distant kin. However there are many examples of co-operation and altruism among unrelated individuals. How can this evolve? We now consider three other hypotheses.

Mutualism

It may sometimes pay two or more individuals to co-operate simply because each gains a net benefit from doing so. Two pied wagtails, *Motacilla alba*, who join together to defend a winter feeding territory can each enjoy a greater feeding rate than they would by being alone because the benefits of the association in terms of improved territory defence outweigh the costs of having to share the food (Davies & Houston 1981, chp. 5).

In other cases, both kin selection and mutualism may be relevant. A single lioness is not very effective at capturing a zebra, but by hunting with another female she improves her capture success so much that the benefits outweigh the costs of having to share the meat once the prey is caught (Caraco & Wolf 1975). Lionesses in a pride are related (chp. 1) and so individuals will get kin selected benefits from the association, but in principle it would also pay two unrelated females to hunt together. Similarly, a single male lion has little chance of taking over a pride. By co-operating with another male he gains an individual advantage in terms of increasing his own reproductive success. Again, male lions who co-operate are often relatives (full or half-brothers) so kin selection will favour the relationship. However it would pay even unrelated males to form a coalition and there are in fact observations of them doing so and holding a pride together (Bygott *et al.* 1979; Packer & Pusey 1982).

Our third hypothesis is that what looks like altruism on the part of the donor may in fact be manipulation by the recipient (Dawkins 1982). This is most obvious, for example, when a parent bird feeds the offspring of a brood parasite, such as a cuckoo. The host gains nothing from its altruism; it has simply been tricked by the cuckoo into feeding the wrong species (see chp. 4). Manipulation also occurs within a species. For example, some female birds lay eggs in the nests of conspecifics, thus avoiding the costs of incubation and parental care (Yom-Tov 1980). Female starlings, *Sturnus vulgaris*, who 'dump' eggs in other female's nests first remove a host egg before laying their own. They then deposit the host egg on the ground nearby. It was at first thought that these eggs which appeared on the ground were laid by females who were unable to get back to their nests in time to lay normally! Then, when eggs were marked in nests as they were laid, it was found that the eggs which appeared on the ground were often marked ones, in other words eggs which had been removed from nests (Feare 1984). In this case, just as with interspecific brood parasitism, the host is being tricked into behaving for another individual's benefit.

Reciprocity

Whenever the benefit of an altruistic act to the recipient is greater than the cost to the actor, then as long as the help is reciprocated at some later date, both participants will gain (Trivers 1971). For example, A helps B today and then B helps A tomorrow. Reciprocation is common in human society and we use money and laws to regulate its use. The problem for its evolution in animal populations is the possibility of cheating. Because of the time delay between one individual gaining and the other doing so, B may accept help from A today but refuse to repay the favour tomorrow (cf. mutualism, where both individuals get the benefit at the same time). We will explore the conditions under which reciprocity can be evolutionarily stable by using a simple model.

THE PRISONER'S DILEMMA

The Prisoner's dilemma model was originally developed to help us think about human behaviour, but it provides an elegant model to illustrate the problems of achieving co-operation in animal societies (Axelrod & Hamilton 1981). Imagine two players in a game who have the choice of co-

Table 11.3. The Prisoner's dilemma game (from Axelrod & Hamilton 1981). The pay-off to player A is shown with illustrative numerical values

		Player B	
		Co-operate	Defect
Player A	Co-operate	R = 3 Reward for mutual co-operation	S = 0 Sucker's pay-off
	Defect	T = 5 Temptation to defect	P = 1 Punishment for mutual defection

operating or defecting (being selfish). The pay-off matrix is given in Table 11.3 with some illustrative numerical values. These values represent the gain in fitness from the interaction (e.g. number of offspring gained).

Imagine player A finds another individual B who always co-operates. If A co-operates too it gets a reward of 3, whereas if it defects it gets 5. Therefore if B co-operates, it pays A to defect. Now imagine player A discovers that B always defects. If A co-operates it gains nothing (the sucker's pay-off) whereas if it defects it gets 1. Therefore if B defects, it pays A to defect. The conclusion is that irrespective of the other player's choice, it pays to defect even though with both players defecting they get less (1) than they would have got if they had both co-operated (3). Hence the dilemma!

In other words, co-operation is not an ESS because in a population of 'all co-operators' a mutant who defected would spread. Defect, however, is an ESS; in a population of 'all defect' a mutant co-operator does not gain an advantage. Any population with a mixture of heritable strategies will therefore evolve to 'all defect'. More generally the conditions for this conclusion to hold in the matrix in Table 11.3 are, $T > R > P > S$ and $R > \dfrac{(S + T)}{2}$ which define the Prisoner's dilemma game. The problem is essentially that while an individual can benefit from mutual co-operation, it can do even better by exploiting the co-operative efforts of others.

Is there any way in which individuals can escape this dilemma and come to stable co-operation? The answer is no if the two players only meet once; defect is the only stable strategy in Table 11.3. Defect is likewise the stable strategy if the total number of interactions is precisely known in advance because defection will be optimal on the last encounter and therefore also on the next-to-last and so on back to the first encounter. However, if the series of encounters goes on indefinitely or, more realistically, if there is always a finite probability, w, that the two players will meet again, then we have to consider the possibility of more complex strategies involving mixtures of co-operation and defection in various sequences.

Axelrod (1984) ran a computer tournament in which 62 different strategies, submitted by scientists from all over the world, were paired against each other in a round robin tournament. Each entry played every other entry and also played against itself and against 'random', a program that randomly co-operates and defects with equal probability. The pay-offs for each contest were as in Table 11.3 and the game was iterated between the pair of contestants with w = 0.99654. This game can be thought of as a contest between intelligent individuals (the scientists who submitted the various strategies) competing to win a prize, or it can be thought of as a model of an evolutionary game, like the Hawk-Dove game (chp. 7), in which the different strategies represent different genetic mutants competing for representation in the gene pool.

Some of the strategies were very complex, for example 'on each move model the behaviour of the other player as a Markov process and then use a Bayesian inference to select what seems to be the best choice in the long run'. Some were very unforgiving, for example 'never be the first to defect, but once the other player defects, always defect for the rest of the game'. Others were sneaky and tried to get away with occasional exploitation, for example 'if the other defects then defect too, but on 10 per cent of the occasions the other co-operates, defect'. The winning strategy was the simplest one of all those submitted, called 'tit for tat'. This co-operated on the first move and thereafter did whatever its opponent did on the previous move. Thus 'tit for tat' is a strategy of co-operation based on reciprocity. The success of 'tit for tat' against the variety of contenders was due to two features.

1 *It was retaliatory*, which discouraged the other side from persisting whenever a defection was tried.

2 *It was forgiving* after just one act of retaliation, which helped restore mutual co-operation and the greater rewards this gave.

On the other hand, 'tit for tat' did not enjoy the possible advantages of exploiting other rules to any great extent. While such exploitation could sometimes be fruitful, it often led to problems. First, a rule which defects to see what it can get away with risks retaliation from rules which are provocable.

Second, once mutual recriminations have set in it can be difficult to escape from long sequences of unprofitable defection. Finally, the attempt to give up on unresponsive rules (e.g. 'random') often leads to giving up on rules which are in fact profitable to co-operate with by a more patient rule such as 'tit for tat'.

Axelrod's computer tournament showed that 'tit for tat' wins in an environment consisting of a variety of strategies. A further analysis examined what would happen if the tournament continued with the frequencies of the different strategies in the next round being proportional to the score they achieved in the previous one. We can imagine, for example, that the scores represent number of offspring, or strategy copies, produced in the previous round. A round of a tournament can therefore be regarded as equivalent to a single generation and a continuation of the game over successive rounds as a simulation of 'survival of the fittest' over successive generations. The results showed that as the less successful rules were displaced, 'tit for tat' continued to do well and eventually displaced all the other rules and went to fixation.

Once a strategy has gone to fixation, we have to consider whether it will be evolutionarily stable, in other words resist invasion by a mutant strategy (chp. 7). Axelrod and Hamilton (1981) showed that 'tit for tat' is an ESS provided that w, the probability that two contestants meet again, is sufficiently large (Box 11.3). Thus, once 'tit for tat' is established, a mutant defector cannot gain an advantage. Therefore, things now look good for the evolution of co-operation based on reciprocity. Unfortunately, however, this is not the end of our analysis because 'all defect' is also an ESS, whatever the value of w. A population of 'all defect' results in pay-off P per move (Table 11.3). A mutant 'tit for tat', which co-operated on the first move would get the sucker's pay-off, S, with no chance of future compensation.

Our problem, therefore, is that although 'tit for tat' is stable once established, how can it ever get going in the first place in a primaeval world full of selfish individuals who always defect? There are two possibilities. One evolutionary scenario is as follows. Co-operation could emerge at first between pairs of relatives, evolving by kin selection. Selection would favour the recognition of cues which gave indications of relatedness. One cue that could be used is simply the fact of reciprocation of co-operation. Then individuals could use rules such as 'co-operate if the other co-operates (i.e. tit for tat) and co-operation could now emerge even between unrelated individuals (Axelrod & Hamilton 1981).

Another mechanism that can get co-operation started when

Box 11.3. *'Tit for tat' is an ESS provided w is sufficiently large (from Axelrod & Hamilton 1981).*
'Tit for tat' has a memory of only one move. An effective challenger will take maximum advantage of it by repeating whichever sequence of co-operate and defect results in greatest pay-off. But because of the short memory of 'tit for tat', we need only consider repeated sequences of two moves. Thus, the most effective challengers will be repeated sequences of 'defect: defect' or 'defect: co-operate'. If neither of these strategies can invade 'tit for tat', then no strategy can. We now examine the iterated Prisoner's dilemma game in Table 11.3.

When 'tit for tat' plays another 'tit for tat' it gets a pay-off of R each move for a total of $R + wR + w^2R \ldots$ moves. The sequence of moves which reiterate with probability w sum to $1/(1 - w)$, so the total pay-off will be $R/(1 - w)$. 'All defect' playing with 'tit for tat' gets T on the first move and then P thereafter, so it cannot invade 'tit for tat' if

$$R/(1 - w) \geq T + wP/(1 - w)$$

The alternation 'defect: co-operate', when playing 'tit for tat' gets a pay-off of

$$T + wS + w^2T + w^3S \ldots = (T + wS)/(1 - w^2)$$

An alteration of 'defect and co-operate' cannot invade 'tit for tat' if

$$R/(1 - w) \geq (T + wS/(1 - w^2)$$

In conclusion, neither strategy (and hence no strategy at all) can invade 'tit for tat' if both of the following conditions hold: $w \geq (T - R)/(T - P)$ and $w \geq (T - R)/(R - S)$.

most of the population is using 'always defect' is clustering. If 'tit for tat' individuals occur in clusters then they can interact with each other more than expected from random encounters in the population at large, and hence enjoy the increased benefits of mutual co-operation. Clustering is often associated with kinship and the two mechanisms can enforce each other in promoting the initial spread of reciprocal co-operation.

Predictions of the model

We have now shown that co-operation based on reciprocity can spread and be stable, provided two conditions hold:
1 Individuals must not be able to get away with defecting

without the other individual being able to retaliate effectively. Some animals may be able to recognize individuals and so spot the cheats. Another possibility is for two individuals to have a fixed place of meeting so they can be more or less certain of interacting again and again with the same individual.

2   For reciprocity to be stable, the probability, w, of the same two individuals meeting again must be high. If w falls, then defection may pay. For example if your partner became ill or old then there would be an incentive to defect and make a one-off gain when the probability of future interactions drops.

We now turn to some examples in nature where individuals do indeed play 'tit for tat' co-operation.

EXAMPLES OF RECIPROCITY

(a) Spawning in the black hamlet fish

The black hamlet fish, *Hypoplectrus nigricans*, is a simultaneous hermaphrodite (all individuals have both male and female gonads) which lives in the Caribbean. Fish come together in pairs to spawn in the last two hours of the day before sunset. Fischer (1980) discovered that spawning occurred in bouts, in which each fish assumed alternate male and female roles ('egg trading'). Thus, individual A released a few eggs which B fertilized (fertilization is external), and then B would release some eggs which A fertilized, and so on. Why does this lengthy process occur? Surely it would be more efficient, and simpler, for A to release all its eggs and B to fertilize them and then for B to release all its eggs and A to fertilize them?

Fischer suggests that egg trading has evolved because it is stable against cheating. The problem is as follows. Eggs are more costly to produce than sperm, so on the first move A is in effect an altruist; B gains the same number of zygotes for less investment. On the second move, B returns the favour. Cheating is a possibility because hamlets do not have mature eggs to release every day, but they can probably produce sperm every day. It would therefore pay an individual to pair up on days when it had no eggs to trade. If A gave up all its eggs at once, B could then get away with cheating. However because A releases just a few eggs at a time it can then easily spot whether B is a cheat and has no eggs to pay back in turn. In fact Fischer discovered that in pairs where the partner failed to reciprocate, the donor refused to give up any more eggs and left. Egg trading is therefore somewhat reminiscent of the 'tit for tat' co-operation envisaged in Axelrod and Hamilton's model.

(b) Regurgitation of blood by vampire bats

Wilkinson (1984) studied a population of individually marked vampire bats, *Desmodus rotundus*, in Costa Rica. Individuals quite often failed to obtain a blood meal during the night and they then begged for blood from other individuals in the day-time roosts. In an experiment, it was found that 5 out of 8 bats captured in the evening before feeding and released into the roosts at dawn, were subsequently given blood by a well-fed individual in the roost. By contrast, none of 6 bats captured and released after feeding successfully were given a meal.

Wilkinson discovered that regurgitation occurred only be-tween close relatives or between unrelated individuals who were frequent roost-mates. For reciprocity to evolve, the fol-lowing conditions are needed:

1 *Donors must be able to recognize cheats, and refuse to feed previous recipients who fail to reciprocate.* Wilkinson per-formed some clever experiments in the laboratory where he formed a group of bats, some individuals (all unrelated) com-ing from one roost and others (also unrelated) from another roost. In a series of trials, one bat, chosen at random, was removed and kept hungry while all the others had access to blood. The hungry bat was then reintroduced. It was found that 12 of the 13 regurgitations occurred between individuals from the same roost in the field, in other words individuals which were familiar with each other. Furthermore, the starved bats which received blood later reciprocated the donation signi-ficantly more often than expected had the exchanges occurred randomly.

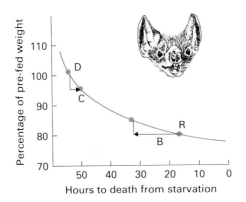

Fig. 11.4. In vampire bats, weight loss after feeding follows a negative exponential decline, with death from starvation occurring at 75 per cent of pre-fed weight at dusk. Therefore a donation of 5 per cent of pre-fed weight when at weight D should cause a donor to lose C hours but will provide B hours to a recipient at weight R. From Wilkinson (1984).

2 *Sufficient repeated pairwise interactions so that there are interchanges of roles and therefore net benefits to all donors.* In the field it was found that some unrelated individuals were constant companions in the roosts, sometimes for several years.

3 *The benefit of receiving aid must outweigh the cost of donating it* (R > S in Table 11.3). Figure 11.4 shows that the bats lost weight with increasing time since their last meal. The decelerating curve means that a donation of a small amount of blood by a well-fed individual results in little cost in terms of time moved along the bottom axis towards the threshold of death. However this same amount can bring an enormous benefit to a starving individual, moving its position considerably to the left along the time axis. The blood meal therefore has little cost to the donor and great benefit to the recipient; indeed it may save the recipient's life and enable it to survive until it has the chance to forage again for itself the next night.

Fig. 11.5. Two unrelated female vervet monkeys; one is grooming the other. Each female is nursing a ten-week old infant. Unrelated females co-operate in territory defence, alarm calls at the approach of predators and in interactions against other females in the group. Photo by Phyllis Lee.

(a)

(b)

Fig. 11.6. (a) A male olive baboon guarding an oestrus female. (b) Two male olive baboons about to separate a consorting male from an oestrus female. The consorting male is closest to the female in the centre of the photo. The male on the right turning his head, is soliciting aid from the male on the far left. Photo by © Craig Packer.

(c) Alliances in primates

Grooming (Fig. 11.5) is the most common form of affinitive behaviour in primates and it is also common for two individuals to join together to fight a third. Most grooming and alliances involve close relatives but sometimes friendships form between non-relatives. Seyfarth and Cheney (1984) performed field experiments which showed that grooming between unrelated vervet monkeys, *Cercopithecus aethiops*, increased the probability that they subsequently attended to each other's solitations for aid. The experiments were done in Amboseli National Park, Kenya, and involved playbacks of vocalizations used by individuals which sought support from others in an alliance. It was found that individual A would look towards the speaker broadcasting B's vocalization for a longer time if B had recently groomed A. Interestingly, this effect worked only between unrelated individuals. Relatives attended to the vocalizations equally whether the caller had recently groomed them or not. These results suggest that vervet monkeys are more willing to aid an unrelated individual if it has behaved affinitively to them in the recent past.

Another example comes from Packer's (1977) study of the olive baboon, *Papio anubis*. When a female comes into oestrus a male forms a consort relationship with her, following her around wherever she goes, awaiting the opportunity to mate (Fig. 11.6a). Sometimes a male who does not have a female enlists the help of another, unrelated male (Fig. 11.6b). This solicited male engages the consort male in a fight and while they are busy doing battle, the male who enlisted help goes off with the female! On a later occasion, the roles are reversed; the male who gave help is now assisted by the one who received help previously and so true reciprocation is involved.

Summary

Altruism is acting to increase the lifetime number of offspring of another individual at a cost to one's own chances of survival and reproduction. Co-operation can evolve in four ways.
1 *By kin selection.* An individual can increase its genetic representation in future generations by helping close relatives, who share copies of its genes identical by descent. The conditions for the spread of an altruistic allele are given by Hamilton's rule. Examples of traits favoured by kin selection may include alarm calling in prairie dogs and wife sharing in Tasmanian native hens. Experiments show that individuals can distinguish close kin from others (e.g. Belding's ground squirrel).

2 *Mutualism.* Both individuals gain a net benefit from co-operating.
3 *Manipulation.* One individual is 'tricked' by another into behaving altruistically.
4 *Reciprocity.* The iterated Prisoner's dilemma model explores the conditions under which 'tit for tat' co-operation involving related or unrelated individuals can spread and be stable. Examples in nature include blood sharing in vampire bats, egg trading in black hamlet fish and alliances in primates.

Further reading

Grafen (1984) gives an excellent account of the logic and assumptions underlying the idea of kin selection. Sherman (1981a,b) considers whether there is a limit to nepotism (care for relatives) in ground squirrels. Experiments on kin recognition include Greenberg (1979) and Buckle & Greenberg (1981) on sweat bees, Waldman (1981) on tadpoles, and Bateson (1980) on quail. Grosberg & Quinn (1986) demonstrate a genetic basis for kin recognition in a colonial tunicate. Dawkins (1982) gives a thought-provoking view of the interaction between genes and phenotypes. The book by Axelrod (1984) explores the lessons of the prisoner's dilemma model for human society.

Topics for discussion

1 All altruism is genetically selfish. Discuss.
2 Under what conditions would co-operation be expected to evolve (a) by kin selection, (b) by reciprocity?
3 What are the mechanisms that can be used for recognition of kin?

Chapter 12. Co-operation and Helping in Birds, Mammals and Fish

In chapter 11 we showed that it is possible to estimate the probability of a parent sharing a particular gene with its offspring. The answer for a sexually reproducing species was shown to be 0.5; in other words any gene in a parent's body has a 50 per cent chance of being duplicated in an offspring. Hence to make a genetic profit by sacrificing its life (as for example in certain arthropods such as Cecidomyian gall midges in which a mother offers herself as food for her offspring (Gould 1978) a parent must produce more than two offspring. But in chapter 11 we also showed that full siblings have a 50 per cent probability of sharing genes, exactly the same degree of relatedness as for mother and child. Therefore if we view helping and self-sacrifice purely from the point of view of making a profit for selfish genes, our arguments for parental care should apply with equal force to caring for siblings (and also with slightly less force to caring for grandchildren and cousins). The reasons why helping offspring is commoner than helping brothers and sisters are therefore likely to be ecological and practical considerations such as the ease of recognizing offspring and the degree to which offspring or siblings benefit from a given amount of aid rather than genetic factors. Some of these are summarized in Box 12.1.

Genetic predispositions and ecological constraints

Although helping in animals usually consists of parents caring for offspring, there are many examples of helping directed at other individuals. In more than 150 species of birds, for example, some individuals spend part or all of their lives helping others to reproduce: they help to feed or protect the offspring of others while apparently foregoing the chance to breed themselves. This behaviour appears to be *altruistic*, since the helpers give benefit to others while incurring costs. These costs might be measurable as, for example, energy expended by the helper, but ultimately they must be scored as lost opportunities for the helper to have its own offspring.

In this chapter we will discuss how this kind of helping might be favoured by selection in birds, mammals, and fish. The problem can be considered as having two parts.

1 *The genetic predispositions for helping.* In the last chapter

269

*Chapter 12
Co-operation
and Helping
in Birds,
Mammals
and Fish*

Box 12.1. *Some hypotheses about the ecological and proximate constraints that favour care of offspring over care of siblings in most species.*

1 Young benefit more from a given amount of aid. One insect fed to a helpless nestling contributes more to its survival than the same insect would to the survival of a healthy sibling. In other words young are 'superbeneficiaries' (West Eberhard 1975). In species with overlap of generations this constraint may not apply since newly born younger siblings might benefit just as much from a helper's aid as would an offspring of the helper.

2 Young are generally easy to identify as relatives of the helper, while siblings may be harder to recognize as kin.

3 While young are often born close to the parents, siblings may disperse from their birth site and therefore not be available for receiving help.

4 Young, especially in slow or continuously growing species (e.g. fish), are very different in size from the helper and are therefore not likely to compete with it for food in the future. In contrast, if the helper increases the survival of one of its siblings from the same brood, it may suffer in the future from increased competition.

5 Young may be more valuable in terms of expected future reproduction. If a youngster is successfully reared to maturity it has a higher expected future reproductive output than does a sibling of similar age to the helper.

6 In a species in which both parents care for the young, suppose that an average individual can raise x youngsters. By helping to raise sibs at home an individual could add up to x offspring-equivalents (full sibs) to the fledged brood, but by mating and having its own offspring it could produce 2x offspring, because of the contribution of its mate. In other words the mate's contribution to rearing offspring generally favours parenting over helping (Charnov 1981).

we saw how altruistic behaviour might arise in several ways. Perhaps helpers substitute sibling care for parental care, in which case helping could be viewed as a form of aid to close relatives or *nepotism* favoured by kin selection. As an alternative, or in addition, helpers might make genetic gains in other ways, such as mutualism and reciprocity, discussed in chapter 11.

2 *Ecological constraints which favour helping in some*

270

Chapter 12
Co-operation
and Helping
in Birds,
Mammals
and Fish

species *but not in others*. Even if the potential genetic gains
for helping and parenting are similar, practical considerations
usually favour parenting (Box 12.1) Perhaps in some species
these contraints are reversed in their effects. Sometimes, the
option of becoming a parent is simply not available.

We will start by describing one of the best studied examples
of helping in birds and then go on to consider some of the
complications which arise in trying to extend the results of
this study to other species.

An example of helping in birds — the Florida scrub jay

The Florida scrub jay (*Aphelocoma coerulescens*), as its name
implies, lives in oak scrub in Florida. Suitable habitat for
birds is patchy and scarce, so there are isolated pockets of
breeding jays with totally unoccupied and unsuitable areas
between. The breeding birds live as pairs in year-round territories. Just over half the pairs have helpers, the average number being about 1.8 per pair. These birds help the breeding
pair to feed their young and to defend the nest (Fig. 12.1); they
are generally 1−2 year-old birds and almost without exception are related to the breeders. Out of 165 helpers, 64 per cent
helped their parents, 24 per cent helped one parent (the other
having died) and only 4 per cent helped breeders that were
not their relatives. Both males and females help, but males
generally stay at home longer than females. When helpers
leave home they disperse and attempt to set up their own
territory. The scrub jay has been studied for 15 years by Glen
Woolfenden and John Fitzpatrick (1984) and their beautifully
detailed records show how genetic and ecological factors contribute to the explanation of why birds stay at home and help.

(a) Breeders benefit from the presence of helpers

Breeding groups of scrub jays with helpers rear more young
than those without help (Table 12.1). Helpers make two kinds
of contribution.
1 They help to defend the nest against predators, such as
snakes, by mobbing and giving alarm calls to warn the chicks.
This antipredator behaviour is primarily responsible for the
increase in survival of chicks when helpers are present in a
group.
2 They help to feed the nestlings, providing up to 30 per
cent of all food brought to the chicks in some groups. While
this apparently does not result in an increase in chick survival,
since the total amount of food brought to the nest is not
increased by the efforts of helpers, it does relieve the burden
on parents so that they survive better from one season to the
next. Breeders with helpers have to put less effort into feeding

271

Chapter 12
Co-operation
and Helping
in Birds,
Mammals
and Fish

Fig. 12.1. The Florida scrub jay. Helpers provide food for the young in the nest and defend the nest against predators such as snakes. Photo by Glen Woolfenden.

Table 12.1. Breeding pairs of Florida scrub jays benefit from the presence of helpers. Both experienced and inexperienced (first time) breeders rear more young when they have helpers (Woolfenden & Fitzpatrick 1984)

	No helpers	With helpers	Average number of helpers in territories with at least one helper
Inexperienced pairs	1.24	2.20	1.7
Experienced pairs	1.80	2.38	1.9

their chicks and probably as a consequence of this their annual survival rate is 85 per cent, as opposed to 77 per cent for breeders without helpers. Since the helpers and parents are related to one another by about 0.5, this increase in parental survival represents a genetic gain for helpers in addition to that made by rearing younger siblings.

(b) Habitat saturation is an ecological constraint

Do helpers benefit from staying at home and helping? Clearly they make some genetic profit because they help to rear their own younger siblings which, as we said at the beginning of the chapter, is approximately equivalent to rearing offspring. But the crucial question is whether helpers could in theory do better by setting up their own territory and breeding themselves right from the start, or by staying at home and helping.

272

Chapter 12
Co-operation
and Helping
in Birds,
Mammals
and Fish

Table 12.2. A scrub jay helper would do better by setting up its own breeding territory if it could find one, than by helping its parents at home (Woolfenden & Fitzpatrick 1984)

Option		Result
1. Stay at home and help	Young produced by experienced parents with no help	1.80
	Young produced by pair with helpers	2.38
	Extra young due to presence of helpers	0.58
	Average number of helpers	1.78
	Genetic equivalents to a helper (r to nestlings = 0.43)	0.14
2. Go off and rear own young	Young reared by first time breeders	1.24
	Genetic equivalents (r to own offspring r = 0.5)	0.62

One helper adds on average 0.33 offspring to its parent's production of young, and a novice pair of breeders rears about 1.24 young (a generous estimate of what a helper would do if it set off and bred instead of helping). So a helper could make 0.62 genetic equivalents by going off to breed instead of only 0.14 equivalents by staying at home (Table 12.2). These calculations are greatly oversimplified. They do not, for example, take into account the following: the mortality of young birds is increased by setting off too early in life to seek a territory; by staying at home a youngster can increase the survival of its parents, which as mentioned earlier is part of the genetic payoff for helping; helping may act as a form of training and increase the youngster's success when it breeds on its own; and the best strategy for a young bird may depend on what all the others do (the more that decide to leave home the more competition there will be for breeding sites). But these complications are not likely to alter the general conclusion that a young scrub jay would do better to leave home if it could find a territory, in spite of the fact that it rears siblings when it helps at home. The most plausible interpretation is that helpers stay at home because there are no spaces for them to set up a breeding territory: the habitat is saturated. This interpretation is supported by Woolfenden and Fitzpatrick's observation that as soon as vacancies arise helpers usually leave home to set up on their own. In 1979–1980, for example, almost half the adults in the study population died of an unidentified disease. The following year most young birds had set up territories in the empty spaces and very few remained at home as helpers.

1970

1971

1973

1976

1975b

1975a

300 m

Fig. 12.2. An illustration of how male scrub jays may inherit part of their parents' territory. This is one of the benefits of helping. The diagrams show how three helpers (B,C,D) acquired territories. Shaded areas represent land inherited by two sons (B,C) and one grandson (D) of the pair occupying territory A. Broken lines show the incipient point of budding off and small arrows show how territorial boundaries expanded. From Woolfenden and Fitzpatrick (1978).

(c) Males benefit by inheriting a breeding space

About half of the surviving male helpers eventually acquire a breeding space by inheriting part or all of their parental territory (Woolfenden & Fitzpatrick 1978; Fig. 12.2). The mortality rate of adult males is low and therefore very few vacant territories appear each year; so one way for a young male to acquire a territory is to help his parents to increase the size of their territory until it is large enough for him to split off part of it. By helping his parent the young male increases their production of young and therefore the size of group occupying the territory. Larger groups can expand their territories at the expense of small groups, and eventually the young male buds off to set up his own territory at the edge of the parental area (Fig. 12.2). When there are several male helpers in a territory there is a dominance hierarchy and the oldest dominant male is the first to inherit his own territory.

273

274

Chapter 12
Co-operation
and Helping
in Birds,
Mammals
and Fish

As with most birds, females do not breed in or near their parents' territory, but instead they disperse to find a vacant breeding space elsewhere. As discussed in chapter 9 this sex difference in dispersal might be a mechanism to avoid inbreeding, since females usually end up mating with an unrelated male as a consequence of dispersal. The female's benefit from helping is therefore less than that of the male: she rears close relatives and has a safe base from which to explore, but she does not inherit her own breeding territory as a result of helping. As one might predict from these observations, males tend to stay at home longer than females. In other words help is given to the parents in proportion to the future expected gain for the helper. In addition to the ecological factors predisposing males to help more than females, there may possibly be genetic considerations (Box 12.2).

Box 12.2. *Genetic predispositions for males to become helpers.*

Ric Charnov (1981) has pointed out that there may be a genetic predisposition for males to become helpers. This arises from uncertainty of paternity. Suppose that a male does not father all his mate's offspring (because he is cuckolded). A daughter of this male is more closely related to her own children than to her siblings. Her relatedness to her children is always 0.5, but if she and her siblings are not all fathered by the same male the average relatedness between her and her siblings is less than 0.5. In the extreme example where each offspring has a different father, the degree of relatedness between brothers and sisters is 0.25.

Now consider the effect of the same situation on a male. If he never fathers his mate's children his relatedness to them is zero, while he is always related to his siblings by at least 0.25 (since all siblings share the same mother). If males father half their mate's children their average relatedness to offspring is $0.5/2 = 0.25$. Their relatedness to siblings in this case is 0.25 (genes shared via the mother) + 0.25/2 (genes shared by siblings through the father). This gives a total of 0.375: again males are more closely related to their siblings than to their own mate's offspring.

These calculations serve to illustrate the point that any degree of uncertainty of paternity might predispose males, more than females, to become helpers and rear siblings instead of offspring.

To summarize, the study of the scrub jay shows that the question 'why do some birds stay at home and help?' has quite a complicated answer. Birds stay at home because (a) they increase their own chance of survival, (b) the chance of successful dispersal to a breeding site is low (c) if they are males they may inherit all or part of their father's territory. Helping is advantageous because (a) it increases the survival of helper's parents, (b) it increases the production of young relatives of the helper and (c) it may may eventually result in an increase in territory size which in turn gives the young male helper a chance to acquire part of the territory for himself (Fig. 12.3).

The scrub jay example illustrates the point that it is difficult to resolve the issue of whether the benefits of rearing siblings or the various other benefits are the crucial ones that promote helping. A considerable controversy has arisen in the literature about this (Woolfenden & Fitzpatrick 1984; Brown 1987) but the resolution of the debate cannot be in terms of either one or the other. Both factors are important and the emphasis one should place on them depends on the question one asks. If the question is 'why stay at home?', the calculations in Table 12.2 show that benefits from raising siblings are not enough to account for this, but if the question is 'why

275

Chapter 12
Co-operation
and Helping
in Birds,
Mammals
and Fish

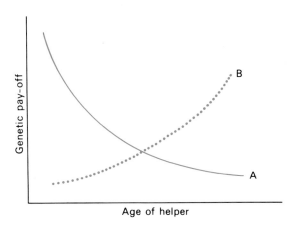

Fig. 12.3. A graphical summary of helping in the scrub jay and many other similar species. The genetic pay-off for helping at home (line A) decreases with increasing age because it is likely that eventually one or both parents will be replaced and the degree of relatedness between helpers and young is therefore decreased. The benefit from breeding (line B) increases with age because as birds get older they are more likely to take over their own territory. The position of the curves is influenced by the ecology of species. For example when territories are plentiful, curve B is shifted far to the left and helping may not occur at all, but when territories are scarce the period of helping is prolonged because curve B moves to the right. The two curves are not necessarily independent, since helping might increase the experience of young birds and therefore their future breeding success.

276

Chapter 12
Co-operation
and Helping
in Birds,
Mammals
and Fish

help?' the fact that helpers are related to the nestlings is important.

HELPING IN OTHER SPECIES AND SOME COMPLICATIONS

The pattern found by Woolfenden and Fitzpatrick in scrub jays seems to apply to many bird, mammal, and perhaps fish species with helpers. Helpers are usually offspring of the breeding pair, their presence usually increases the breeding success of the group, and they usually have little opportunity to breed themselves because of ecological constraints. The black-backed jackal (*Canis mesomelas*) in the Serengeti fits this pattern. Monogamous breeding pairs have 1−3 young from previous litters that act as helpers by regurgitating food for the pups and lactating female as well as grooming, guarding and playing with the pups (Fig. 12.4, Fig. 12.5). Similarly in the Princess of Burundi cichlid fish (*Lamprologus brichardi*) helpers are young of the breeding pair from previous broods. These small fish live in the shallows at the edge of Lake Tanganyika where they defend a territory around a nesting hole. Young stay on their parent's territory for 2−3 breeding cycles (up to 12 months) where they both guard the nest and clean the eggs or larvae. As a result of this help the mother is

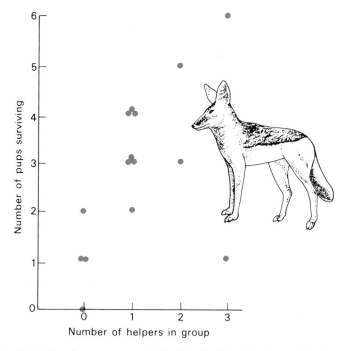

Fig. 12.4. Breeding success of black-backed jackals is positively correlated with the number of helpers in a group (Moehlman 1979).

able to lay more eggs, but at the same time, the energetic expense of helping causes the young fish to grow more slowly (Taborsky & Limberger 1981; Taborsky 1984).

277

Chapter 12
Co-operation
and Helping
in Birds,
Mammals
and Fish

DO HELPERS REALLY HELP?

A correlation between the presence of helpers and breeding success of the group does not necessarily show that helpers help. The correlation could arise, for example, because good quality parents produce many young each year and therefore have both a high annual success and a large number of helpers from previous seasons. Alternatively some pairs might live in good territories and have enough food or shelter from predators to rear many young each year and therefore acquire a large number of helpers. When Woolfenden and Fitzpatrick (1984) compared the number of young raised by the same pair in the same territory with and without helpers, the difference was much smaller than in the overall comparisons shown in Table 12.2. In order to disentangle cause and effect it is necessary to do an experiment such as the one carried out by Jerram Brown *et al.* (1978, 1982) on grey-crowned babblers (*Pomatostomus temporalis*). The grey-crowned babbler lives in year round territorial groups of 2−13 birds in open woodland in Queensland, Australia. As with the scrub jays, each group consists of a breeding pair and a variable number of helpers which are young from previous broods. The breeding success of a group is positively correlated with the number of helpers present, and in order to show that helpers actually cause the increase in success Brown did a removal experiment. He removed all the helpers but one from 10 breeding groups and left another 10 groups as controls with 4−6 helpers per group. The control groups reared more young per season than the experimentals, mainly because the helpers feed the young and relieve the pressure on the breeding female so that she can recover from one breeding attempt and start again sooner.

HELPERS ARE NOT ALWAYS RELATIVES

(a) Dwarf mongoose

The scrub jay pattern does not, however, apply to all species with helpers. In particular, helpers may be quite unrelated to the offspring they help to rear. The dwarf mongoose (*Helogale parvula*) is an example. It is a small diurnal carnivore that lives in groups of about 10 individuals made up of a breeding pair and helpers; the helpers are either offspring of the breeding pair or they are unrelated immigrants. Females tend to be

(a)

(b)

Fig. 12.5. Black-backed jackals (*Canis mesomelas*) are monogamous and have helpers who are offspring from a previous litter. (a) A mated pair grooming their year-old son (centre) who remains on the parental territory as a helper. Helpers contribute to pup survival by regurgitating food (b), chasing off predators, in this case a spotted hyena (c), and by helping to defend the territory (d), the intruder is on the right. Photos by © Patricia D. Moehlman (1980).

278

(c)

(d)

more active as helpers and they bring food such as beetles, termites and millipedes to the young in their den as well as guarding the den against predators such as other species of mongoose (Rood 1978). The striking observation made by Rood is that unrelated helpers do as much helping as those which are siblings of the young ones in the den. In one troop an immigrant female did more work than any other individual, including the father and mother! In contrast to related

279

280

Chapter 12
Co-operation
and Helping
in Birds,
Mammals
and Fish

helpers, unrelated individuals gain no genetic profit through the young ones they help to raise, so the only possible gain they can make is to increase their own success in the future. In the mongoose this comes about because unrelated helpers, especially females, sometimes take over as breeders when one of the resident parents dies. As with the scrub jays there is a shortage of breeding opportunities and helping is part of the strategy of eventually acquiring a slot as a breeder. But why should the unrelated mongoose help instead of just waiting on the sidelines for a chance to start breeding? There are three factors which might be important.

1 Helping may be a form of 'payment' for permission to stay on a breeding territory while waiting to take over. If the resident breeding pair gain nothing from having an unrelated helper close at hand they may simply evict it.

2 Helping keeps the group and therefore the territory intact, which is vital to the helper's future success when it becomes a breeder itself.

3 Finally, some of the young reared by the helper will in turn help it when it becomes a breeder, regardless of whether or not they are relatives. These points can be summarized by saying that helping is a long term investment for selfish future breeding success.

(b) Anemone fish

A similar interpretation may be applicable to helpers in the anemone fish *Amphiprion akallopisos*. Breeding pairs of fish defend anemones, which are an essential scarce resource for successful reproduction. Sometimes the pair are helped in defence by a non-breeder, which is most unlikely to be a relative of the breeding pair since the young fish disperse in the plankton before settling on a territory. Instead the helpers are probably unrelated and as hypothesized for mongoose helpers, they may be 'hopeful reproductives' waiting for a chance to take over when one of the breeders dies (Fricke 1979). Helping may again be a form of payment for permission to stay in the territory.

(c) Pied kingfisher

The idea of helping as a payment for permission to remain in a group is clearly illustrated by the pied kingfisher (*Ceryle rudis*) (Fig. 12.6) in which the breeding pair apparently accept unrelated helpers only when the payment is sufficiently advan-

281

*Chapter 12
Co-operation
and Helping
in Birds,
Mammals
and Fish*

tageous to them. The pied kingfisher, unlike most of the species referred to so far, is a colonial nester and does not live in a year round territory. Uli Reyer (1980) compared two colonies, one at Lake Victoria and the other at Lake Naivasha in East Africa. At Naivasha breeding pairs have at most one helper which is nearly always a 1–3 year old male offspring of at least one of the breeding pair (a 'primary helper'). He helps by feeding fish to the female, defending the nest against predators, and bringing food to the nestlings. At Lake Victoria there are sometimes secondary as well as primary helpers; these birds are also males but are unrelated to the breeding pair. They show up at both colonies just after the young hatch out and try to feed the breeding female of a pair; they are persistently chased away by the breeding male at Naivasha, but at Victoria they are eventually tolerated and allowed to stay and help to feed the young. The reason why secondary helpers are accepted at one lake but not the other is probably related to feeding conditions. Victoria is not as good for fishing as Naivasha: the water is rougher so that it takes longer for a kingfisher to catch a fish, the fish are smaller, and the feeding grounds are further away from the colony than at Naivasha. As one might expect, the parents expend more energy per day feeding their chicks at Victoria than at Naivasha (Reyer & Westerterp 1985). This means that parents with only one helper have difficulty in bringing enough food for their chicks and the secondary helpers make a big difference to breeding success (Table 12.3). At Lake Naivasha one helper is enough to rear all the young successfully and a secondary helper could therefore contribute little to the success of a breeding pair. Reyer and Westerterp (1985) confirmed this interpretation by the experiment of adding chicks to broods at Naivasha and

Table 12.3. Helpers in the pied kingfisher. At Lake Naivasha nests have no helpers or one helper, feeding conditions are good and one helper is therefore enough to ensure that all the chicks get enough to eat. At Lake Victoria there are sometimes second (unrelated) helpers, feeding conditions are poor, and the second helper makes a big difference to chick survival (Reyer 1980, 1984)

	Per cent nests with helpers			Feeding conditions		Breeding success				
	0	1	2	Time to catch a fish (min)	Per cent dives successful	Clutch size	Young hatched	Young fledged with helpers		
								0	1	2
Lake Naivasha	72	28	0	5.9	79	5.0	4.5	3.7	4.3	—
Lake Victoria	37	43	20	13.0	24	4.9	4.6	1.8	3.6	4.6

282

Chapter 12
Co-operation
and Helping
in Birds,
Mammals
and Fish

taking them away at Victoria. The Naivasha parents, now facing the same degree of energy expenditure as birds would normally have at Victoria, began to accept secondary helpers while the Victoria birds with reduced broods rejected them.

The conclusion is that unrelated helpers are accepted only when their help is effective at increasing the breeder's success or easing their heavy energetic burden. But why do secondary helpers help at all? Adult females are in short supply, probably because they are eaten while on the nest by snakes, mongooses or lizards, and secondary helpers sometimes acquire a wife as a result of helping. Seven out of 17 helpers that survived to breed mated with the female they had helped and 3 of them actually displaced the old breeding male. This is why the male only accepts a helper if there is an immediate large benefit, which has to be weighed against a possible future cost. In addition, secondary helpers have a slightly higher chance of surviving than do birds that simply wait for a year and do nothing.

Fig. 12.6. A pied kingfisher bringing food to the nest. Photo by Uli Reyer.

283

*Chapter 12
Co-operation
and Helping
in Birds,
Mammals
and Fish*

Primary helpers obviously make some genetic gain by helping relatives. At Lake Victoria their degree of relatedness to
the nestlings is 0.35, and one primary helper on average increases the production of chicks by 2.46 (note that this is not
apparent in Table 12.3 where the two kinds of helper are
combined). Another advantage is that they increase the survival of their parents (especially the mother) by helping. To be
weighed against these genetic gains are the facts that primary
helpers, who work harder in feeding the chicks than do secondary helpers, have a lower annual survival than they would by
doing nothing, and of course they forgo the possibility of
breeding themselves. When Reyer (1984) compared the genetic gains over the first two years of life for four strategies he
found that, as in the scrub jay, the highest pay-off would be
from breeding in the first year if a mate could be found, while
the lowest pay-off of all was for doing nothing in the first year.
Helpers achieve an intermediate pay-off. Although the pay-off
is similar for the two kinds of helper it comes from different
sources: primary helpers gain mainly through helping kin,
while secondary helpers gain almost entirely from increasing
their own chances of mating.

We can now extend the conclusions of the scrub jay study
by saying that helping in birds, mammals and fish may lead to
genetic gain through aid to relatives or through future reproduction of the helper (relatedness is not essential for helping
to be favoured by selection) and the environmental constraint
which leads to helping might be a shortage of territories or
mates.

Conflict in breeding groups

We have already seen in the pied kingfisher that helpers and
breeders may have conflicting interests. It has even been suggested that helpers in some species may try to disrupt breeding in order to decrease the number of future competitors and
therefore increase their own chances of success at getting a
territory (Zahavi 1974), although there is as yet no firm evidence for this view. However, conflict of interest is very
clearly demonstrated in birds where co-operative breeding involves several females sharing the same nest (as opposed to
the species described so far in which just one female breeds).

(a) Groove-billed anis

The groove-billed ani (*Crotophaga sulcirostris*) lives in a
group territory defended by 1–4 monogamous breeding pairs.
The group builds a communal nest in which all the females

284

Chapter 12
Co-operation
and Helping
in Birds,
Mammals
and Fish

lay some eggs and to which all the group members contribute parental care. Not all the eggs laid in the nest survive to be incubated and females compete to make sure that their eggs are the ones which survive. Competition takes the form of females rolling each other's eggs out of the nest, so that by the time incubation starts there are usually broken eggs strewn over the ground below the nest. Female anis are apparently unable to recognize their own eggs so each female rolls eggs out of the nest only up to the time when she herself starts to lay, otherwise she would run the risk of evicting her own eggs. Sandy Vehrencamp (1977) found that there is a dominance hierarchy among the females in a group and that the most dominant female ends up with the largest number of eggs in the nest at the time of incubation. She achieves this by starting to lay after the others, who are therefore unable to evict her eggs. The subordinate females start to lay earlier than the dominant, but they have several tactics for increasing the chances of survival of their eggs: they lay more eggs than dominants, usually wait two or three days between eggs, often produce a late egg after incubation has started, and initiate incubation earlier than the dominant. Once incubation starts the embryos begin to develop and the dominant female is forced to stop laying since otherwise her young would be late hatchers and at a disadvantage in competition between nest-lings for food. In spite of these tactics, nearly all the eggs of dominant females survive to be incubated while less than half those of subordinates survive (Table 12.4). Although the exis-tence of conflict between females is clear it is not yet known how dominant females succeed in laying last.

There is also an asymmetry in the extent to which female anis care for the eggs and young. Surprisingly the dominant female, with the biggest stake in the nest, does the least care (although the dominant male is very attentive as a parent). Again it is not known how the asymmetry is brought about, although presumably the subordinate females would not care for young if the nest contained none of their own offspring.

The ecological condition which favours communal nesting in anis seems to be predation pressure. In open pasture habitats where the nests are conspicuous to diurnal predators, groups are better able to defend the nest against predators at least in the egg stage. In dense marsh habitats the incubating anis suffer a great deal of predation at night, and when the incu-bation is shared between the individuals of a group each one has a higher chance of survival (Vehrencamp 1978). However both these advantages are benefits to the average group mem-ber and as we have seen, dominant birds gain much more out

285

Chapter 12
Co-operation
and Helping
in Birds,
Mammals
and Fish

Table 12.4. In groups of anis the first female to start laying lays more eggs but has fewer that survive to be incubated than the last female (Vehrencamp 1977)

	Order of laying		
	1st ♀	2nd ♀	3rd ♀
1. Eggs laid	7.0	6.3	5.8
2. Eggs tossed out	4.0	2.5	0
Incubated (= (1.) − (2.))	3.0	3.8	5.8

of being in a group than do subordinates. Subordinates must still gain more from group nesting in terms of predator defence than they lose from interactions with dominants, or they would leave the group and nest as solitary pairs (see p. 132)

(b) Acorn woodpeckers

Conflict between breeding females in a communal nester may occur even if the females are close relatives, as illustrated by the acorn woodpeckers (*Melanerpes formicivorus*) (Fig. 12.7) studied by Walter Koenig, Ron Mumme and Frank Pitelka in

Fig. 12.7. The acorn woodpecker lives in groups, including several breeders of each sex and non-breeding helpers which are offspring from previous years. Acorns and other hard-shelled nuts are stored in granary trees, some of which have as many as 30 000 holes, drilled by successive generations of the group. The nuts are stored in the autumn and used during the winter. On the right are two males and a female (lower right, with darker head band).

286

Chapter 12
Co-operation
and Helping
in Birds,
Mammals
and Fish

central coastal California (Koenig *et al.* 1984). These birds live in groups of up to 15 individuals typically consisting of about two breeders of each sex and their one-year old offspring. Groups are often formed by a band of brothers and a band of sisters from a different family coming together. All group members help to defend a communal territory and to feed the young. The breeders share the same nest and may even share paternity: genetic analysis of protein polymorphisms has shown that sometimes both males in a breeding group may contribute genes to the same female's clutch of eggs. However the close genetic relationship between females and the sharing of so many aspects of daily life does not mean that there is no conflict within the group. Females sharing a nest sometimes remove each other's eggs and as with Sandy Vehrencamp's anis, the first female to start laying is susceptible to having her eggs tossed out by the other female. Once the second one has started to lay she stops removing eggs, presumably because she is unable to distinguish between her own and the other bird's. Unlike the anis, an acorn woodpecker female does not simply chuck her nestmate's eggs on the floor; instead she carries them to a suitable nook in a branch and proceeds to peck them open. The whole group (including the female that laid the egg the in first place) then joins in an ovicidal feast!

A distinctive feature of acorn woodpecker natural history is that in many places they build enormous storehouses of acorns and pine seeds to keep them going in the winter and through the following spring breeding season. These stores, called granaries, consist of thousands of holes drilled into oak or pine trees (or even occasionally into telegraph poles). In the early autumn the group collects seeds and carefully stores them in the granary holes. It is the difficulty of constructing and maintaining granaries (combined with the high adult survival that results from the granary) that probably constitutes the crucial ecological constraint that both prevents young woodpeckers from leaving home and leads to the evolution of communal nesting. In southeastern Arizona, acorns are much less abundant and here the woodpeckers do not live in groups, nor do they construct granaries or live year-round in the same territory. The population migrates south each winter and new territories are set up each spring (Stacey & Koenig 1984).

Division of labour and specialized helpers

We have seen in this chapter that helping in vertebrates is usually a stepping stone for young individuals to become a breeder. Although many helpers must die before they ever have the chance to breed, helping is not a way of life for

Fig. 12.8. The nest chamber of a colony of naked mole rats. The large individual lying on its back across the middle of the picture is the breeding female suckling some two-week old young. Other individuals are sleeping non-workers who feed the breeding female and young with faeces. Photo by Michael Lyster © Zoological Society of London.

which individuals are permanently specialized. This, as we shall see in the next chapter, is in contrast to the social insects, where individuals develop specialized morphology and behaviour for a permanent life as a helper. Is there anything analogous to this in the vertebrates? A tantalizing example which seems to fit the bill has been described by Jenny Jarvis (1981). Naked mole rats (*Heterocephalus glaber*) are unprepossessing, hairless, small, blind, pink mammals that feed on roots and tubers, and live in colonies in underground burrows in East Africa (Fig. 12.8). A colony may contain up to 80 individuals but only one pair in the group breeds. The other females have undeveloped ovaries and the males, while they may have active sperm, are apparently non-breeders. The most striking aspect of mole rat life for our present purposes is that non-breeders come in at least two sorts that are different in size and do different jobs. Small ones dig burrows and search for food, while large ones remain close to the reproductive female in the nest, perhaps guarding or brooding the young. The difference between the two roles may be related to age;

288

Chapter 12
Co-operation
and Helping
in Birds,
Mammals
and Fish

young rats start off as small foraging helpers and grow into the larger nest-based helpers. The mechanisms maintaining different roles are yet to be fully elucidated, but it seems probable that chemical cues are involved.

Although mole rats may be more extreme in their development of division of labour than are other vertebrate species with helpers, the ecological and genetic factors that favour helping are probably quite similar to those in other species. The colony members are almost certainly all closely related to one another, and the difficultly of constructing an elaborate burrow system essential for survival severely limits the chances of successful dispersal away from home.

Conclusions

The main points to emerge from this chapter on helping in birds, mammals and fish are as follows:

1 Most but not all helpers aid close relatives (often they help their parents to rear siblings). In this respect helping can be seen as a way of promoting genetic representation in the next generation, analogous to being a parent.

2 Individuals become helpers because the alternative of leaving to become a breeder, although in principle a better option, is prevented by limitations such as a shortage of territories or mates.

3 In addition to the genetic gains from helping relatives, helpers may increase their own chance of survival or future reproductive success by staying at home. This has led to disagreements about whether helping is favoured because of the gains from helping kin or because of the direct gains to the individual helper. These are not, however, mutually exclusive and the relative importance of the two kinds of genetic gain may vary from species to species. Where helpers are unrelated to the individuals they help, the second kind of advantage must be the crucial one.

4 Conflicts of interest may arise between helpers and breeders, for example helpers may be potential sexual rivals of one of the breeders. These conflicts of interest are most marked in species where more than one female uses the same nest.

Summary

In some species of birds, mammals and fish there are individuals that do not breed, but help others to rear offspring. These helpers are usually, but not invariably, close relatives of the breeders that they help, for example they might be young

289

*Chapter 13
Co-operation
and Helping
in Birds,
Mammals
and Fish*

from a previous season. Helping close relatives is a way of promoting genetic representation in the next generation, analogous to having children. However helpers could in theory usually do better in terms of genetic representation in the next generation by breeding instead of helping. They are prevented from doing this by environmental limitations such as a shortage of suitable breeding sites. Where helpers are unrelated to the breeders, their benefit from helping is that they increase their chances of surviving and breeding in the future.

Further reading

Emlen (1984) gives a good general account of helping in birds and mammals. Brown's (1987) book is the most extensive review of helping in birds and includes discussions of the relative importance of kinship and individual gains in the evolution of helping. The last chapter of Woolfenden and Fitzpatrick (1984) discusses this issue and reaches a different conclusion. Kerry Rabenold's (1984) work on the stripe-backed wren is an exceptionally good example of a careful field study.

Topics for discussion

1 Why are there often conflicts of interest in breeding groups and how are they resolved?
2 Is kin selection necessary to account for the evolution of helpers in birds and mammals?
3 Discuss the relative merits of long term field studies and field experiments for evaluating the costs and benefits of helping.

Chapter 13. Co-operation and Altruism in the Social Insects

The social insects

THE PROBLEM

Co-operation and helping in vertebrates pales into insignificance beside what happens in the social insects. In these insects apparent self-sacrifice reaches the point where some individuals are completely sterile; they never reproduce themselves but instead spend their whole adult lives devoted to rearing the young of others. As Darwin himself and many other biologists since his time realized, this presents a real paradox, for if natural selection favours maximum genetic contribution to future generations, how can it lead to the development of totally sterile individuals that never reproduce? In the last chapter we saw that self-sacrifice and helping others to breed may be favoured by selection in vertebrates because the help is usually given to close relatives. Some individuals are prevented from having their own offspring by ecological constraints such as lack of a territory, so instead they help to rear their younger siblings. We also saw that helping non-relatives or relatives is sometimes part of a long-term strategy by which an individual eventually gains a mate or a territory. In this chapter we will consider to what extent similar ideas can be used to understand how sterile castes and helping have evolved in the social insects.

THE DEFINITION OF 'SOCIAL INSECT'

What exactly is meant by the term 'social insect'? To be more precise, this chapter is largely about the *eusocial insects*, which are characterized by three features: they have co-operative care of the young involving more individuals than just the mother; they have sterile castes; they have overlap of generations so that mother, adult offspring and young offspring are all alive at the same time. This is important because it provides the opportunity for the young to rear their younger siblings instead of having their own offspring, as in the scrub jays described in the last chapter. All three of these traits are necessary for a species to qualify as eusocial, but there are many species with intermediate stages, such as co-operative nest building without sterile castes. Eusociality occurs in three insect orders: Hymenoptera (ants, bees and wasps), Isoptera (termites) and Homoptera (aphids). The existence of eusociality in the first two orders has been well known for a long time, but in aphids it has been discovered only fairly recently (Aoki 1977). As we shall see later on, there may be special genetic predispositions towards the evolution of sterile castes in Hymenoptera and aphids.

291

Chapter 13
Co-operation
and Altruism
in the Social
Insects

The social insects are not only important because of their central role in the attempts of evolutionary theorists to understand the origin of altruism, but they are also extremely impressive in terms of their natural history. In a persuasive piece of salesmanship E.O. Wilson (1975) advertises that there are

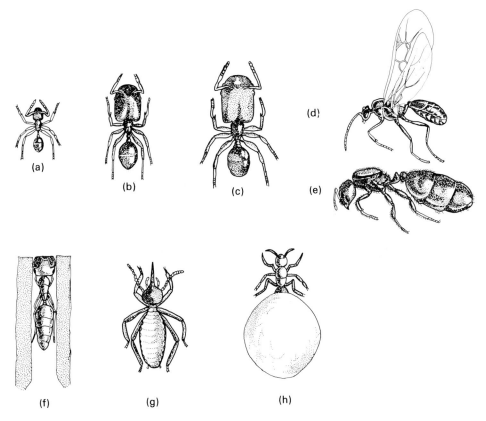

Fig. 13.1. Examples of castes in social insects. The top row are the female castes and male of the myrmicine ant, *Pheidole kingi instabilis*. (a) Minor worker. (b) Media worker. (c) Major worker. (d) Male. (e) Queen. The bottom row are various specialized castes in other species. (f) Soldier of the ant *Camponotus truncatus* blocking a nest entrance with its plug-like head which serves as a 'living entrance' to the nest. (g) A sterile caste of the nasute termite, *Nasutitermes exitiosus*, which has a head shaped like a water pistol to spray noxious substances at an approaching enemy. (h) Replete worker of a *Myrmecocystus* ant, which lives permanently in the nest as a 'living storage cask'. From Wilson (1971).

292

Chapter 13
Co-operation
and Altruism
in the Social
Insects

more than 12 000 species of social insects in the world, which is approximately equivalent to all the known species of birds and mammals. The staggering natural history of social insects can be illustrated by the following small sample of facts. In terms of size, a colony of African driver ants (*Dorylus wilverthi*) may contain up to 22 million individuals weighing a total of 20 kg. In terms of communication, the honeybee dance language, in which successful foragers tell other worker bees about the direction and distance to a food source, is one of the only known communication systems in wild animals in which an abstract code (the speed and orientation of the dance) is used to transmit information about remote objects. In terms of feeding ecology, the diets of social insects include seeds, animal prey, fungus grown in special gardens on collections of leaves or caterpillar faeces, and the excreta ('honeydew') of tended herds of aphids. Social insect colonies are often populated by individuals specialized to perform different tasks (so called castes). Sometimes castes have bizarre morphological modifications to help them carry out their jobs (Fig. 13.1). For example the head of soldier termites of the species *Nasutitermes exitiosus* is modified into a 'water pistol' used for squirting defensive sticky droplets at enemies, while the head of soliders in ant species such as *Camponotus truncatus* is shaped like a plug which fits neatly into the nest entrance to keep out intruders, rather like the operculum of a snail.

In the following sections, we will first describe the life history of one example of a eusocial insect to provide a background, then consider two theories which have been put forward to account for the evolutionary origins of sterile castes, before going on to discuss the special features of hymenopteran genetics, which are thought to predispose this group to evolve sterile castes.

The life cycle and natural history of a social insect

Myrmica rubra is a species of ant commonly found in Europe in woodlands, farmland and gardens. It builds a nest, the chambers of which are excavated in the ground underneath flat stones or sometimes in rotten tree stumps or in open soil. The nest is started by a single fertile queen. She is fertilized during a 'nuptial flight' in August or September in which very large numbers of winged reproductive females and males swarm around in the air and copulate (only the sexual forms can fly, and then only at this stage in their life cycle). The queen loses her wings after the nuptial flight and spends her first winter sealed inside the nest chamber which she has built by excavating a hole in the ground or in a tree stump. During the

293

Chapter 13
Co-operation
and Altruism
in the Social
Insects

following summer, eggs which she has laid develop into larvae and may mature into adult workers (the typical ants one sees scurrying around near ants' nests) before the autumn or in the following spring. Up to the time of maturity of the first worker, the queen feeds herself and her brood entirely off her own reserves of fat and protein, but when the workers mature they start to care for their younger siblings and collect food for them. The workers are female, but they are sterile. They never develop wings, their ovaries do not mature and they never take part in a nuptial flight. In successive years the colony and its nest grow slowly, until after about nine years it contains roughly 1000 workers and still a single queen who has laid all the eggs. At this stage the colony starts to produce a new generation of reproductives, winged females and males that eventually leave the colony on their nuptial flights. The old colony may then continue for a few more years, but as soon as the old queen dies and stops laying eggs to replenish the worker force, the colony dwindles in size and dies off.

This picture of the life cycle is typical of many temperate zone ants, although of course the details vary greatly from species to species. The details of worker behaviour are also very variable, but the following generalizations are typical of many species. Workers usually spend the first few weeks of their lives inside the colony handling dead prey which have been brought back by foragers, feeding the larvae and the queen with regurgitated food, cleaning the nest, and guarding the entrance. Later in life (this change takes place at an age of about 40 days in *Formica polyctena*, a species which has been studied in detail) workers begin to do jobs outside the colony, mainly foraging and defence against enemies. The total length of life of ant workers is not very well known but probably ranges from a few weeks to a few years. In wasps and bees, workers usually live for about 3 – 10 weeks. In addition to changes in worker behaviour with age, in some species of ants there are two major castes of worker (both sterile females): soldiers and normal workers. Soldiers are usually larger and have large heads with jaws or glands for producing defensive secretions. As their name implies they are specialists in colony defence.

The females belonging to different castes (queen, worker, soldier) do not usually differ genetically; the determination of caste depends on environmental conditions during larval development. In *Myrmica*, for example, whether a larva develops into a queen or a worker depends on factors such as nutrition, temperature and age of the queen who laid the egg. In honeybees the queen can suppress the development of new queens by chemical signals which prevent the workers from

294

*Chapter 13
Co-operation
and Altruism
in the Social
Insects*

feeding larvae the special diet ('Royal Jelly') needed to make them grow into queens.

How eusociality evolved: two theories

In this section we will reconstruct two possible pathways along which the evolution of sterile castes proceeded. The theories described refer to evolutionary history and therefore cannot be tested by direct experiments. Instead they have to be assessed in terms of the feasibility of the selective pressures they propose and by how well they account for the variety of stages of development of eusociality seen in present day insects.

In trying to understand how eusociality and in particular sterile castes evolved, it is useful to remember the distinction made in the last chapter between *ecological constraints* and *genetic predispositions*. The ecological constraints are the features of the environment which may either favour group living and co-operative breeding and/or reduce the chances for young individuals to breed themselves. The genetic predispositions refer to the degree of gene sharing between helpers and those they help: the greater the degree of relatedness, the smaller the benefit: cost ratio that is necessary for helping to be favoured by selection (chapter 11). In the social insects members of a colony are very often part of the same family, so that relatedness is high and workers are genetically predisposed to help. There are two theories about how sterile castes have evolved in insects, both of which postulate certain ecological constraints and genetic predispositions (Fig. 13.2).

HYPOTHESIS ONE: STAYING AT HOME TO HELP

(a) Ecological constraints

The ancestors of present day social insects were probably parasitoid wasps which laid their eggs inside or on the surface of a host, on which the larvae fed as they grew up (Evans 1977). A simple form of parental care in present-day parasitoids is to anaesthetize a host by stinging it, and then to carry it back to a safe site such as a hole in the ground or crevice in a tree before laying an egg on it. A slightly more advanced form of parental behaviour, seen for example in the digger wasp (chp. 10) is to create a safe place by burrowing into the ground. The burrow is then provisioned with one or more paralysed insect prey and an egg laid on each before sealing up the burrow entrance.

Parental care of this type could lead to the formation of

295

Chapter 13
Co-operation
and Altruism
in the Social
Insects

(a) **Families at home**

Solitary parasitoids

↓

Nest guarding by ♀

↓

Young stay at home
and help defend nest
or enlarge it

↓

Young permanently at
home and never breed

(b) **Shared nests**

Several reproductive females
(often sisters) build nests
close together

↓

Co-operation in nest building
and defence, but each female
reproduces herself

↓

Domination by one ♀ in group.
Others lose chance to reproduce

↓

Overlap of generations and
young females becoming workers

Fig. 13.2. Two hypotheses about the origin of eusociality in insects. According to hypothesis (a), sterile castes originated as daughters staying at home to help their mothers (called 'the subsocial route' to eusociality); according to (b), they originated from groups of reproductives nesting together in which one female dominated the rest (the 'parasocial route').

larger groups if there was sufficient ecological pressure for the offspring to stay near their birth site after reaching maturity and for mothers to tend their broods. Two main ecological factors are likely to have favoured staying at home.

1 *Defence of the eggs and larvae against parasites.* A major cause of mortality of young in provisioning species such as digger wasps is often parasitism by other insects, and in eusocial species one of the important tasks of workers is to defend the nest against parasites and other enemies.

2 *Nest building.* Although the ancestral parasitoids probably carried prey to naturally occurring refuges such as cracks and holes (or even burrows dug by their hosts), this soon gave rise to the construction of artificial refuges known as nests. In habitats where natural sites were in short supply, there would have been strong selection for the ability to build nests out of mud, chewed vegetation, and so on. Nest building is a laborious time-consuming business, so it is easy to imagine that newly matured adults might have done better to stay at home and help their mother to enlarge or repair her nest rather than go off to start on their own. At first they may have simply used their mother's nest for laying their own eggs, but eventually they may have evolved towards rearing younger siblings instead in their own offspring.

296

Chapter 13
Co-operation
and Altruism
in the Social
Insects

In the termites there may have been an additional ecological pressure for staying at home. These insects digest cellulose by means of protozoans living in their intestine, and the protozoa have to be passed literally from anus to mouth between generations. Thus ancestral young termites had to stay at home long enough to become infected with cellulose-digesting protozoa. As a thought provoking aside, it is worth mentioning Richard Dawkins' (1979) suggestion that termite eusociality evolved because the protozoans manipulate termites into staying at home to build an ideal environment in which protozoans can grow and replicate!

(b) Genetic predispositions

By helping to rear siblings, the workers pass on their genes to the next generation, so there is an obvious genetic predisposition towards helping. But Ric Charnov (1978) has pointed out that there is more to it than this: a mother makes a genetic gain by 'persuading' her children to stay at home and rear younger siblings while the children themselves make no genetic loss and therefore are 'willing' victims of maternal persuasion. The argument, which is a development of Alexander's hypothesis of 'parental manipulation' (Alexander 1974), goes like this. Suppose there is enough time to rear two broods in a season and the queen could either rear the first brood, die, and let her daughters lay eggs and rear the second brood, or she could lay both lots of eggs and persuade her first batch of daughters to stay at home and rear the second brood. In the first case, the young emerging at the end of the season will be grandchildren of the original queen, related to her by 0.25 (chp. 11). In the second case they will be children of the queen, related to her by the usual 0.5. Therefore if the numbers of offspring at the end of the season are the same in the two cases, the queen will have doubled her genetic representation by persuading her daughters to stay at home and help. If the mother has a very large capacity to lay eggs, it is possible that she could supply each of her daughters with as many eggs as they themselves could lay and rear.

There is therefore strong selection on the queen to 'persuade' her daughters to help. But, and here we come to the crucial part of the argument, the daughters lose nothing by staying at home. They rear younger siblings ($r = 0.5$) instead of children ($r = 0.5$) (assuming the original queen mated once at the beginning of the season). As long as the queen can lay enough eggs, the daughters make the same genetic gain from rearing siblings as from rearing offspring (Fig. 13.3).

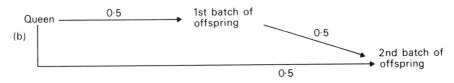

Fig. 13.3. An illustration of Charnov's (1978) idea that offspring may loose nothing by staying at home to help their mother, while the mother makes a large genetic gain. In (a) the queen starts the season by having children and ends with grandchildren (related by 0.25). In (b) she has two successive batches of children and ends with offspring related by 0.5. In (a) the offspring care for their own children and in (b) for their younger siblings, both related by 0.5.

HYPOTHESIS TWO: SHARING A NEST

In many tropical and some temperate zone wasps, nests are founded not by a single queen but by a group of co-operating individuals. Very often the founders are sisters (e.g. *Trigonopsis cameronii*) but sometimes they are not sisters and may possibly be unrelated (e.g. *Ceceris hortivaga*) (West Eberhard 1978a). In primitively social wasps each queen lays her own eggs and rears her own young, but if one queen succeeded in dominating the others and preventing their eggs from developing, the stage would be set for the evolution of workers.

(a) Ecological constraints

The ecological pressures favouring nest sharing are probably similar to those suggested earlier as favouring young that stay at home to help: defence against parasites and nest construction. For example in the wasp *Trigonopsis cameronii*, 1−4 sisters build a communal nest out of mud. Although each female builds her own brood cell, lays her own eggs, and provisions her own young, the sisters co-operate in building shared walls of the nest and in driving away ants or other enemies (Eberhard 1972). The way in which this kind of co-operation between sisters might lead to an unequal division of

297

298

Chapter 13
Co-operation
and Altruism
in the Social
Insects

reproductive labour is illustrated by *Metapolybia aztecoides*, another neotropical wasp, this time with a sterile worker caste. Here a nest is built by several queens, all of which lay eggs which develop into workers, that in turn help to enlarge the nest. Co-operation between queens is essential at the start in order to produce enough workers to get the colony and nest going, but once the colony is well established, one queen fights with and evicts the rest (that is, if more than one queen is still surviving) before they have a chance to produce any reproductive offspring. Each queen therefore starts the colony with a certain chance of ending up having reproductive off-spring, and the losers are 'hopeful reproductives' whose only chance at the start was to co-operate with the other queens. In *Metapolybia* the founders are probably sisters, so even the losers make a genetic gain from the colony, but if the chances of reproduction without co-operation were small enough, one could imagine co-operation even between unrelated founders (West Eberhard 1978b).

Nest sharing might also arise by accident. In the great golden digger wasp (*Sphex ichneumoneus*) two unrelated females sometimes end up using the same burrow because the second female was attempting to take over an apparently abandoned nest hole which happened to be occupied (chp. 10). Although in *Sphex* it is disadvantageous for females to share a burrow because they fight with each other and steal each other's prey, it is possible that accidental sharing could be the start of communal nesting if ecological pressures from parasites, for example, were very great.

(b) Genetic predispositions

If, as in *Metapolybia*, the founders are sisters, even those that fail to reproduce will still gain some genetic representation in the next generation. This is nicely illustrated by Bob Metcalf's study of paper wasps (*Polistes metricus*) (Metcalf & Whitt 1977). These wasps sometimes found nests as solitary queens and sometimes two sisters share a nest. When two queens share, one (the α female) does nearly all the egg laying while the other (the β female) passes on her genes largely through her sister's offspring. Metcalf calculated the degree of related-ness between sisters sharing a nest (by means of electrophore-tically detectable enzyme polymorphisms) and also estimated the number of offspring produced by α, β and solitary females. These calculations showed that the α female does better than the β, while the β does about as well as a solitary queen in terms of genetic representation in the next generation, and

that the β female's genetic contribution is made almost entirely through nephews and nieces. Nests of pairs of females
produce more young than those of solitary females because
they are better guarded against predators and parasites. Metcalf's study shows, therefore, that a *Polistes* female can do as
well by joining her sister as by setting up a nest on her own,
even if she produces hardly any children in the former case
(Table 13.1).

299
Chapter 13
Co-operation
and Altruism
in the Social
Insects

Table 13.1. Metcalf's calculations of relative gene contribution to future
generations in *Polistes metricus*. The table shows a comparison between
the gene contributions of solitary queens and α and β queens that share a
nest. The shared nests have a higher probability of success than solitary
nests. α females produce most of the young in a shared nest, while β
females make most of their gene contribution through helping their sister's
young. The measures of success are expressed relative to that of a solitary
queen

	Solitary queens	Joint nesters α	β
Relative probability of nest success (± S.E.)	1	1.38 ± 0.02	1.38 ± 0.02
Average relatedness to offspring	0.47	0.45	0.34
Relative gene contribution	1	1.83 ± 0.57	1.39 ± 0.44

The fact that social insect colonies are often founded by
sisters raises the interesting question of how siblings recognize each other. If dispersal is very limited a simple rule such
as 'co-operate with the first queen you meet' might be ade-
quate, but in at least some species there is a genetically based
ability to recognize relatives. Greenberg (1979) has shown that
workers of the sweat bee *Lasioglossum zephyrum* selectively
prevent unrelated bees from entering their nest. There is a
linear correlation between degree of relatedness and the tendency of workers to allow intruders into the colony. Greenberg suggests that the recognition is based on genetically
determined family odours and discrimination by workers
against bees with an unfamiliar smell (see chp. 11 for a more
detailed discussion of kin recognition).

To summarize the discussion of how sterile castes might
have originated in the social insects, the two main hypotheses
are that they arose by offspring staying at home to help their
mother, or by sisters (and perhaps unrelated queens) sharing a
communal nest. In both cases the ecological pressures favouring sociality were probably nest defence and nest construction, and the genetic predisposition towards helping was that

300

Chapter 13
Co-operation
and Altruism
in the Social
Insects

the recipients of aid were close relatives of the altruists. The two hypotheses are not mutually exclusive, and most authors agree that the first is probably correct for wasps, termites, ants and some bees, while the second applies to the remaining eusocial bees. The two hypotheses are equally applicable to any sexually reproducing diploid organism, but we now turn to a very important special feature of the Hymenoptera which may give them an additional predisposition towards eusociality.

Haplodiploidy and altruism

W.D. Hamilton (1964) was the first to fully appreciate the significance of the special genetic predisposition of Hymenoptera to form sterile castes. The special feature is *haplodiploidy*: males develop from unfertilized eggs and are haploid, while females develop from normally fertilized eggs and are therefore diploid.

A haploid male forms gametes without meiosis, so that every one of his sperm is genetically identical. This means that each of his daughters receives an identical set of genes to make up half her total diploid genome. With a diploid father, a female would stand a 50 per cent chance of sharing any particular one of his genes with her sisters, but with a haploid father she is certain to share all of them. The other half of a female hymenopteran's genes come from her diploid mother, so she has a 50 per cent chance of sharing one of her mother's genes with a sister. If we now think about the total degree of relatedness between sisters we come to a remarkable conclusion. Half their genome is always identical, and the other half has a 50 per cent chance of being shared, so the total relatedness is $0.5 + (0.5 \times 0.5) = 0.75$. In other words, because of haplodiploidy full sisters are more closely related to one another than are parents and offspring in a normal diploid species. Hymenopteran queens are diploid and are therefore related to their sons and daughters by the usual 0.5 (Box 13.1, Table 13.2). A sterile female worker can therefore make a greater genetic profit by rearing a reproductive sister than she could if she suddenly became fertile and produced a daughter! This extraordinary state of affairs also suggests why in the Hymenoptera only females help to rear sisters. Males are related to their sisters by 0.5 instead of the 0.75 value calculated above for females. The relatedness of males to their sisters is calculated as follows: a haploid male inherits all his genes from his mother and there is a 50 per cent chance that a given sister will have inherited the same maternal genes. (Note, however that a female is related to her brother by only 0.25, since the 50 per cent of her genes that come from her father

301

Chapter 13
Co-operation
and Altruism
in the Social
Insects

Table 13.2. Degrees of relatedness between close relatives in a
haplodiploid species

	Mother	Father	Sister	Brother	Son	Daughter	Niece or nephew (via sister)
Female	0.5	0.5	0.75	0.25	0.5	0.5	0.375
Male	1	0	0.5	0.5	0	1	0.25

have no chance of being shared with a brother, and the other
half of her genes have a 50 per cent chance of being shared:
$0.5 \times 0.5 = 0.25$.)

By contrast, in the termites in which both males and
females are equally related to their siblings, both sexes may
become sterile workers.

Haplodiploidy might also help to explain why sterile castes
have evolved more often in the Hymenoptera than elsewhere
in the insects. Wilson (1975) gives a figure of 11 independent
evolutionary origins for sterile castes in the social ants, bees
and wasps, which constitute only 6 per cent of all insect
species, and only one origin of sterility in all the rest of the
insects, namely the termites. Since Wilson wrote this, non-
reproductive soldiers have been discovered in a Japanese
aphid (Aoki 1977) but aphids are genetically even more prone
to sterility than the Hymenoptera. They reproduce asexually
(for at least part of the year), which means that the members of
an aphid colony are genetically identical just like the cells of a
body. So the existence of sterile workers in aphids is perhaps
no more of an evolutionary suprise than the fact that our nose
cells do not produce sperms or eggs.

Although haplodiploidy in the Hymenoptera may predispose
them to become eusocial, it does not *cause* eusociality to
evolve. This is easy to see because not all haplodiploid insects
have sterile castes, and in the termites sterile castes have
evolved in a normal diploid species. As we have stressed
earlier in the chapter, ecological pressures and genetic predispositions work together to determine whether or not sterility will evolve. We will come back to the question of whether
or not haplodiploidy paves the way for the evolution of eusociality on p. 309.

A second cautionary note is that the simple calculations
of relatedness in Table 13.2 hold only if the colony is formed
by a single queen who has mated once. If the queen mates
twice and sperm mixes at random the average relatedness
between sisters is only 0.5. If a colony is founded by several
sisters as in *Trigona*, helping is still more likely to evolve in a

302

*Chapter 13
Co-operation
and Altruism
in the Social
Insects*

Box 13.1. *Calculating coefficients of relatedness, r, in haplodiploid species.*

General point. Males develop from unfertilized eggs and so are haploid; all of a male's sperm are genetically identical so the probability of sharing a copy of a gene via the father is 1. Females develop from fertilized eggs and so are diploid; the probability of sharing a copy of a gene via the mother is 0.5, because of meiosis.

Method. Draw out a pedigree, linking the two individuals through their recent common ancestors. To determine the coefficient of relatedness between individual A and individual B, draw arrows along the pathways, pointing from A to B. Indicate on each link in the pathway the probability that a copy of a gene will be shared.

Examples.

(a) Sister–sister

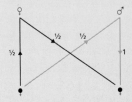

Half of a female's genes come from the father; the probability that a copy of one of these is shared with the sister is 1. The other half come from the mother; the probability that a copy of one of these is shared is 0.5.

Via mother = (0.5 × 0.5) + Via father = (0.5 × 1)
r = 0.75

(b) Sister–brother

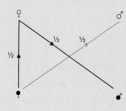

A female is linked to her brother only via her mother, as her brother develops from an unfertilized egg. Half of her genes come from her mother; the probability that a copy of one of these is shared is 0.5. The other half come from her father; the probability that a copy of one of these is shared is zero.

Via mother (0.5 × 0.5) + Via father (0.5 × 0) r = 0.25

(c) Brother–brother

All of a male's genes come from his mother. There is a 0.5 chance of sharing a copy of a particular gene with his brother.

Via mother (1 × 0.5) r = 0.5

303

Chapter 13
Co-operation
and Altruism
in the Social
Insects

(d) Brother–sister

All of a male's genes come from his mother. There is a 0.5 chance of sharing a copy of a particular gene with his sister.

Via mother (1×0.5) $r = 0.5$

Note the asymmetry in relatedness between brothers and sisters [cf. (b) above].

haplodiploid than in a diploid species because in the former a female is related to her sister's children by 0.375 and in the latter by 0.25. Regardless of whether or not the queen is monogamous, the coefficient of relatedness is not always exactly as calculated by working out pedigrees (Craig 1979; Grafen 1984). An hypothetical example will serve to illustrate this point. Suppose that in a social insect there are two strategies for females controlled by a single allele difference: stay at home and help, or set up a nest and accept help. If a female with the helping allele finds a sister with a nest willing to accept help, she can be sure that the sister does not share the helping allele, even though her relatedness by pedigree is 0.5 (Grafen 1986a).

CONFLICT BETWEEN WORKERS AND QUEEN

Our discussion so far has referred primarily to evolutionary history; we have described the ecological pressures and genetic predispositions that might have been important in leading to the evolution of sterile castes. In this section we are not going to discuss the origin of eusociality, but the selection pressures that act within present-day hymenopteran colonies. Our question will be: 'Given that there are sterile castes, how do workers and queens maximize their genetic profit?'. The answer to this question will probably tell us something about the selective forces *maintaining* eusociality today, but can be used only as indirect evidence for its origin.

Hamilton's theory, which we described in the last section, can be used to analyse how workers and queens might maximize their genetic profit. As we shall show, the theory predicts a conflict of interest between workers and queen over the sex ratio of reproductives in the colony.

304

*Chapter 13
Co-operation
and Altruism
in the Social
Insects*

CONFLICT OVER THE SEX RATIO

Our account of Hamilton's theory can be summarized as follows. Imagine a young female with the hypothetical choice of going off to rear her own daughters or staying at home to rear a new generation of younger reproductive sisters. Since she is more closely related to sisters than daughters, she would do better to stay at home and have sisters rather than the same number of daughters. In fact the queen seems to be the loser since she is condemned to have offspring!

However there is a further twist to the story that we have not considered. It may well be better for a young female to stay at home and rear sisters, but this presupposes that the queen is going to produce sisters for her to rear. Obviously what the daughters can do will depend on what the queen is doing. As we saw earlier in the book, whenever we encounter situations like this we need to analyse the problem in terms of what will be the stable strategy or ESS.

Let us consider the queen first of all. She is equally related to her sons and daughters ($r = 0.5$ in each case) and so just like diploid females of any sexual species she is expected to produce equal numbers of male and female reproductive offspring. To be more precise she should *invest equally* in the two sexes (chp. 8). It is important to emphasize that we are referring to equal investment in *reproductive* offspring, not sterile workers. Recall that in chapter 8 the argument was that a 50:50 sex ratio was stable because the expected *reproductive success* of a male and a female is the same. Hence the discussion of sex ratios is only pertinent to reproductives.

Now for the twist: if the queen produces an equal sex ratio, workers will spend their lives rearing equal numbers of brothers (to which they are related by 0.25) and sisters (related by 0.75). Their average relatedness to reproductive siblings will therefore be only 0.5, exactly the same as they would have to their own progeny if they had decided to leave home and have their own offspring!

In order for female workers to gain the full genetic benefit from staying at home and rearing sisters, they must rear more queens than drones. But how much bias in favour of reproductive sisters should they show? Once again we search for the ESS sex ratio, this time from the workers' point of view. The workers are more closely related to their sisters and so should rear more of them than brothers. But if they rear too many sisters then the sex ratio in the population will become so female-biassed that a drone will have very much greater reproductive success than a queen. The stable sex ratio for the workers is 3:1 in favour of reproductive females. When female

305

Chapter 13
Co-operation
and Altruism
in the Social
Insects

reproductives are exactly three times as common as males, drones have three times the expected success of queens because on average each drone can mate with three females. From the workers' point of view this would exactly compensate for the fact that brothers are only one third as closely related as are sisters: a worker expects to get three nieces or nephews from her brothers for every one she gets from her sisters. Nieces and nephews on her sister's side are three times as closely related to her, so the total gain per unit investment via brothers and sisters is the same.

To summarize this rather complicated argument, the queen prefers an equal investment in male and female reproductive offspring, but the workers prefer a ratio biassed 3:1 in favour of females. There is a direct conflict of interest over the sex ratio between workers and the queen. Who wins?

TESTS OF WORKER−QUEEN CONFLICT

Bob Trivers and Hope Hare (1976) attempted to test whether the queen or workers win by analysing the ratio of investment (more accurate than simply looking at numbers) in male and female offspring in 21 species of ants. The ant species were chosen because they were ones in which the conditions for the hypothesis apparently hold (one queen, one mating). Despite a considerable amount of scatter in their data, Trivers and Hare found that on average, the ratio of investment was much closer to 3:1 than to 1:1 (Fig. 13.4). They concluded that the workers win the conflict and successfully manipulate the sex ratio towards their own optimum and away from that of the queen. To put it baldly, the workers are successfully farming the queen as a producer of nieces and nephews: a far cry from the idea of workers as subordinate females making the best of a bad job! Trivers and Hare suggest that the workers win simply because they have practical power; they provision the young and are in a position to selectively kill off males and nurture queens. The queen presumably retaliates during evolution by attempting to control worker behaviour with pheromones or direct aggression.

Although Trivers and Hare considered two extreme divisions of practical power, total queen control and total worker control of the sex ratio, a more plausible alternative is that power is shared. The queen can choose the sex of the eggs she lays (by whether or not she fertilizes them with stored sperm) and the workers can choose whether or not to rear the larvae. With shared power the problem becomes 'given that the workers control the sex ratio after egg laying, what ratio should

306

Chapter 13
Co-operation
and Altruism
in the Social
Insects

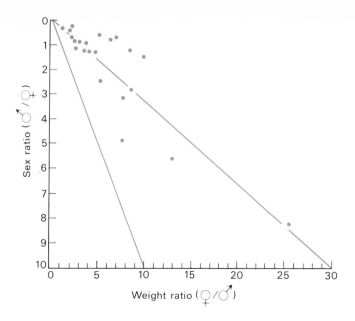

Fig. 13.4. Ratio of investment (measured by weight) in 21 species of ants. The x-axis is the ratio of female: male weight and the y-axis is the ratio of numbers of males: females in the colony. The lower line is the prediction if the investment ratio is 1:1 and the upper line is 3:1 in favour of females. The data are closer to the 3:1 line, as predicted if workers control the sex ratio. (To understand how the lines are drawn take the example of a ♀:♂ weight ratio of 6:1. Equal investment would mean six ♂ per ♀, and a 3:1 investment ratio in favour of ♀ would mean a ratio of 2 ♂ per ♀.) Trivers and Hare (1976).

the queen lay', and 'given that the queen lays a certain ratio of eggs, how should the workers manipulate it'. In one analysis of the problem, Michael Bulmer and Peter Taylor (1981) suggested that the queen can in general prevent a female-biassed investment ratio by limiting the supply of diploid eggs to the workers, although the workers may be able to respond by investing more in each queen egg that is made available to them. The general message of these more complicated analyses of conflict over the sex ratio is that quantitative predictions are not easy to make, although the qualitative prediction of Trivers and Hare generally holds. Indeed it could be argued that the scatter in Trivers and Hare's data reflects the fact that the balance of power varies between species.

Richard Alexander and Paul Sherman (1977) criticized Trivers and Hare's paper on the following grounds. (a) The queen often mates more than once, altering the bias in relatedness from the 3:1 assumed by Trivers and Hare. (b) Workers often lay male eggs (in one study of bumblebees, for example, 39% of male eggs were laid by workers (Owen & Plowright 1982)) and with laying workers the 3:1 prediction no longer holds.

307
Chapter 13
Co-operation
and Altruism
in the Social
Insects

More important, they offer an alternative explanation for the female biassed sex ratio. In chapter 8 we mentioned that Fisher's theory of equal investment in males and females no longer holds when there is competition between brothers for mates (local mate competition). If a mother 'knows' that her daughters will all be fertilized by her sons, she should produce a brood made up mostly of daughters, with just enough sons to do the fertilizing. Alexander and Sherman suggest that there may be some local mate competition which accounts for the biassed sex ratio in the ants studied by Trivers and Hare. This hypothesis is not incompatible with the idea of worker manipulation. If the queen favours a biassed sex ratio because of the effects of local mate competition the workers will favour an even more biassed ratio because of the additional impact of 0.75 relatedness. Without knowing the exact degree of local mate competition it is not possible for Alexander and Sherman to make a quantitative prediction of the optimal sex ratio for the queen, and the 3:1 bias observed by Trivers and Hare must be taken as suggestive, but not conclusive, evidence for the idea of worker manipulation in ants.

It will be apparent from the discussion of Trivers and Hare's paper that in order to test whether workers or queens control the investment ratio we need to know about the extent of local mate competition, whether or not workers lay eggs, and the degree of relatedness of workers and reproductives. A few studies have succeeded in collecting all this information, and the pattern that emerges is that when the conditions assumed by Trivers and Hare are met, sometimes the investment ratio is at the queen's optimum of 1:1, sometimes it is at the workers' optimum and sometimes it is in between. This probably serves to emphasize that the outcome of worker – queen conflict depends on how the life history of the species influences the division of power. In *Polistes metricus*, studied by Bob Metcalf (1980) the queen appears to win. In this species, as we have already described, nests are founded by one or more mated sisters in the spring. The nests produce reproductive males relatively early in the summer, followed by queens at a later stage; workers are produced continually through the season. The young queens mate in the late summer and overwinter before starting up their own nests the following spring.

Genetic analyses of enzyme polymorphisms showed that workers are related to their reproductive sisters by an average of 0.65 (slightly less than the theoretical maximum of 0.75 because the queen sometimes mates more than once). Also, workers do not normally lay eggs. There was little evidence of local mate competition: this was inferred from the fact that there is little or no inbreeding and the observation that fe-

308

Chapter 13
Co-operation
and Altruism
in the Social
Insects

Fig. 13.5. *Polistes apachus* workers on the nest. This species has a female biassed sex ratio of reproductives, perhaps indicating that workers benefit more than the queen. Photo by permission of Bob Metcalf.

males do not disperse far. Males must therefore disperse rather a long way and so brothers are not likely to compete. Thus, workers should prefer a female biassed ratio and the queen a 1:1 ratio of investment in reproductives.

Why do the queens win? Metcalf suggests that they win because of practical considerations. How might the workers manipulate the sex ratio? The obvious way would be by killing off male eggs or larvae (given that the queen lays eggs in a

309

Chapter 13
Co-operation
and Altruism
in the Social
Insects

1:1 ratio and that workers cannot lay female eggs themselves this seems to be the only obvious possibility). In *P. metricus* the queen produces male offspring early in the season. At this stage few of the workers have emerged and the few that have emerged can be effectively controlled by the queen. By this ploy it appears that queens have removed the opportunity for workers to control the sex ratio. The workers lose, in that they rear reproductives with an average degree of relatedness of 0.45 (0.25 for brothers and 0.65 for sisters) while the queen is related to her offspring by 0.5. An example of a study in which strongly female-biassed sex ratios were found and in which the various conditions of Trivers and Hare's argument were fulfilled (no local mate competion, no worker laying, mongamous queens) is Phil Ward's (1983) of *Rhytidoponera* ants in Australia (see also Fig. 13.5).

The evolutionary battle between workers and queen over the sex ratio has taken place in spite of the fact that queen and workers are not genetically differentiated. It is the diet and not the genes of a female that determine whether she becomes sterile or reproductive. If one thought in terms of genes for controlling the sex ratio, these genes must produce a conditional strategy 'if in a worker body favour 3:1, if in a queen favour 1:1'.

Haplodiploidy and the origin of eusociality

Now let us return to our earlier question (p. 301) of how hymenopteran helpers might have evolved in the first place. Was haplodiploidy important in the origin of helping? It might seem from what we have said so far that the answer should be yes. Let us summarize the argument beginning with 'Hamilton's rule' which we introduced in chapter 11. The rule is that helping pays whenever the ratio of benefits to costs (K) exceeds $1/r$. Now for our present calculations K is the ratio of (extra siblings raised as a result of helping/offspring lost by helping) and r is ratio (relatedness to siblings reared/relatedness to offspring lost).

In diploid animals a full sibling is genetically equivalent to an offspring ($r = \frac{1}{2}$ for both) so it would pay to help whenever K is greater than $(\frac{1}{2} / \frac{1}{2}) = 1$. In other words it pays to help if you can replace one lost offspring with just over one sibling. For a female hymenopteran, however, with a bias towards rearing sisters, the average relatedness of a worker to her reproductive siblings is greater than $\frac{1}{2}$; therefore the critical value of K to favour helping is less than 1. For example

310

Chapter 13
Co-operation
and Altruism
in the Social
Insects

with the 3:1 bias discussed earlier, the average relatedness to reproductives is $(\frac{3}{4} \times \frac{3}{4}) + (\frac{1}{4} \times \frac{1}{4}) = \frac{5}{8}$ and the critical value of K is $\frac{1}{2} / \frac{5}{8} = \frac{4}{5}$. In other words instead of having to rear just over one sibling for every potential offspring lost, the worker has to raise just over 4 siblings for every five offspring sacrificed. It would appear that haplodiploidy makes it easier for helping to evolve.

This line of reasoning is, however, too simple. To see why we have to introduce the concept of the *value* of males and females, the proportion they will contribute to the gene pool of future generations. Recall that when the sex ratio is 3:1 in favour of females, the expected reproductive success of a male, or in other words his value as a machine for making grand-children, is three times that of a female. If you include this factor, the pay-off for raising siblings and offspring must be calculated as (number raised × value × relatedness). Taking the example where the sex ratio of the population as a whole is 3:1 (this applies therefore to the average ratio of offspring and of siblings) we get the following answers:

For helping to raise a sibling: $(\frac{3}{4} \times 1 \times \frac{3}{4}) + (\frac{1}{4} \times 3 \times \frac{1}{4})$ $= \frac{12}{16}$

For raising an offspring: $(\frac{3}{4} \times 1 \times \frac{1}{2}) + (\frac{1}{4} \times 3 \times \frac{1}{2})$ $= \frac{12}{16}$

(In each line the first bracket is for females, the second for males. So on the top line the first bracket means 'three-quarters of sibs are females, with a value of one and a relatedness of three-quarters'). Therefore with a population sex ratio set at 3 reproductive females for every male, the critical value of K for helping to pay is $K > \frac{12}{16} / \frac{12}{16} = 1$, the same as for a diploid species! In other words *when the sex ratio in the population as a whole* is 3:1 haplodiploidy gives no advantage to helping: the extra relatedness to females is counter-balanced by the higher value of males (Grafen pers. comm.).

So can a female-biassed sex ratio combined with haplodi-ploidy ever tip the balance in favour of helping, or is it totally irrelevant? The answer is that a female bias might favour helping if it does not lower the value of females to offset their greater relatedness. This could happen if the sex ratio within the nest is female biassed while the overall population sex ratio is not. If the bias in the nest was 3:1 and the population sex ratio was 1:1, the value of males and females would be equal and we would be back to the simple calculation of K based on relatedness which gave $K = \frac{4}{5}$. But how could this situation ever arise?

Jon Seger (1983), in an important paper, has pointed out that it could arise as a result of a particular kind of life cycle found in some hymenoptera. In this life cycle, which is called

311

Chapter 13
Co-operation
and Altruism
in the Social
Insects

a 'partially bivoltine' cycle, there are two generations each year, but the generations are not completely distinct; individuals from generation 1 may survive to take part in generation 2. One kind of partially bivoltine cycle is found in halictine bees (Fig. 13.6a). The females overwinter as fertilized adults; in the spring they produce a first generation of offspring. The spring generation then gives rise to the summer generation and at the end of the summer this second generation mates and the fertilized females overwinter. The important fact to add is that some of the spring males survive to mate with second generation females (Fig. 13.6a). These males have the opportunity to mate with two generations of females, so they are more valuable to their mother (as grandchild-makers) than are spring females. The overwintering female should therefore produce a male-biassed spring generation of reproductives. Some of these spring males survive to mate with summer females, in effect pre-empting the chances of summer males. It therefore pays a mother in the summer to have a female-biassed sex ratio among her offspring, since males, their job have been partly pre-empted by survivors from the spring, are less valuable than females.

Now we have just the conditions needed for haplodiploidy to favour helping and it has arisen as a consequence not of selection for eusociality but as a 'preadaptation' resulting from the halictine life history pattern. Imagine an early summer female with the option of staying at home to help her mother rear siblings or going off to breed herself. In either case she will be contributing to the generation that overwinters as fertilized females, a generation that is female-biassed relative to the population as a whole (don't forget that the population includes males surviving from the spring).

Let us take an hypothetical numerical example to look at the effect of this. Suppose the population sex ratio of reproductives was 1:1 and the ratio amongst the overwintering generation was biassed 2:1 in favour of females. If the value of males and females was equal, the pay-off for raising a sibling would be $(\frac{3}{4} \times \frac{2}{3}) + (\frac{1}{4} \times \frac{1}{3}) = \frac{7}{12}$, while the pay-off for an offspring would be the usual $\frac{1}{2}$. The critical value of K is therefore $\frac{1}{2} / \frac{7}{12} = \frac{6}{7}$. The female halictine has to rear only 6 siblings for every 7 offspring lost, so haplodiploidy combined with a biassed sex ratio has shifted the balance slightly in favour of helping. If you try working out examples with different sex ratios, you will find that the general result holds even when the sex ratio in the overwintering generation is not female biassed in absolute terms, as long as it is female biassed *relative* to the population as a whole.

But this is not quite the end of the argument. Remember that

312

Chapter 13
Co-operation
and Altruism
in the Social
Insects

summer males are less valuable than females because of pre-emptive matings made by males of the previous (spring) generation. Even if the sex ratio in the summer was not at all female biassed relative to the rest of the population the higher value of females would tip the balance in favour of helping. Suppose that as a result of pre-emptive matings females were twice as valuable as males and the sex ratio was 1:1, the pay-off for helping is (value × relatedness) = ($\frac{2}{3}$ × $\frac{3}{4}$) for females + ($\frac{1}{3}$ × $\frac{1}{4}$) for males = $\frac{7}{12}$, while the pay-off for having offspring would be ($\frac{2}{3}$ × $\frac{1}{2}$) + ($\frac{1}{3}$ × $\frac{1}{2}$) = $\frac{1}{2}$ and the value of K would therefore be $\frac{6}{7}$. (Note that 'value' has been scaled to add up to one.) (Grafen 1986b)

To summarize this complicated argument, the halictine life cycle favours helping by females in the summer because (a) the local sex ratio is more female biassed than that of the population and (b) the value of males is decreased because some females are mated by spring males.

There is one final twist to the story. Some partially bivoltine hymenoptera have a life cycle in which the events are shifted around one step; this is the pattern found, for example, in almost all sphecid wasps (Fig. 13.6b). Here the population overwinters as male and female larvae. The larvae complete their development in the spring and mate to give rise to the summer generation. The summer generation then matures and mates to form a second, autumn generation. The autumn generation does not mature but develops into the overwintering last instar larvae of the next year. As in the halictine life cycle, males of the first generation (the overwintering larvae from the previous year) may survive to mate with those of the later generation, so here a male-biassed sex ratio would be expected in the autumn and a compensatory female biassed ratio in the summer. Now think of the young summer female with the hypothetical choice of staying at home to help her mother or going off to mate. In either case, the youngsters she rears will contribute to the autumn generation and hence be male-biassed. There is no special genetic incentive for becoming a helper arising from haplodiploidy in the sphecid life cycle: haplodiploidy actually works against helping here.

Seger's ingenious idea that the partially bivoltine life cycle of halictines might favour the evolution of sociality is supported by three lines of evidence.

1 There are, as predicted, seasonal changes in the sex ratio: in 10 species of sphecid for which data are available the overwintering generation tended to be male-biassed and the summer generation female-biassed.

2 As predicted, the early stages of helping are found more often in species with the halictine than with the sphecid life cycle.

3 Both the primitive stages of sociality (representing evolutionary beginnings of helping) and the partially bivoltine life cycle are more common in temperate than tropical regions, suggesting that the former may be facilitated by the latter.

313

*Chapter 13
Co-operation
and Altruism
in the Social
Insects*

Parental manipulation or daughter advantage?

According to Seger's hypothesis, offspring originally stayed at home to help because the sex ratio of the brood was female-biassed relative to the population as a whole and this shifted the offspring's critical ratio of benefits to costs for helping from 1 to something less than 1 (e.g. ⅘ with a 3:1 bias).

From the mother's point of view, the critical value of K is 0.5. This is because by persuading her offspring to help the mother gets more of her own offspring to whom she is related

(a) The Halictine life cycle

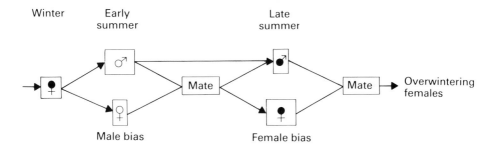

(b) The Sphecid life cycle

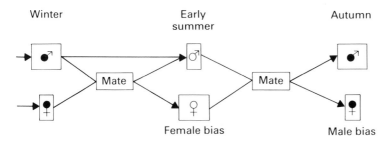

Fig. 13.6. Simplified diagrams of two partially bivoltine life cycles. In (a) summer females (open symbols), if they stay at homes would rear a female-biassed brood while in (b) they would rear a male-biassed brood. The relative size of the male and female boxes indicate a male or female bias in the offspring. After Seger (1983).

314

Chapter 13
Co-operation
and Altruism
in the Social
Insects

by ½ instead of grandchildren, to whom she is related by only ¼ (Fig. 13.3). It appears, therefore, that ecological conditions (values of K greater than the critical threshold) under which parental manipulation was favoured at the origin of eusociality were likely to be reached before those for which helping was of advantage to the daughter. Does this mean that parental manipulation is more likely as a route for the evolution of eusociality by staying at home, as Alexander (1974) has argued? Robin Craig (1979) has pointed out that if the value of K was below the daughters' threshold, the daughters would have resisted manipulation, so a likely way for the queen to have encouraged workers to help was to alter the value of K to move it above the daughters' critical value. She could have done this, for example, by giving less food to the daughters and thereby stunting their growth and reducing their capacity to lay eggs if they left home (egg-laying capacity in insects is very much size-dependent). With a reduced ability to lay eggs, the cost of staying at home would have been lowered, in other words the value of K increased. If we accept this kind of reasoning, although parental manipulation and daughter advantage are theoretically distinct, one would soon have merged into the other in evolutionary time.

Another line of evidence suggesting that parental manipulation played a role in the origin of eusociality is its taxonomic distribution. All the eusocial hymenoptera belong to the group known as the 'aculeate' hymenoptera, and none of the remaining, much larger, group of haplodiploid insects (including the parasitic hymenoptera) have evolved eusociality. The special feature of the aculeates is that they have elaborate parental care, which places the mother in a strong position to exert parental manipulation (Stubblefield & Charnov 1986).

Comparison of vertebrates and insects

With the possible exception of naked mole rats (chp. 12), there are no known examples of sterile castes in vertebrates, but in other respects there are some close parallels between the conclusions of this chapter and those of chapter 12. The two hypotheses about the evolutionary origin of helping in insects are similar to the two kinds of co-operative nesting described for birds in chapter 12. The first resembles scrub jays, in which young birds stay at home to help their parents rear younger siblings, while the second is parallel to anis where several females share the same nest and one female is dominant over the rest and produces more young.

The ecological pressures thought to favour helping in vertebrates and insects are also quite similar. For example both

315

*Chapter 13
Co-operation
and Altruism
in the Social
Insects*

scrub jay helpers and workers in social insects play a similar role in defence of the nest against predators or parasites. Nest sharing by reproductive females is also thought to be a response to predator pressure in both vertebrates and insects. Another major ecological factor which leads to helping in birds is lack of opportunity for young birds to find a nesting territory and this again is paralleled by the importance of co-operative nest building in social insects: a lone female has little chance of building a nest alone, so she is forced to co-operate.

In spite of these similarities there must also be important differences which have led insects but not vertebrates to evolve sterile castes. The differences could either be ecological or genetic.

First, the ecological and life history constraints are different in insects and vertebrates. Vertebrate helpers such as young scrub jays help as part of a long term strategy of acquiring a breeding territory or mate in the future. In short lived insects, such long term gains are less (although there are exceptions such as *Metapolybia*), and the emphasis is shifted to gains from helping siblings. If a young scrub jay, for example, stays at home to help, it does less well than one that succeeds in setting up on its own (chp. 12). This is because the helper is capable of rearing more young than the mother is able to provide, the constraint on the mother being her inability to incubate a very large clutch. An insect queen, in contrast, can often provide her daughters with as many eggs as they themselves could lay, because the constraint on brood size is the availability of nest chambers and not the need for incubation. For both insects and vertebrates one constraint which prevents helpers from setting up on their own is the problem of obtaining a nest or territory. In insects nest construction takes relatively longer than in vertebrates, so this constraint is more severe. In other words, a nest is a scarcer resource for an insect than is a territory for a vertebrate.

The differences between birds and insects in genetic predispositions towards sterility have already been dealt with in depth. Hymenopteran haplodiploidy and aphid asexuality are features that are not shared by any of the vertebrates discussed in chapter 12. The termites appear to have no particular genetic mechanisms that could predispose them towards eusociality (but see Box 13.2) and may thus be a closer parallel to the vertebrates.

Summary

In the social insects there are sterile workers that never have

316

*Chapter 13
Co-operation
and Altruism
in the Social
Insects*

Box 13.2. *Termites: genetic factors that might favour eusociality.*

1 *Cycles of inbreeding and outbreeding.*

Stephen Bartz (1979) has suggested that alternation between inbreeding and outbreeding might favour the evolution of sterile castes in termites. To understand the argument, take an extreme case. Suppose the king and queen of a termite colony are unrelated to one another but both were born in highly inbred colonies and are therefore effectively homozygous. The two parents will be related to their offspring by the usual ½, but the offspring will be related to each other by 1! They will all be heterozygotes like the F1 generation in a Mendelian cross between two homozygotes with alternative alleles. Thus offspring will be genetically predisposed to help their siblings rather than their own offspring. Although this is an extreme example it makes the general point that an alternation between inbreeding and outbreeding could favour helping.

How might these events actually come about? Bartz points out that in termite colonies, winged reproductives are often produced not by the founding king and queen but by secondary reproductives that are offspring of the original founder and foundress. In other words the reproductives that disperse to find a mate and start a new colony are the product of brother–sister matings. Given that the winged dispersers are unlikely to end up mated to a sibling, the alternating pattern of in- and outbreeding seems to hold.

2 *Linked genome.*

An additional possible genetic predisposition for eusociality in termites was discovered by Syren and Luykx (1977). In some species an appreciable proportion of the genome is linked to the sex chromosomes. As a result of this, relatedness among siblings of the same sex is higher than 0.5 (Lacy 1980). The general importance of this is hard to assess without more information on its prevalence in termites (Crozier & Luykx 1985).

offspring but instead help to rear younger siblings. This appears at first sight to go against the idea of natural selection favouring maximum efficiency at passing on genes. However the fact that sterile workers rear close relatives genetically predisposes them to be altruistic.

There are two theories about how sterile helpers arose in evolution: by young staying at home to help their mother and by sisters sharing a nest with one producing most of the offspring. According to both theories the ecological advantages of communal nesting are defence against parasites or predators and construction of an elaborate nest. The genetic predisposition is that help is given to close relatives. In the case of the first theory there may be strong selection for mothers to persuade their daughters to stay at home and little selection for daughters to resist.

317
Chapter 13
Co-operation
and Altruism
in the Social
Insects

In the Hymenoptera there may be an additional genetic predisposition towards helping: haplodiploidy. Sterile female helpers in the Hymenoptera could pass on their genes even more efficiently than the reproductive queen if they bias the sex ratio of the reproductive siblings they help to rear. This is because sisters are more closely related to one another than are mothers and offspring. Haplodiploidy acts as a predisposition for the evolution of helping only when the sex ratio in the nest is more female biassed than in the population at large.

Further reading

Hamilton (1972) summarizes the importance of haplodiploidy in Hymenoptera as a predisposition towards eusociality. Brockmann (1984) is a very useful overview of social hymenoptera. Wilson (1975) in the chapter on social insects gives an excellent review of the biology of this group. Michener (1974) is the definitive work on social bees.

Dawkins (1979) enumerates an entertaining and salutory list of twelve common misunderstandings in the literature on kin selection. For example the statement that 'all individuals in a species share a high proportion of their genes (shown by DNA hybridization studies) and they should therefore be altruistic to one another is fallacious. To see why, look at Dawkins' paper.

Wilson (1980) discusses an aspect of social insect behaviour not covered in this chapter, namely the optimal allocation of workers in a colony to different tasks. He takes the example of leaf cutting ants (*Atta*) and shows that one size of worker is more efficient than the others at cutting and carrying leaves for the colony's store. This is the size at which workers normally change from nest maintenance to foraging duties. A general introduction to the economics of caste in social insects is given by Oster and Wilson (1978). Andersson (1984) reviews the origin of eusociality.

318

*Chapter 13
Co-operation
and Altruism
in the Social
Insects*

Topics for discussion

1 What is the importance of haplodiploidy for the evolution of eusociality?

2 What ecological factors are important for the evolution of helping in vertebrates and insects?

3 What data would you need to collect to evaluate the outcome of worker—queen conflict in hymenopteran colonies?

Chapter 14. The Design of Signals: Ecology and Evolution

Most of the interactions between individuals described in this book involve communication. Males attract females or repel rivals, offspring beg from their parents, and poisonous caterpillars warn their predators by using special *signals* or *displays* — behaviours and structures designed for use in communication. This chapter is about how signals are designed by natural selection for effective communication. We will discuss the influence of two kinds of selection pressure on signals: ecological constraints imposed by the environment and the response of reactors at whom the signals are directed. But first let us clarify what we mean by communication.

The most obvious characteristic is usually that a signal or display of one individual (actor) in some way modifies the behaviour of another (reactor). The reactor's response may be immediate and obvious (a male firefly rapidly flies towards a flashing female of the correct species); it may be subtle and difficult to detect (a male antelope slightly alters its direction of walking to avoid crossing a territorial boundary when it detects the scent marks of a resident); it may be delayed in time (the ovaries of a female budgerigar gradually develop as the result of the stimulus of male song); or it may not occur all the time (a territorial male blackbird sings for several hours during which time only one or two intruders hear the song and retreat). If, in spite of these difficulties, we can detect the response of a reactor, we could characterize communication as *the process in which actors use specially designed signals or displays to modify the behaviour of reactors*. The qualification 'specially designed signals' saves our definition of communication from becoming too wide. When we see a drunkard lurching down the street towards us late at night we may cross the road to get out of the way, but since there is no reason to suspect that the drunkard's lurch has been developed by natural selection to signify advanced inebriation we would not include lurching within the definition of communication. In contrast, the grasshopper's incessant buzz, which is created by rubbing the legs together and probably evolved from simple walking movements, is an example of communication. The ancestral male grasshopper may have lurched towards a female accidentally producing a chirrup at the same time, but natural selection has acted on the sound to make it into a loud and conspicuous signal for attracting females.

320

*Chapter 14
The Design
of Signals:
Ecology and
Evolution*

As an aside it is worth pointing out that our definition of communication would not satisfy all students of behaviour. Our concern in this chapter is with how signals evolve, and we therefore stress the importance of specially adapted signals in communication. But for someone interested in human or animal social interaction a broader definition 'any aspect of A's presence or behaviour which influences B' might be appropriate. This would encompass observations such as those of Michael Argyle (1972) who found that many subtle aspects of posture (leaning back in a chair, crossing legs, etc.) play a major role in human communication, even though they may have not evolved as special signals.

Ecological constraints and communication

Different groups of animals rely on different sensory channels for communication. Small mammals live in a world of smells, birds in a world of music, and coral reef fish in a world of brilliant poster colours. Why are there such differences? Part of the answer is that the utility of different channels depends on constraints imposed by the habits and habitats of a species. It is obvious, for example, that sound or scent are more useful than visual signals for nocturnal small mammals and that birds living in dense bushes can more readily hear than see each other. Roe deer living in dense forest mark their territories with loud calls and scents deposited on the vegetation, while in open habitats they use primarily visual signals (Loudon pers. comm.). But differences in the effectiveness of transmission are not the only considerations in assessing the costs and benefits of various communication channels (Table 14.1). Sound is very flexible: enormous numbers of signals can be fitted into a short space of time by rapid changes in pitch, loudness and harmonic structure. Scent may be less flexible but is energetically cheap to produce and can last for a very long time, an advantage for an animal such as a fox with a very large territory in which it can announce its presence at any particular site only once every few hours, or even every few days. Brilliant body colours are permanently on display (at least seasonally); while this may be advantageous for attracting females and deterring rivals, it may be a considerable disadvantage in attracting unwelcome predators!

COMMUNICATION IN ANTS

The use of different communication channels in different ecological conditions is nicely illustrated by Bert Hölldobler's (1977) study of recruitment signals in ants. When ant workers

Table 14.1. Advantages of different sensory channels of communication. From Alcock (1984)

321

Chapter 14
The Design
of Signals:
Ecology and
Evolution

Feature of channel	Type of signal			
	Chemical	Auditory	Visual	Tactile
Range	Long	Long	Medium	Short
Rate of change of signal	Slow	Fast	Fast	Fast
Ability to go past obstacles	Good	Good	Poor	Poor
Locatability	Variable	Medium	High	High
Energetic cost	Low	High	Low	Low

return from a foraging trip to the colony they often recruit others to take part in harvesting the supply of food. Hölldobler describes three kinds of recruitment, two of which are illustrated in Fig. 14.1.

1 Species such as *Leptothorax* feed on single, immobile prey items (e.g. dead beetles) which are too big for one worker to carry but which can be brought back to the nest by two individuals pulling together. After finding a prey, *Leptothorax* workers return to the nest, regurgitate some of the food and secrete a chemical signal from their abdomen to attract other workers. One worker is recruited to help bring back the prey and it is 'led by hand' to the foraging site — so-called 'tandem running' in which the recruit follows the leader by keeping contact with its body: the recruit's antennae rest on the leader's abdomen (Fig. 14.1a).

2 Fire ants (*Solenopsis*) feed on large mobile prey (large

(a)

(b)

Fig. 14.1. Two types of recruitment communication in ants. (a) Tactile (b) Chemical.

322

*Chapter 14
The Design
of Signals:
Ecology and
Evolution*

insects and so on) which need several individuals to carry them back to the nest and they use odour trails as a means of recruitment. After finding a suitable prey, a worker returns to the nest laying a trail of scent secreted from a special abdominal gland. The odour trail excites other workers to join in, running along the scent trail to the prey, and returning to the colony laying their own trail if they have found the prey (Fig. 14.1b). The scent trail builds up very rapidly as large numbers of workers add to it, but it also decays rapidly as soon as workers stop renewing it because the scent is very volatile and lasts only for a few minutes. This means that the trail persists only as long as the prey is available, and it can change position to keep up with the moving prey.

3 A third kind of ant foraging trail is characteristic of leaf cutters (*Atta*) and seed eaters such as *Pogonomyrmex*. These ants feed on long lasting or renewing patches of food, so that workers use the same trails for days, weeks or years on end. There are two ways of marking trails, by long lasting scents and by cutting a path through the vegetation. Both of these provide durable signals.

These examples illustrate how different kinds of signals using different sensory channels are used according to the feeding ecology of the ants: tactile signals for recruiting a single companion to stationary food, rapidly decaying odour signals for recruiting large numbers of workers to mobile prey, and visual cues or long lasting scents for large patches of renewing food.

BIRD AND PRIMATE CALLS

The way in which habitat structure and meteorological conditions affect signal transmission has been studied in detail for sound signals, especially bird songs. Differences between species and between populations within a species can sometimes be explained in these terms.

Gene Morton (1975) and Claude Chappuis (1971) were the first to show that the structure of bird songs is correlated with habitat. Morton found that the songs of species living below the canopy in tropical forests in Panama were characterized by lower frequencies, a larger proportion of pure tones and a narrower range of frequencies than songs of grassland birds in the same country (Table 14.2). Typically the songs of tropical forest birds contain many low pitched pure whistles and those of grassland species sound like buzzy trills.

Several subsequent studies have shown that geographical variation within a species can sometimes be correlated with habitat type. For example Fernando Nottebohm (1975) found

323

*Chapter 14
The Design
of Signals:
Ecology and
Evolution*

Table 14.2. Differences between the songs of forest and grassland birds in Panama (Morton 1975)

Habitat	Emphasized Frequency (kHz)	Per cent pure tones	Frequency range (kHz)
Forest (below canopy)	2.2	87	1.5
Grassland	4.4	33	3.5

the rufous-collared sparrows (*Zonotrichia capensis*) living in South American forests sing slower trills than their grassland cousins (Fig. 14.2a). The great tit, like the rufous-collared sparrow, lives in many different kinds of habitat over a wide geographical area. Its distribution stretches from Ireland to Japan and from Finnish birch forests to Malaysian mangrove swamps. Mac Hunter and John Krebs (1979) recorded the territorial songs of great tits in two contrasting kinds of habitat, open woodland or parkland and dense forest. Regardless of geographical location (recordings were made in various countries stretching from Norway to Iran) there were consistent differences between songs from the two habitat types (Fig. 14.2b). The birds in more open habitats have songs with higher maximum frequency, more rapidly repeated notes, and a wider range of frequencies than those of forest birds. So striking are the correlations with habitat type that the songs of birds in a park in southern England are more similar to those from the same sort of habitat in Iran (5000 km away) than they are to birds in a dense forest 100 km away in another part of southern England.

What is the survival value of these differences in song structure between habitats? The answer is still not certain. Morton hypothesized that songs are designed to carry the maximum possible distance and suggested that the differences he observed between habitats might be related to differences in the attenuation of sound. He tested this by measuring the attenuation of pure tones of different frequency that he broadcast in the two habitat types from a loudspeaker. He found that in both habitats high frequencies attenuated more rapidly than lower ones. This is not surprising, since on theoretical grounds one would predict that high frequencies would be more readily attenuated by obstacles such as leaves and branches, by air turbulence, and by the viscosity of the air itself. The unexpected aspect of Morton's results was that sounds of about 2kHz carried particularly well, better than higher or lower pitched tones, in forests but not grassland. Since forest birds sing at about 2kHz, Morton suggested that their songs are

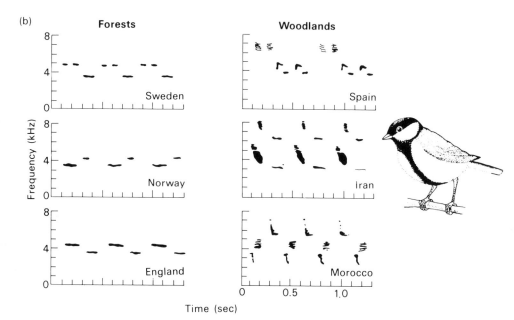

Fig. 14.2. (a) The song of the rufous-collared sparrow (chingolo) has slower trills in forests than in open country. (b) Great tits in dense forests sing songs with a narrower range of frequencies, lower maximum frequency and fewer notes than the songs of open country birds.

tuned into the 'frequency window' of minimum attenuation. The reason for higher pitched, wider frequency band songs in grassland was less clear.

However, more recent studies of sound attenuation have shown that there are probably only fairly small differences between habitat types (Marten & Marler 1977a,b; Wiley & Richards 1978). While in forests branches and leaves cause attenuation of high frequencies, in more open habitats local air turbulence often has a similar effect. Since open areas are often less sheltered than forests, the overall pattern of attenuation of different frequencies tends to be similar in the two habitats. Furthermore, Morton's frequency window seems to be more related to attenuation of lower (below 2kHz) frequencies by the ground than to differences between habitats.

325

Chapter 14
The Design
of Signals:
Ecology and
Evolution

Wiley and Richards (1978) propose an alternative explanation for the differences between forest and open habitats. They point out that there are two problems facing a singing bird. One is *attenuation* of sound: if the song attenuates too much the receiver will not be able to detect it because it is lost in the background noise. The other is *degradation*: if the song is degraded or distorted in its passage through the environment it may be confused by the receiver with other sounds. They suggest that there are much greater differences in degradation than in attenuation between habitats. In forests, the major source of degradation is echoes or reverberations from branches and leaves, while in open habitats the major source is irregular fluctuations in amplitude caused by gusts of wind that mask the song. These two kinds of degradation can be reduced by different design features. Reverberations are more severe for high frequency than low frequency sounds (because high frequencies are deflected by small objects such as leaves and branches). They also cause a problem if the song contains rapidly repeated notes as are found in the trills of many bird songs, because the echoes will become confused with the original notes. Therefore songs designed to overcome reverberations in forests should be of low frequency and contain either pure notes as opposed to trills, or trills with widely spaced notes. These are precisely the patterns observed by Morton and in the studies of chingolos and great tits. The irregular amplitude fluctuations of open habitats, in contrast, favour rapid trills. Since the song is masked by wind at irregular intervals it has to contain notes that are short enough and sufficiently rapidly repeated to be detected in a short space of time. The same need to fit the songs into short intervals between gusts may account for the high frequencies of open habitat songs, since a high frequency sound requires a shorter time for a given number of wavelengths. Again the patterns predicted by Wiley and Richard's hypothesis fit well with those observed in the field.

An additional complication is that in at least two of the studies referred to (Morton's and the great tit study), forest birds occupied larger territories than those in open habitats.

326

Chapter 14
The Design
of Signals:
Ecology and
Evolution

The songs in the two habitats may be adapted not for maximum detection distance, but for an optimum distance which differs between habitats because of the difference in bird density. (An optimum distance could arise if, for example, it was disadvantageous to signal over too great a distance — see below.) Forests songs with their low frequencies and energy concentrated into a narrow band are likely to carry further than the higher pitched, wider frequency band songs of open country birds.

There is a striking parallel between these results of studies of bird song and the interpretation by Busnel and Klasse (1976) of a peculiar kind of human language. In four parts of the world, Andorra, Turkey, Mexico and the Canary Islands, local peasants have developed extraordinary whistling languages with elaborate vocabularies. Although the details of the languages and the methods of sound production vary from place to place, all are designed for long distance communication. The four places are all mountainous with steep sided, rocky valleys; distances are not very great as the crow flies but are large in terms of the effort needed to cross from one side of the valley to the other. The whistling languages are used for communicating across the steep valleys. They have all their sound energy concentrated into a narrow band of frequencies and are therefore well designed, like forest bird songs, to be detected and interpreted over relatively large distances.

Not all sounds are used for long distance signalling, and short distance sounds may be designed not to carry further than necessary. The grey-cheeked mangabey (*Cercocebus albigena*) (Fig. 14.3) is a monkey which lives in the forests of East Africa; it lives in troops and defends a group territory. Peter and Mary Sue Waser (1977) studied two vocalizations of the mangabey: the 'whoop-gobble' which is used in intergroup signalling, and the 'scream' which is used during agonistic encounters within a group. The two sounds are produced at the same volume by the monkeys, but because the whoop-gobble is lower-pitched and has a narrower range of frequencies it carries much further than the scream (Table 14.3). The differences in frequency structure and therefore carrying power of the two sounds reflect their design for communication over different distances.

Probably the reason why the mangabey uses a song which does not carry far for signalling within the troop is to avoid attracting predators. Another way to make sounds less likely to attract a predator is to produce unlocatable sounds, as was first suggested by Peter Marler (1955) in a classic paper about alarm calls of small birds. He pointed out that the calls given by many species of small birds when a hawk flies over are

327

Chapter 14
The Design
of Signals:
Ecology and
Evolution

Fig. 14.3. The grey-cheeked mangabey. Photo by P.M. Waser.

Table 14.3. Comparison between two calls used by the grey-cheeked mangabey in tropical forests of Uganda (Waser & Waser 1977)

Call	Function	Distance from which heard by human ear	Sound pressure 5 m from monkey	Frequency (Hz)
Whoop-gobble	Spacing between troops	1000 m	75 dB	300–400
Scream	Within troop agonistic encounters	300 m	78 dB	1000–3000

remarkably similar. All are thin high-pitched 'seet' sounds which are hard for the human ear to locate. Marler suggested that the calls have evolved a structure which makes them hard for predators to locate and therefore reduces the risk to the caller of attracting the hawk to itself. (Why the caller should warn his flock mates in the first place is a different matter — see chp. 11.) Elegant and appealing though Marler's hypothesis is, attempts to test its assumption that the calls are hard to locate have produced ambiguous results. In one study pygmy

328

Chapter 14
The Design
of Signals:
Ecology and
Evolution

owls (*Glaucidium* spp.) and Goshawks (*Accipiter gentilis*), both of which prey on birds, were tested to see if they could locate alarm calls. The predators sat on a perch between two loudspeakers and their head movements were used to test if they could locate which speaker played the call. They seemed to be as good at locating alarm calls as they were at locating other, supposedly more locatable calls (Shalter 1978). Brown (1982), on the other hand, found in a somewhat similar experiment that red-tailed hawks (*Buteo jamaicensis*) and great-horned owls (*Bubo virginianus*) had more difficulty in locating alarm calls than other calls.

Reactors and the design of signals

The constraints of the environment impose broad limits on the design of signals, but within these limits the way in which signals evolve results from selection to increase their effectiveness in altering the behaviour of reactors. Reactors play an important role both in the evolutionary origin of signals and in their subsequent evolution towards increased effectiveness.

HOW SIGNALS ORIGINATE

To the casual observer animal displays often seem to be inexplicably bizarre and absurd. Why should male ducks perform sham drinking and preening movements as part of their courtship ritual? Why do wolves mark their territories by urinating? Why do rhesus monkeys grin as a signal of fear and appeasement?

A great step forward in understanding the answers to these questions came when ethologists such as Lorenz and Tinbergen realized that many signals have evolved from incidental movements or responses of actors which happened to be informative to reactors. Selection favoured reactors who were able to anticipate the future behaviour of actors by responding to slight movements which predicted an important action to follow. If a dog always bares his teeth before biting, reactors who are able to anticipate and escape from an aggressive attack by observing bared teeth will be favoured by selection. Once this happens selection will favour actors who bare their teeth as a means of rapidly deterring reactors, and teeth baring will begin to evolve into a threat display.

We should expect then, that the incidental movements and responses from which signals have evolved are those which were originally most informative about future actions. This is borne out by many studies of the detailed form of displays in birds, fish and mammals. Many signals in these animals have

329

Chapter 14
The Design
of Signals:
Ecology and
Evolution

apparently evolved from *intention movements* such as are made by a bird when it crouches and tenses its muscles for take-off: it is not hard to imagine that intention take-off movements were originally good predictors of a lunging attack by a rival, approach by a prospective mate and so on. Other movements performed at moments of transition from one major activity to another are also frequently the raw material from which displays have evolved. Often these movements reflect motivational conflict or indecision as the animal vacillates between, for example, attack and running away (Table 14.4).

Our examples of the duck, wolf and monkey can be interpreted in a similar way. The drinking and preening movements of courting ducks are probably derived from *displacement activities* — seemingly irrelevant actions which tend to occur at moments of balance between incompatible motivational states such as aggression and sex. Urinating is like blushing and hair erection, which are also the basis for displays in some species, a consequence of autonomic nervous activity during moments of stress. The ancestral wolf may have urinated uncontrollably when confronted with a rival at its territory boundary (anyone who has nursed a nervous dog will know all about this sort of thing!), and the response has subsequently evolved into a way of signalling 'keep out'. The grin of the terrified monkey is very similar to the reflex response with which a monkey protects the most vulnerable parts of his face such as eyes and mouth from the onslaught of an attacker. As with urinating, an autonomic reaction to stress has become a signal during evolution.

Most of the conclusions described in the above two paragraphs are inferences based on studies of the detailed structure of displays and the contexts in which they occur. For example the observation that threat displays occur at territory boundaries and in sequences of behaviour which include transitions between attack, threat and retreat suggests that threat signals arise when the animal is in motivational conflict. Similarly the structure of displays can be viewed as reflecting motivational conflict. The 'zig-zag' display of the three-spined stickle back (*Gasterosteus aculeatus*) involves a strange movement in which the courting male approaches the female in a series of short arcs, as if he is in a conflict between approach and avoidance.

There is also some more direct evidence for the conflict hypothesis from experiments in which the tendencies of an animal to attack and flee are manipulated independently. Nick Blurton Jones (1968) studied the threat displays of captive great tits in this way. He found that the birds would attack a pencil pushed through the bars of their cage and that they

330

Chapter 14
The Design
of Signals:
Ecology and
Evolution

Table 14.4. Examples of the kinds of behaviour patterns and other responses from which displays in birds, fish and primates are thought to have evolved (Hinde 1970)

Behaviour or response	Example of display	
1. Intention movement	Sky pointing in the gannet	
2. Ambivalent behaviour	Oblique threat posture of black headed gull	
3. Protective response	Primate facial expressions	
4. Autonomic response (e.g. sweating, urinating, rapid breathing)	Vocalizations (from rapid breathing). Scent marking	
5. Displacement activities	Preening in duck courtship	
6. Redirected attack	Ground pecking in herring gulls	

would flee from a bright light. When the light and the pencil were presented at the same time the birds tended to perform threat displays. Other threat displays could be elicited by presenting the pencil outside the cage so that direct attack was thwarted. Thus motivational conflict and thwarting of attack seem to generate threat displays.

HOW SIGNALS ARE MODIFIED DURING EVOLUTION:
RITUALIZATION

331

Chapter 14
The Design
of Signals:
Ecology and
Evolution

Although signals started off as incidental movements or responses such as those shown in motivational conflict, during their evolution they have become modified by selection to improve their effectiveness as signals. For example the courtship movement in which male ducks preen their wings has been emphasized in some species by the evolution of conspicuous bright coloured feathers on the wing, towards which the male points his bill during courtship preening. An extreme case is the mandarin duck (*Aix galericulata*) which has some of its wing feathers modified to form a bright orange 'sail' which is erected during courtship preening. The ancestral preening movement is reduced to a quick turn of the head so that the bill points to the orange sail.

The term *ritualization* is used to refer to the evolutionary modification of movements and structures to improve their signal function. Thus the ancestral preening movement of ducks has become ritualized in species such as the mandarin duck. The changes that occur during ritualization include the following: the movements tend to become highly stereotyped, repetitive and exaggerated; often the movements are emphasized, as in the mandarin duck, by the development of bright colours on the body. It is of course not possible to observe these evolutionary changes taking place and the evidence that they have occurred comes from comparative studies of displays in closely related species. In the example shown in Fig. 14.4 the probable ancestral movement is pecking at pieces of food on the ground during courtship (perhaps originally a displacement activity). This pattern is seen in present-day jungle fowl (Fig. 14.4). In other species of galliform bird the display is ritualized, for example in pheasants and peacocks it is emphasized by the evolution of a large tail and the original ground pecking movement is reduced to a stereotyped bobbing of the head or pointing of the beak.

We have said that ritualization occurs because it improves the signal function of a display. What exactly does this mean? As we saw in chapter 8 some displays used by males during courtship may be exaggerated and very elaborate as a result of sexual selection: females may exhibit a preference for elaborate displays. While this hypothesis may account for the ritualization of some courtship signals, it cannot be a general explanation of ritualization, since other displays such as those used in threat also appear to have undergone ritualization.

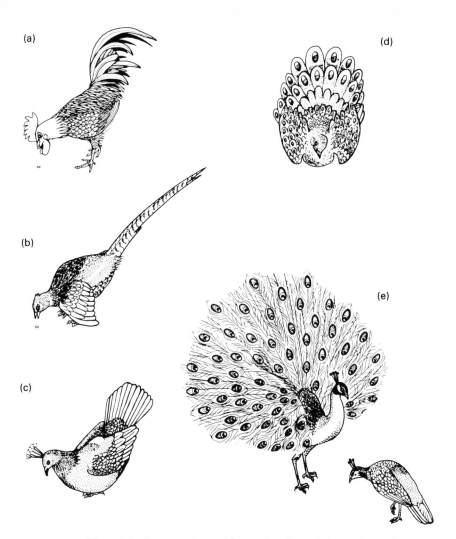

Fig. 14.4. Comparative evidence for the origin and ritualization of a display. The ground pecking display of phasianid birds. (a) The least ritualized form is shown in the male domestic fowl (*Gallus*). It scratches the ground with its feet and pecks at food or small stones (perhaps originally a displacement activity). This attracts the female. (b) The male ring-necked pheasant (*Phasianus colchicus*) attracts females by means of a similar display. (c) The impeyan pheasant (*Lophophorus impejanus*) and (d) the peacock pheasant (*Polyplectron bicalcaratum*) both emphasize the pecking display with rhythmic bobbing of the tail or the head. (e) The peacock (*Pavo*) shows little of the ancestral movement. The male spreads his enormous tail and points his beak towards the ground. Redrawn from Cullen (1972).

(a) Reduction of ambiguity

According to the traditional view, ritualization is the result of the selective advantage to signallers in reducing the chance of confusion between their various displays (Cullen 1966). Thus

333

Chapter 14
The Design
of Signals:
Ecology and
Evolution

a ritualized threat signal, by its exaggerated, stereotyped and repetitive nature clearly states 'I am about to attack' and not 'I am very frightened' or 'I am a sexy male'. In fact (as Darwin first pointed out) threat and appeasement signals are often extreme opposites: a threatening dog stands erect while a fearful dog crouches near the ground in appeasement. The principle of reduction of ambiguity can also be illustrated with reference to courtship signals. Often the displays of closely related species are clearly distinct so as to minimize the chance of confusion between species (Fig. 14.5).

Increasing the clarity of signals by increasing their stereotypy may at the same time reduce the amount of information they convey about the actor. The exact structure and movement pattern of an ancestral threat signal may have reflected the precise balance between aggression and fear in the actor, but a stereotyped, ritualized signal probably conveys less information about the actor's state. Morris (1957) called this stereotypy 'typical intensity' and he viewed the loss of accuracy of information as the price paid by the actor for reduction of ambiguity. This assumes that it is advantageous for the actor to communicate information about its internal state, but in fact just the opposite may often be the case. As pointed out in chapter 7, if two animals are contesting a resource by means of ritualized signals, it will not pay either to reveal its exact balance between attack and fleeing until the last possible moment. In fact stereotypy of displays may have evolved precisely *because* it reduces the information available to reactors about the actor's internal state. This leads us to the second hypothesis about ritualization, which emphasizes the use of signals by actors to control the behaviour of reactors.

(b) Ritualization and manipulation

A rather different view of ritualization emerges if we start to think about costs and benefits of communication to the individual. The basic principles of natural selection that we have used in previous chapters would lead us to expect that animals would only use signals when signalling is to their advantage because of its effect on the behaviour of others. We can put this in more formal and more precise terms as follows. Signals are specially evolved pieces of behaviour with which one individual, the actor, alters or manipulates the behaviour of another, the reactor, to the advantage of the actor. If signals really evolved in this way, how is ritualization to be interpreted? Imagine that individual A manipulated the behaviour of B in a way that was advantageous to A but not to B; it would then have been of advantage to B to resist responding

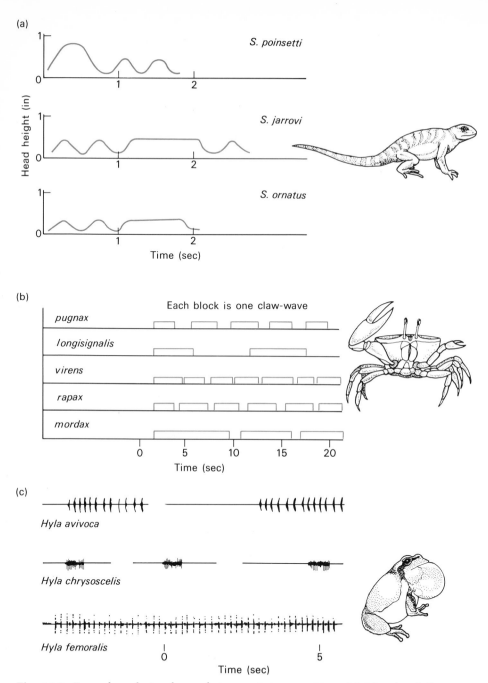

Fig. 14.5. Examples of signals used in species recognition. (a) Lizards of the genus *Scleroporus* have species specific head bobbing patterns used in courtship and threat. The graph shows the head height as a function of time in the bobbing displays of three species. (b) Male fiddler crabs (*Uca*) attract females by waving their enlarged claw. Each species waves in a different way. The graph to the left shows the waving patterns of four species. A block represents a complete wave of the claw. (c) Tree frogs (*Hyla*) have species characteristic calls. Females are attracted only to males of their own species. The oscillogram traces to the left show the gross temporal characteristics of each of three species' calls. There are also species characteristic differences in the fine structure of calls that are important in species recognition.

335

*Chapter 14
The Design
of Signals:
Ecology and
Evolution*

to A's signal, or in other words to develop 'sales-resistance'. This in turn would have generated pressure on A to develop a more effective signal to overcome the resistance of B, for example by increasing the amplitude of the signal, repeating it or exaggerating it. Of course A and B would not have been single individuals, but *roles* played through evolutionary time, and the changes would have taken place over many generations. To summarize this argument, if signals arose as manipulative devices, there would have followed a coevolutionary 'arms race' (chp. 4) between sales-resistance of reactors and manipulative power of actors. It is the outcome of this coevolutionary race that we see as ritualization. Ritualization, according to this view, did not occur because of selection for more accurate or more unambiguous transfer of information by signalling, but simply to overcome sales-resistance. In overcoming sales-resistance the concepts of information and ambiguity are irrelevant. All that counts is persuasion or manipulation.

Perhaps a useful analogy is with advertising. Most advertisements work because they exploit the psychological susceptibilities of the public. When you see an advertisement with a bronzed, tough, healthy cowboy smoking a cigarette, you do not start to believe that smoking will improve your health or your horse-riding ability, you are (or at least some of you must be) simply persuaded, against your better judgement, by psychological manipulation, that smoking is a 'good thing'. Just as the concept of information is of little relevance to persuasion, so is deceit. Advertisements do not (at least where there is legislation to prevent it) actually tell lies, they simply persuade.

Although the 'manipulation' view offers a different explanation of why signals became ritualized, it takes the same starting point as that proposed by Lorenz, Tinbergen and others for the origin of signals. Recall that the probable starting point for many signals was the opposite of manipulation by actors, namely the advantage gained by reactors through responding to subtle cues that allowed them to anticipate the actor's future behaviour. Given this starting point, reactors were tuned to the signals of actors and primed for actors to turn the situation to their own advantage.

The term 'manipulation' may seem to be rather emotive, but it is a good description of the effect of ritualized signals on reactors in many kinds of communication. A striking example is the effect of a young cuckoo on its foster parents. By begging in the nest, the cuckoo manipulates the parental behaviour of its foster parents to the advantage of cuckoo genes and the manifest disadvantage of foster parent genes. A similar kind of manipulation in normal offspring−parent communication has been proposed by Bob Trivers (1974). He pointed out that a

336

Chapter 14
The Design
of Signals:
Ecology and
Evolution

young animal is more closely related to itself than it is to its siblings. Therefore it ought to be more interested in its own survival than in the survival of its brothers and sisters. The parents, however, are usually related equally to all their off-spring and are therefore expected to allocate their resources equally between their children. (There are obvious exceptions to this rule, for example some birds of prey in which the brood hatches asynchronously and the parent only feeds the smallest young if food is plentiful.) Trivers therefore predicted that there would be conflicts of interest over allocation of parental resources. Each young would try to acquire more than its 'fair share' of resources while the parents would try to allocate resources equally. (This is analogous to the parent-offspring conflict described in chp. 13.) It is not surprising, therefore, that once the offspring have grown parents often play an active role in encouraging their independence. For example mother rhesus monkeys actively resist their offspring's attempts to gain access to the nipple (Hinde 1974).

(c) Co-operative and non-cooperative signals

So far in our discussion of manipulation we have assumed that communication is not co-operative, that the interests of actor and reactor are not the same. But as we have seen in chapter 11, co-operative behaviour of many kinds is by no means uncommon, especially where individuals are closely related or where they interact with each other repeatedly and exchange benefits in a reciprocal way. Communication between the members of an ant colony, in a family of jackals, or even between repeatedly interacting territorial neighbours, might well be co-operative. If ritualization of signals is the product of the manipulation—sales-resistance arms race we have de-scribed, what about co-operative signals? Should they have become ritualized in the same way? If the ancestral reactor *benefited* from receiving the signal, rather than evolving sales-resistance, it would have evolved heightened sensitivity to the signal. This in turn would have paved the way for actors to decrease the cost of signalling by reducing the amplitude of their signals. All signals have a cost, for example some cicadas spend 20 times their resting energy while they are calling (MacNally & Young 1981), while other animals attract lethal enemies through their conspicuous signals, so that if reactors are highly sensitive to the signals, actors would evolve towards minimizing their cost. The minimum might be set by the sensory limits of the reactors or the noise levels in the environment. In short, co-operative signals should be quite dif-ferent from the typical ritualized non-cooperative signals

which ethologists have studied: they should be quiet or in-
conspicuous, in fact they may be barely detectable. Unfortunately no one has undertaken a systematic comparison of
co-operative and non-cooperative signals, so the predicted
difference remains to be tested.

337

Chapter 14
The Design
of Signals:
Ecology and
Evolution

(d) Honesty

Before leaving the subject of ritualization, it is worth mentioning yet another hypothesis. Amotz Zahavi (1979) has suggested that ritualization is the result of selection by reactors for
reliable signals. During evolution, he argues, reactors are
selected not just to become sales-resistant but to discriminate
between signals on the basis of their reliability as indicators of
an individual's strength, size, motivation to fight, parental
ability and so on. One way to select for reliability is to select
for stereotypy and repetition of the display. In the same way
that human judges assess the difference between athletes by
asking them all to perform exactly the same task under the
same conditions, reactors will be better able to judge differences between actors by selecting for stereotyped displays.
The crucial feature of ritualization according to Zahavi is not
just that displays become more stereotyped and repetitive, but
that the stereotypy provides a uniform background against
which subtle differences between individuals are emphasized.
Zahavi might argue that when the male zebra finch hops towards a potential partner in his stereotyped courtship dance,
the female is assessing the male's agility, as an indicator of his
ability to gather food for the young.

These various ideas to account for the evolutionary ritualization of signals, reduction in ambiguity, manipulation, and
honesty are not necessarily mutually exclusive; they could all
play a role at different stages or in different instances of ritualization. Signals may initially become ritualized because of the
benefit of reduced ambiguity, then selection on actors to manipulate reactors may have led to greater exaggeration and
stereotypy, and finally sales-resistance and discrimination by
reactors may have led to displays, particularly those used in
assessment, being performed in a way which allows assessment of differences between actors. Signals which do not
allow reliable assessment may gradually lose their effectiveness, fall into disuse and be replaced by new ones.

Variability of signals and information

We now turn from the question of how signals became ritualized during evolution to that of how they are used in com-

338

Chapter 14
The Design
of Signals:
Ecology and
Evolution

munication on a moment to moment basis. A major difference between animal and human communication is that animals do not, on the whole, communicate about abstract ideas or about remote objects in their environment. Instead they communicate about immediate events connected with the actor and its surrounds; things such as ownership of a territory, readiness to mate, and approach of a predator. Exceptions to this generalization include the well-known dance language of bees, in which information about the direction, distance and quality of remote food sources is communicated. Given that animal signals carry simple messages such as 'keep out' and 'I am hungry' it is surprising how variable they are. Often there is not just one signal for each kind of message, but a repertoire with several threat, mating, and danger signals (Box 14.1). We will illustrate our discussion of this by referring to threat signals.

As we saw in chapter 7, variation *between* individuals in their threat signals is often associated with variation in strength or fighting ability. Therefore when animals engage in threat, one of the things they are doing is gauging their opponent's size or strength from its signals. There is, however, often variation *within* an individual in its threat signals. Magnus Enquist (Enquist *et al.* 1985; Enquist 1985) has studied variable threat signals in the fulmar (*Fulmaris glacialis*), a seabird, competing for pieces for fish off the shore in Iceland. The birds use a range of displays (Fig. 14.6) which vary both in their effectiveness in causing the rival to retreat and in their cost, measured as the likelihood that the rival will retaliate with physical attack. The central conclusion of Enquist's study is that cost and effectiveness are correlated. The fulmar has a choice of using a low cost display such as wing-raising (the probability of eliciting physical attack is only 0.017) or a high cost display such as rushing from behind (cost = 0.28). The former causes the opponent to retreat on only 12 per cent of occasions while the latter has this effect on 28 per cent of times it is used. Probably an individual chooses which display to use on the basis of how hungry it is, and therefore how much it values the piece of fish.

The interpretation of variable threat displays is not, however, always so straightforward. In another seabird, the Arctic skua (*Stercorarius skua*) there are 11 different kinds of display used as threats in contests, and careful analysis revealed little difference between them with regard either to their association with the actor's future behaviour or in their effect on reactors. Although the different displays are associated with slightly different probabilities of ensuing attack, the differences are not very great and furthermore, the differences vary

Fig. 14.6. Aggressive displays by fulmar petrels. (a) Rushing (left bird) and wing-raising (right bird). (b) Breast-to-breast, which may lead to pecking and wing-beating fights. From Enquist *et al.* (1985).

between locations and years: there is no consistency about which display gives the best prediction of future attack (Paton & Caryl 1986). This is, of course, exactly what would have been predicted from game theoretic analyses of contests (the animal should conceal its hand, see chp. 7), but it leaves the puzzle of what the various displays are for. Robert Hinde (1981) has argued that to look simply at the probability of attack (or any other behaviour) following a display may be too simple. His idea is that in contests, and perhaps in other kinds of social interaction as well, signals are often used as a form of interrogation, so that the actor's behaviour following a parti-cular signal may depend on the reactor's response. If this view is correct, there may be very complex relationships between signals and the ensuing behaviours. Contrary to the prediction of this hypothesis, however, Paton (1986) did not find any consistent differences in the response of reactors to the dif-ferent threat displays. It may in fact be inappropriate to think of this species as having different displays since no differ-ences can be found in their effect. Perhaps the birds simply use a range of postures in contests because they are attacked or start an attack from different physical positions.

VERY COMPLEX SIGNALS: BIRD SONGS

Perhaps the most puzzling signals from the point of view of design are the very complex song repertoires of passerine birds. A male nightingale or mockingbird may produce hun-dreds of different phrases in a seemingly endlessly varying

339

340

Chapter 14
The Design
of Signals:
Ecology and
Evolution

Box 14.1. *Animal signals and information.*

In the literature on animal signals, the word 'information' is used in two distinct ways.

(a) The technical meaning. The reduction in uncertainty of an observer about the actor's future behaviour following a signal. If the actor was equally likely to do one of two activities (say attack and retreat) before giving a signal, but afterwards was certain to do one of them (say attack), then the observer's uncertainty about the future behaviour of the actor has been reduced by one 'bit' as a result of seeing the display. There is no doubt that most animal signals convey information in this sense.

(b) The colloquial meaning. In everyday terms when we talk of information we mean information 'about something'. In animal communication information might be about the species, size, age, intentions, mating status and so on of the actor. In chapter 7 and again in the present chapter we have referred to the question of whether or not signals convey information in this second sense. The main points are as follows.

1 Animals are not expected to convey information about their intentions, except perhaps in co-operative interactions. Some evidence supports this view.

2 Signals may convey information about size, age and strength, and they tend to do so by variations that are not open to bluff or cheating (so-called reliable cues) (chp. 7).

3 Signals may convey information about the environment, for example the kind of predator that is in the vicinity (Seyfarth *et al.* 1980) or the whereabouts of food (von Frisch 1967).

outburst of song. If signals convey messages such as 'keep out', 'I am a male of species A' and so on, why the need for such extraordinary elaboration? As E.O. Wilson (1975) neatly put it, the signal seems redundant to the point of inanity. Many questions remain unanswered, however evidence suggests that although sometimes the different songs in a repertoire convey different messages, often they do not. In these cases complexity may be important either for mate attraction or territory defence.

(a) Attracting mates

As we saw in chapter 8, the very complex song of the sedge warbler is a product of sexual selection: females are more attracted to males with the largest repertoires. A similar interpretation of complex songs is suggested by Don Kroodsma's (1976) observation that female canaries are more readily stimulated to build nests when they hear a complex song repertoire than when they are given an artificially simplified repertoire. Older male canaries have larger repertoires, so females may use song complexity as a way of assessing the age and experience of potential mates.

341
Chapter 14
The Design
of Signals:
Ecology and
Evolution

(b) Territory defence

Many species use song not only for attracting mates but also for proclaiming territory ownership, and here complex repertoires may also play a role. This was shown in an experiment in which great tits were removed from their territories and the empty spaces were occupied with loudspeakers broadcasting song. Territories occupied with song repertoires were invaded more slowly than those occupied with loudspeakers playing a single song type (control territories with no song were invaded fastest of all) (Krebs *et al.* 1978). It is not yet known why repertoires are better at keeping out intruders, but one suggestion is that they are a form of deception. Intruders are looking for an empty territory, or at least a sparsely populated piece of habitat. Residents deter intruders by singing, and a variety of songs may be a more effective deterrent because it creates the impression of a habitat crowded with many singing males. Great tits tend to change from one song to another at the same time as changing perches, which enhances the illusion of many birds. Selection will act on intruders to spot the deception and the result will be an evolutionary race. The outcome of the race may sometimes be very elaborate song repertoires.

In summary, the influence of reactors on the design of signals leads to ritualization — an increase during evolution of exaggeration, repetition and stereotypy. Although ritualization may sometimes occur because of the benefits of reducing ambiguity in signals, it may often be the product of an evolutionary chase in which actors are selected to increase the effectiveness of their signals and reactors are selected to become sales-resistant.

342

*Chapter 14
The Design
of Signals:
Ecology and
Evolution*

Summary

Communication in animals occurs when one individual uses specially designed signals or displays to modify the behaviour of others. The design of signals is influenced both by ecological constraints and by their effectiveness in modifying the behaviour of reactors. The habitat can influence the effectiveness of different sensory channels of communication (e.g. scent versus visual signals) and the exact form of signals within a sensory channel. The latter point is illustrated by differences between the songs of birds living in different kinds of vegetation.

As signals evolve selection improves their effectiveness by making them stereotyped, repetitive and exaggerated. This evolutionary process of ritualization may have come about through a coevolutionary race between signallers and reactors.

A puzzling feature of some animal signals is their variability. Sometimes different variants convey different messages but in other instances they apparently do not.

Further reading

Catchpole (1979) is a good general review of bird song including discussions of song repertoires and song mimicry. Cullen (1972) is a good summary of the older ethological literature on animal signals, and Wiley (1983) covers more recent literature.

Dawkins & Krebs (1978) and Krebs & Dawkins (1984) develop the idea that communication is a matter of manipulation by actors of reactors. They contrast this view with the 'classical ethological view' that signals evolve for efficient transfer of information. Hinde (1981) criticizes Dawkins and Krebs, and questions whether their view is very different from that of earlier ethologists. See also Caryl's (1982) reply.

Andersson (1980) discusses the idea of threat displays falling into disuse because they are mimicked by bluff.

Wiley and Richards (1978) provide a technical review of ecological constraints on sound signals and Bowman (1979) describes how the songs of Darwin's finches are adapted to their habitats.

The book by Hailman (1977) is an excellent review of visual displays in animals.

Topics for discussion

1 Do animals tell lies?
2 How would you test the prediction that co-operative signals

should be less conspicuous than non-cooperative signals?
3 How useful is the analogy between animal signs and human advertising?

343

*Chapter 14
The Design
of Signals:
Ecology and
Evolution*

Chapter 15. Conclusion

The story of behaviour and adaptation that we have told in the last fourteen chapters is inevitably too simple. All the 'ifs' and 'buts' of an impeccably cautious and impregnable account would have made the book twice as long and half as easy to understand. However we do not want to leave the impression that the ideas we have discussed are completely accepted by all evolutionary biologists. Far from it, even our basic assumptions are still very much disputed in the literature.

How plausible are our main premises?

SELFISH GENES

Our discussions of natural selection have always been along the following lines: 'Imagine a gene for such and such behaviour; when would it tend to spread in a population?'. As we saw in chapter 1 this approach does not imply that there are genes 'for' altruism, spite, long tails or whatever, but merely there are some genetic differences between individuals which are correlated with the behaviour or structure in question.

But how plausible is the view that natural selection is a struggle between selfish genes rather than between individuals or groups? Obviously the field biologist sees *individuals* dying, surviving and reproducing; but the evolutionary consequence is that the frequencies of *genes* in the population change. Therefore the field biologist tends to think in terms of individual selection whilst the theorist thinks in terms of selfish genes. It can be very valuable, however, for the field worker to use selfish gene thinking in formulating ideas. This was apparent, for example, in chapter 12 where we saw that from the selfish gene viewpoint there is often no difference between parental care and sibling care, so hypotheses about the reasons for one and not the other occurring in a particular species must be framed in ecological rather than genetic terms. An exception to this rule, also clarified by selfish gene thinking, was in the social hymenoptera described in chapter 13 where females may be genetically predisposed to help their sisters. Sometimes an analysis in terms of selfish genes may unravel a problem that appears puzzling when considered in the context of individuals maximizing inclusive fitness. An example is the phenomenon of *segregation distortion*, in which an allele gets itself into a disproportionately large number of gametes. For a heterozygous parent, each allele is expected to occur in 50 per cent of gametes, but segregation distorters somehow interfere with the process of gamete formation and increase their proportionate representation. In male *Drosophila* it is thought that chromosomes with segregation distorter alleles somehow cause sperm containing the homologous chromosome to become abnormal (e.g. they may have broken tails) (Dawkins 1982). This phenomenon can be understood within the framework of selfish genes competing for representation in future generations but is hard to explain in terms of maximizing inclusive fitness. In fact since the effect of segregation distorter genes is to decrease fecundity (by killing off gametes), the interests of the selfish gene conflict with those of the individual.

GROUP SELECTION

At the beginning of the book we more or less dismissed group selection as a viable alternative to selection acting on individuals or selfish genes. We acknowledged that it could in principle work, but suggested that the conditions for group selection to be a powerful evolutionary force were not likely to be met in nature very often. However this is not a universally accepted point of view. In particular D.S. Wilson (1980) has published a book in which he claims that group selection is a very major evolutionary force indeed and Leigh van Valen (1980), a distinguished evolutionary theorist, has heralded Wilson's book as a major breakthrough and a change of paradigm for evolutionary biologists. How should we treat this claim?

One point to bear in mind from the start is that Wilson's model is more subtle than the simple 'differential extinction of groups' model discussed in chapter 1. The essential feature of Wilson's hypothesis is that populations are divided into groups (so-called 'trait groups') within which selection for or against altruistic traits (or any other trait for that matter) occurs. After selection has operated on the groups, the whole population mixes together before splitting up again into new groups for the next round of selection. There could be several ways in which individuals sort themselves into trait groups, but Wilson takes as the simplest case random assortment.

In Wilson's model the altruists are at a disadvantage *within* a trait group (because of their self-sacrifice) but trait groups

with altruists are more likely to contribute to the next generation than are trait groups with no altruists (Fig. 15.1). If the
population consisted of just one trait group, the altruistic gene
would only spread if the fitness change for the actor (d) is
greater than that for every other member of the population (p).
But, and this is the crux of Wilson's argument, if the population is divided into many trait groups the effect of the altruistic act on the small proportion of non-altruists that happen to
be in the same trait group as the altruist can be more or less
ignored. For example if there were 100 bird flocks in a forest
and one bird in one flock gave a warning call, the effect on the
average fitness of all other birds in the forest would be quite
small. Wilson concludes therefore that with trait groups, the
conditions for an altruistic gene to spread are $d > 0$, instead of
$d > f$ if there are no trait groups. (This means that any slight
advantage to the altruist will cause the gene to spread regardless of the advantage obtained by recipients of the act.) Thus
with selection acting on trait groups the altruistic gene is more
likely to spread.

The issue of group selection is still under active discussion
by theoreticians. In models such as Wilson's, the gene spreads

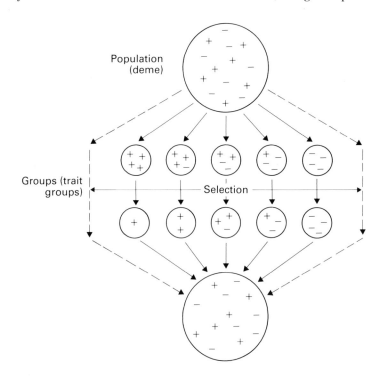

Fig. 15.1. D.S. Wilson's model of group selection. The model works if an
allele ($-$ sign) is selected against in mixed groups but increases in frequency across the population. After Harvey (1985), following Wade
(1978).

because of its advantage to individuals but this advantage only comes about because of the way in which the population is divided into groups. It is not yet clear how important this process will be for understanding the evolution of behaviour.

OPTIMALITY MODELS AND ESSs

In nearly every chapter we have used the ideas of optimality and ESSs. An ESS is the equivalent of an optimal solution when the pay-offs are frequency dependent, so the advantages and limitations of the two kinds of model can be discussed together. In chapter 3 we encountered some of the criticisms which have been levelled at optimality arguments and some of the limitations in putting them into practice. To recap briefly there were three main points.

1 *The idea that animals are optimal cannot be tested.* As we saw in chapter 3 this criticism is based on a mistaken notion. The aim of using an optimality model is not to test whether animals are optimal, but to test whether the particular optimality criterion and constraints used in the model give a good description of the animal's behaviour.

2 *It is hard to tell why the animal's behaviour does not fit the predictions exactly.* Very often the simple models give an approximate but not exact description of the animal's behaviour. This could be because the model makes incorrect assumptions about constraints or goals, or because some component of cost has not been measured. There is no simple way of distinguishing between these possibilities.

3 *Animals are not well enough adapted to optimize.* The main rationale for using optimality and ESS models is the assumption that natural selection produces well adapted animals, the aim of the models being to find out how they are adapted. There are, however, at least three reasons why animals might not be well adapted.

(a) The physical or biological environment may fluctuate too rapidly for the animals to 'catch up' in their adaptations (see chp. 4). For example the Atlantic gannet (*Sula bassana*) lays a clutch of one egg (Fig. 15.2), but when a second chick is experimentally added to the brood, both are reared without difficulty (Nelson 1964). This result appears to show that gannets are not well adapted to their environment, since natural selection should favour individuals that maximize their lifetime reproductive success (chp. 1). The reason that gannets are badly adapted seems to be that the food supply has recently changed. Gannets now feed some of the time on fish offal discarded by fishing boats and this extra food allows them to rear more young than in former times, but selection

Fig. 15.2. The Atlantic gannet, here shown at the nest with a five week old chick, rears only one young per season, although brood manipulations suggest that it could raise two. Photo by J.B. Nelson.

has not yet had time to change the clutch size from one or two. This view is supported by the observation that the southern hemisphere gannet (*Sula capensis*) which feeds in waters relatively unaffected by fishermen can still raise only one chick to full fledging weight (Jarvis 1974).

(b) There may be insufficient genetic variation for new strategies to evolve. If the environment changes or if for some other reasons the optimal phenotype changes, animals can adapt to the new conditions only if there is genetic variation in the population. Although the issue is by no means resolved, it seems likely that in small populations the rate of evolution might well be limited by the rate at which new mutations arise (Maynard Smith 1977b).

(c) There may be coevolutionary arms races (chp. 4) as for example between predator and prey; if one side is ahead in the arms race, the other will appear to be poorly adapted to its environment, for example hosts that are killed or debilitated by pathogens and parasites (see also Rothstein 1986).

Two more criticisms can be added to this list. One is that optimality and ESS models assume that genes exist, but contain no specific genetic mechanisms. For example, the ESS models of fighting strategies in chapter 7 did not allow for sexual reproduction and the mixing of genes that this entails. The attitude of the optimality or ESS theorist is often 'think of the strategies and let the genes look after themselves'. The population geneticist on the other hand would like to know whether the models can really be couched in terms of genes and whether the rules of inheritance will allow the equilibria suggested by the models to actually develop. The answer to this question is largely unexplored at the moment (Grafen 1984).

The second criticism is more specifically directed towards the optimality models used and tested in this book. The critical reader will have noticed that although we stressed the value of making quantitative predictions from optimality and ESS models, most of the tests of these predictions were qualitative. The animals were usually seen to do 'approximately the right thing' the dungflies in chapter 3, for example, copulated for 36 min instead of the predicted 41 min. Some might ask whether it is worth developing quantitative arguments if the tests are only qualitative. The answer is that tests are often qualitative simply because of limitations in the techniques used to carry out the tests. The value of quantitative predictions is still potentially just as great and what is needed is comparable technical developments in ways of testing the models. Once the quantitative predictions can be tested accurately, discrepancies between observed and predicted results help to tell us what is wrong with the models.

It is possible to carry on discussing the pros and cons of optimality models for a long time, but the strongest argument in their favour is that over and over again optimality arguments have helped us to understand adaptations. We have illustrated this point in the proceeding chapters with behavioural examples; foraging, flock size, territory size and so on, but optimality arguments can equally well be used to understand adaptations at the physiological and biochemical level. For example the familiar 'herring bone' arrangement of the swimming muscles of many fish is not merely án incidental design feature. This arrangement allows the muscles to contract at a rate which maximizes their power output (Alexander 1975). At the biochemical level the energy for muscle contraction is generated by oxidation of carbohydrates or fats via the Krebs cycle. It would be chemically feasible to carry out the oxidation by a more direct route, but the advantage of the cycle is that it maximizes the net energy gain per molecule oxidized (Baldwin & Krebs 1981).

The comments discussed so far could apply equally well to

simple optimality and to ESS models. Now let us briefly consider a case where the two kinds of model may lead to different interpretations of field data. Richard Dawkins and Jane Brockmann (1980) demonstrated that digger wasps adopt what appears at first sight to be a suboptimal strategy, but further analysis in terms of an ESS revealed a possible explanation as to why the particular strategy might be used. The behaviour in question was the persistence of fighting by female wasps disputing the ownership of a burrow (chp. 10). The question asked by Dawkins and Brockmann was 'how long should a female persist in a fight?'. They made the assumption that females are designed to maximize the benefit they obtain from the contest, which leads to the prediction that a female's persistence should be related to the value of the burrow. This in turn depends on the number of paralysed insects stored in the burrow. A nest containing four insects is ready for egg laying and the winner of this prize saves herself days of digging and provisioning, while a burrow with only one insect still needs a lot of work before it is ready.

Contrary to expectation, Dawkins and Brockmann found that the persistence of females in a fight was not correlated with the total number of insects in the burrow but with the number put in by the loser. Since the loser is the one that determines the length of the fight (because the fight ends when the loser goes away), Dawkins and Brockmann concluded that wasps fight in proportion to their own past contribution rather than in proportion to the total value of the burrow. It is easy to see that this could lead to females giving up very quickly in a fight over a valuable burrow, just because the other wasp has done more of the provisioning, even though the pay-off goes to the winner regardless of who provisioned the burrow.

Dawkins and Brockmann's first reaction was that the wasps probably cannot 'tell' how many insects are in the burrow, and that the rule 'fight in proportion to own contribution' is a reasonable rule of thumb that approximates the optimal strategy, since there is usually some correlation between the total number and the number put in by each female. This is an example of how the optimal policy depends on assumptions about the constraints. A perfectly knowledgeable wasp should fight in proportion to total burrow contents, but when constrained by ignorance of the total number, the policy 'fight in proportion to own contribution' might be the best option.

However, now consider what would happen if all wasps were endowed with perfect knowledge. If both wasps had the same assessment of the value of the burrow and fought for a time proportional to this value, both would fight equally hard and give up at the same moment! Presumably chance factors would cause one wasp to persist fractionally longer than the other, and if these factors were truly random each wasp would have a 50:50 chance of winning after the long struggle. Imagine now a wasp that decided how long to fight by the toss of a coin and adopted the rule 'if heads give up at once, if tails persist indefinitely' (this is equivalent to the 'bourgeois' strategy in chp. 7). In a population of perfectly knowledgeable wasps this strategy would win half its fights and waste no time in lost contests. Therefore its net pay-off (subtracting total time wasted from the pay-off for winning) would be higher than that of the omniscient wasp. The message of this example is that 'perfect knowledge', which appeared at first sight to be an optimal strategy, is not an ESS. In general terms, whenever the pay-off depends on what others do, the question should be analysed as an ESS rather than a simple optimality problem.

Causal and functional explanations

Behavioural ecology is about functional explanations (the answers to 'why?' questions) of behaviour. As we emphasized in chapter 1 a great deal of misunderstanding can arise if functional and causal ('how?') explanations are confused. A simple illustration of this is an objection that is often raised to labels such as 'selfish', 'spiteful', 'sneak' and 'transvestite', used by behavioural ecologists to describe the behaviours they observe. The objection is that the labels are too anthropomorphic and are loaded with the implication that animals are endowed with human-like motives. The answer to this objection is that the labels are used not to describe the causal mechanisms underlying the behaviour but to describe its functional consequences.

Although it is important to be clear about the distinction between causal and functional explanations, it is equally valuable to realize that the two kinds of question are complementary and that asking 'why?' questions can often help to understand the answers to 'how?' questions, or vice versa. An example of how causal and functional explanations go hand in hand is illustrated in Fig. 15.3. Prairie dogs (Cynomys ludovicianus) are colonial and live in underground tunnels which may be up to 15 m long. The tunnels are usually simple U-shaped passages with an opening to the surface at either end. It has been known for a long time that prairie dogs build little mounds of soil around the two entrances of the tunnel. These mounds were considered to function either as lookout points or to protect the tunnel against floods. However a closer inspection revealed that the two ends of the burrow have different kinds of mounds. At one end there is a high steep

Fig. 15.3. A diagrammatic section of a prairie dog burrow. A typical burrow has two entrances, one with a low, round 'dome' at its entrance and the other with a taller, steeper-sided 'crater'. The different heights and shapes of the burrow entrances cause air to be sucked out of the crater end and therefore in through the dome (Vogel *et al.* 1973).

sided 'crater' mound while at the other end there is a low rounded 'dome' (Fig. 15.3). If the mounds are simple lookouts or flood barriers, why should they be different shapes? The answer to this 'why?' question comes from an understanding of how air is exchanged in the tunnel (Vogel *et al.* 1973). A prairie dog living in the long underground tunnel cannot survive without a regular supply of fresh air, and it appears that the mounds around the two entrances are designed to ensure a continuous flow of air through the tunnel. The crater mound is higher and has steeper sides than the dome; as a consequence, air is sucked out of the crater end of the tunnel and into the dome end.

The forces causing the air flow are viscous sucking and the Bernoulli effect. Viscous sucking refers to the fact that when moving air passes a region of stationary air, the still air is dragged along with the current. The effect is larger at the crater end because the crater is higher than the dome and so it is exposed to faster winds. The Bernoulli effect states that the pressure of a steadily moving fluid decreases when its velocity increases. The velocity of air above the crater is greater than above the mound and the crater has very still air inside be-cause of its steep edges. The pressure drop between the inside and outside of the crater is therefore higher than in the case of the dome so the Bernoulli effect causes air to be sucked out of the crater end of the tunnel. Vogel *et al.* (1973) demonstrated by means of laboratory experiments with miniature model

tunnels and by dropping smoke bombs down real tunnels in
the field that the mound system is so effective that it causes
the air in the tunnel to change once every ten minutes even in
a very light breeze. The rate of air exchange is related to wind
speed, but is unaffected by wind direction since the mounds
are symmetrical. This second feature is important because
wind direction is unpredictable in the prairie dog's natural
habitat. This example illustrates that a functional question
'what are mounds for?' led to a detailed understanding of a
mechanism question 'how do prairie dogs get enough fresh
air?'.

The optimality models we have encountered in this book
bring together mechanisms (in the form of constraints) and
functions (the currency) in building up an explanation of
behaviour.

A final comment

What we have described as behavioural ecology in this book is
the present day equivalent of natural history. It stands in a
lineage which gradually developed from detailed descriptions
of animal behaviour by naturalists such as Gilbert White and
Henri Fabre to experimental studies of natural history by Tinbergen and others. At the moment there is something of a
bandwagon in natural history for inventing functional explanations of behaviour, which gives behavioural ecology and
sociobiology a bad name. We have tried to avoid this as much
as possible (without complete success!) and instead we have
emphasized the idea of making testable predictions about
adaptation. To illustrate how this approach has developed
from studies of natural history let us construct a hypothetical
lineage of studies of mating behaviour in dungflies.

A few hundred years ago naturalists would have been satis-
fied to discover that when two dungflies were seen together,
one riding on the other's back, the top one was a male and the
one underneath was a female and that the two were mating. A
hundred years ago Darwin realized that males in general compete for females. A description of the natural history of dungfly mating at this stage would have included reference to the
fact that males are larger than females and that this may be a
result of sexual selection. Twenty years ago evolutionary biologists would have stressed the idea that males ride on the backs
of females not only to inject sperm, but also for some time
after copulation while the female lays her eggs. By guarding
the female in this way the male guarantees that his sperm are
not displaced by those of another male. In the last ten years
behavioural ecologists have started to try and explain why it is

that the male rides on the female's back for 40 min and not 10, 20 or 60. In developing a theory to explain why, it has become apparent that the same kind of analysis can be used for bumblebees sucking nectar out of flowers, parents investing in their offspring and many other problems. This gradual reductionist progression from broad description to detailed quantitative analysis and simple generalizations is one of the major themes of development of the natural history lineage. But at the same time natural historians are interested in differences between species. Perhaps in the next few years we will see the same quantitative approach used to explain why different species behave in different ways.

Summary

This chapter is in two parts. The first section tries to assess the value and the limitations of the selfish gene and optimality views of evolution. Group selection as an alternative to individual selection is reassessed. The value of optimality arguments can be illustrated by studies of adaptations at the behavioural, physiological and biochemical levels.

In the second part of the chapter, we try to show how different kinds of questions (function and causation) should go hand in hand in studies of behaviour.

Further reading

Maynard Smith (1977b) presents a very clear and thought-provoking summary of unsolved problems in evolutionary theory of relevance to behavioural ecologists using optimality arguments. Particularly valuable is his discussion of what limits the rate of evolution.

Gould (1980) is also a critical and stimulating review of neo-Darwinian evolution. He argues that microevolutionary changes of the type we have discussed in this book may be caused by different mechanisms from those leading to macroevolutionary changes (evolution and extinction of species).

Dawkins (1986) gives an incisive discussion of many current issues in evolutionary theory.

References

Abele, L.G. & Gilchrist, S. 1977. Homosexual rape and sexual selection in acanthocephalan worms. *Science, NY* **197**, 81−3.

Alatalo, R.V., Carlson, A., Lundberg, A. & Ulfstrand, S. 1981. The conflict between male polygamy and female monogamy: the case of the pied flycatcher, *Ficedula hypoleuca*. *Amer. Natur.* **117**, 738−53.

Alcock, J. 1984. *Animal Behaviour: An Evolutionary Approach.* 3rd edn. Sinauer, Sunderland, Massachusetts.

Alcock, J., Jones, C.E. & Buchmann, S.L. 1977. Male mating strategies in the bee *Centris pallida*, Fox (Anthophoridae: Hymenoptera). *Amer. Natur.* **111**, 145−55.

Alexander, R.D. 1974. The evolution of social behaviour. *Ann. Rev. Ecol. Syst.* **5**, 325−83.

Alexander, R.D. 1975. Natural selection and specialized chorusing behaviour in acoustical insects. In D. Pimentel (ed.). *Insects, Science and Society*, pp. 35−77. Academic Press, New York.

Alexander, R.D. & Borgia, G. 1979. On the origin and basis of the male-female phenomenon. In M.S. Blum & N.A. Blum (eds.). *Sexual Selection and Reproductive Competition in Insects*. Academic Press, New York.

Alexander, R.D. & Sherman, P.W. 1977. Local mate competition and parental investment in social insects. *Science, NY* **196**, 494−500.

Alexander, R.McN. 1975. *The Chordates*. Cambridge University Press, Cambridge.

Altmann, S.A. 1974. Baboons, space, time and energy. *Amer. Zool.* **14**, 221−40.

Andersson, M. 1980. Why are there so many threat displays? *J. theor. Biol.* **86**, 773−81.

Andersson, M. 1982. Female choice selects for extreme tail length in a widowbird. *Nature* **299**, 818−20.

Andersson, M. 1984. The evolution of eusociality. *Ann. Rev. Ecol. Syst.* **15**, 165−89.

Andersson, M. & Wicklund, C.G. 1978. Clumping versus spacing out: experiments on nest predation in fieldfares (*Turdus pilaris*). *Anim. Behav.* **26**, 1207−12.

Aoki, S. 1977. *Colophina clematis* (Homoptera: Pemphigidae), an aphid species with 'soldiers'. *Kontyu, Tokyo* **45**, 276−82.

Arak, A. 1983. Sexual selection by male−male competition in natterjack toad choruses. *Nature* **306**, 261−2.

Argyle, M. 1972. Non-verbal communication in human social interaction. In R.A. Hinde (ed.). *Non-verbal Communication*, pp. 243−69. Cambridge University Press, Cambridge.

Arnold, S.J. 1976. Sexual behaviour, sexual interference and sexual defence in the salamanders *Ambystoma maculatum*, *A. tigrinum* and *Plethodon jordani*. *Z. Tierpsychol.* **42**, 247−300.

Arnold, S.J. 1981. Behavioural variation in natural populations. II. The inheritance of a feeding response in crosses between geographical races of the garter snake, *Thamnophis elegans*. *Evolution* **35**, 510−15.

Arnold, S.J. 1983. Sexual selection: the interface of theory and empiricism. In P.P.G. Bateson (ed.). *Mate Choice*, pp. 67−107. Cambridge University Press, Cambridge.

Askenmo, C.E.H. 1984. Polygyny and nest site selection in the pied flycatcher. *Anim. Behav.* **32**, 972−80.

Austad, S.N. 1982. First male sperm priority in the bowl and doily spider, *Frontinella pyramitela. Evolution* **36**, 777−85.

Austad, S.N. 1983. A game theoretical interpretation of male combat in the bowl and doily spider, *Frontinella pyramitela. Anim. Behav.* **31**, 59−73.

Axelrod, R. 1984. *The Evolution of Cooperation*. Basic books, New York.

Axelrod, R. & Hamilton, W.D. 1981. The evolution of cooperation. *Science* **211**, 1390−6.

Baker, R.R. & Parker, G.A. 1979. The evolution of bird coloration. *Phil. Trans. Roy. Soc. Lond. B* **287**, 63−130.

Baldwin, J.E. & Krebs, H.A. 1981. The evolution of metabolic cycles. *Nature, Lond.* **291**, 381−2.

Barnard, C.J. 1984. *Producers and Scroungers*. Croom Helm, Beckenham.

Barnard, C.J. & Sibly, R.M. 1981. Producers and scroungers: a general model and its application to captive flocks of house sparrows. *Anim. Behav.* **29**, 543−50.

Barnard, C.J. & Thompson, D.B.A. (1985) *Gulls and Plovers: The Ecology and Behaviour of Mixed-species Feeding Groups*. Croom Helm, London & Sydney.

Bartz, S.H. 1979. Evolution of eusociality in termites. *Proc. Natn. Acad. Sci. USA* **76**, 5764−8.

Bateman, A.J. 1948. Intra-sexual selection in *Drosophila. Heredity* **2**, 349−68.

Bateson, P.P.G. 1980. Optimal outbreeding and the development of sexual preferences in Japanese quail. *Z. Tierpsychol.* **53**, 231−44.

Bateson, P.P.G. 1983a. Rules for changing the rules. In D.S. Bendall (ed.). *Evolution from Molecules to Men*, pp. 483−507. Cambridge University Press, Cambridge.

Bateson, P.P.G. 1983b. *Mate Choice*. Cambridge University Press, Cambridge.

Bell, G. 1980. The costs of reproduction and their consequences. *Amer. Natur.* **109**, 453−64.

Belovsky, G.E. 1978. Diet optimization in a generalist herbivore: the moose. *Theoret. Pop. Biol.* **14**, 105−34.

Belovsky, G.E. 1984. Herbivore optimal foraging: a comparative test of three models. *Amer. Natur.* **124**, 97−115.

Belovsky, G.E. 1986. Generalist herbivore foraging and its role in competitive interactions. *Amer. Zool.* (in press).

Benzer, S. 1973. Genetic dissection of behavior. *Sci. Am.* **229** (12), 24−37.

Berthold, P. 1984. The endogenous control of bird migration: a survey of experimental evidence. *Bird Study* **31**, 19−27.

Bertram, B.C.R. 1975. Social factors influencing reproduction in wild lions. *J. Zool. Lond.* **177**, 463−82.

Bertram, B.C.R. 1980. Vigilance and group size in ostriches. *Anim. Behav.* **28**, 278−86.

Birkhead, T.R. 1977. The effect of habitat and density on breeding success in common guillemots, *Uria aalge. J. Anim. Ecol.* **46**, 751−64.

Birkhead, T.R. 1979. Mate guarding in the magpie *Pica pica. Anim. Behav.* **27**, 866−74.

Birkhead, T.R. & Clarkson, K. 1980. Mate selection and precopulatory guarding in *Gammarus pulex. Z. Tierpsychol.* **52**, 365−80.

Bischoff, R.J., Gould, J.L. & Rubenstein, D.I. 1985. Tail size and female choice in the guppy *Poecilia reticulata. Behav. Ecol. Sociobiol.* **17**, 253−5.

Blurton Jones, N.G. 1968. Observations and experiments on causation of threat displays of the great tit, *Parus major. Anim. Behav. Monogr.* **1**, 75−158.

Borgia, G. 1979. Sexual selection and the evolution of mating systems. In M.S. Blum & N.A. Blum (eds.). *Sexual Selection and Reproductive Competition in Insects*, pp. 19−80. Academic Press, New York.

Borgia, G. 1985. Bower quality, number of decorations and mating success of male satin bowerbirds (*Ptilinorynchus violaceus*): an experimental analysis. *Anim. Behav.* **33**, 266−71.

Bowman, R.I. 1979. Adaptive morphology of song dialects in Darwin's finches. *J. Orn, Lpz.* **120**, 353−89.

Bradbury, J.W. 1977. Lek mating behaviour in the hammer-headed bat. *Z. Tierpsychol.* **45**, 225−55.

Bradbury, J.W. & Gibson, R.M. 1983. Leks and mate choice. In P. Bateson (ed.). *Mate Choice*, pp. 109−38. Cambridge University Press, Cambridge.

Bradbury, J.W. & Vehrencamp, S.L. 1977. Social organization and foraging in emballonurid bats. III. Mating systems. *Behav. Ecol. Sociobiol.* **2**, 1−17.

Bray, O.E., Kennelly, J.J. & Guarino, J.L. 1975. Fertility of eggs produced on territories of vasectomized red-winged blackbirds. *Wilson Bull.* **87**, 187−95.

Brockmann, H.J. 1984. The evolution of social behaviour in insects. In J.R. Krebs. & N.B. Davies (eds.). *Behavioural Ecology: An Evolutionary Approach*, 2nd edn., pp. 340−61. Blackwell Scientific Publications, Oxford.

Brockmann, H.J. & Dawkins, R. 1979. Joint nesting in a digger wasp as an evolutionarily stable preadaptation to social life. *Behaviour* **71**, 203−45.

Brockmann, H.J., Grafen, A. & Dawkins, R. 1979. Evolutionary stable nesting strategy in a digger wasp. *J. theor. Biol.* **77**, 473−96.

Brooke, M. de L. 1978. Some factors affecting the laying date, incubation and breeding success of the manx shearwater, *Puffinus puffinus. J. Anim. Ecol.* **47**, 477−95.

Brown, C.H. 1982. Ventriloquial and locatable vocalizations in birds. *Z. Tierpsychol.* **59**, 338−50.

Brown, J.L. 1964. The evolution of diversity in avian territorial systems. *Wilson Bull.* **76**, 160−9.

Brown, J.L. 1969. The buffer effect and productivity in tit populations. *Amer. Natur.* **103**, 347−54.

Brown, J.L. 1982. Optimal group size in territorial animals. *J. theor. Biol.* **95**, 793−810.

Brown, J.L. 1987. *Ecology and Evolution of Helping and Communal Breeding in Birds.* Princeton University Press, Princeton, NJ.

Brown, J.L., Brown, E.R., Brown, S.D. & Dow, D.D. 1982. Helpers: effects of experimental removal on reproductive success. *Science* **215**, 421−2.

Brown, J.L., Dow, D.D., Brown, E.R., & Brown, S.D. 1978. Effects of helpers on feeding of nestlings in the grey-crowned babbler, *Pomatostomus temporalis. Behav. Ecol. Sociobiol.* **4**, 43−60.

Buckle, G.R. & Greenberg, L. 1981. Nest mate recognition in sweat bees, *Lasioglossum zephyrum*: does an individual recognise its own odour or only the odours of its nest mates? *Anim. Behav.* **29**, 802−9.

Bull, J.J. 1980. Sex determination in reptiles. *Q. Rev. Biol.* **55**, 3−21.

Bulmer, M.G. & Taylor, P.D. 1981. Worker−queen conflict and sex ratio theory in social Hymenoptera. *Heredity* **47**, 197−207.

Burley, N. (in press). Sex ratio manipulation in color-banded populations of zebra finches. *Evolution.*

Burley, N. & Coopersmith, C.B. (in press). Preferences of adult zebra finches for opposite- and same-sex conspecifics with varying phenotypic attributes. *Evolution*.

Burley, N., Krantzberg, G. & Radman, P. 1982. Influence of colour-banding on the conspecific preferences of zebra finches. *Anim. Behav.* **27**, 686–98.

Busnel, R.G. & Klasse, A. 1976. *Whistled Languages*. Springer-Verlag, Berlin.

Bygott, J.D., Bertram, B.C.R. & Hanby, J.P. 1979. Male lions in large coalitions gain reproductive advantage. *Nature, Lond.* **282**, 839–41.

Cade, W.H. 1979. The evolution of alternative male reproductive strategies in field crickets. In M. Blum & N.A. Blum (eds.). *Sexual Selection and Reproductive Competition in Insects*, pp. 343–79. Academic Press, London.

Cade, W.H. 1981. Alternative mate strategies: genetic differences in crickets. *Science* **212**, 563–64.

Cade, W.H. & Wyatt, D.R. 1984. Factors effecting calling behaviour in field crickets, *Teleogryllus* and *Gryllus* (age, weight, density and parasites). *Behaviour* **88**, 61–75.

Caldwell, R.L. 1985. A test of individual recognition in the stomatopod, *Gonodactylus festae*. *Anim. Behav.* **33**, 101–6.

Calvert, W.H., Hedrick, L.E. & Brower, L.P. 1979. Mortality of the monarch butterfly, *Danaus plexippus*: avian predation at five over-wintering sites in Mexico. *Science, NY* **204**, 847–51.

Capranica, R.R., Frishkopf, L.S. & Nevo, E. 1973. Encoding of geographic dialects in the auditory system of the cricket frog. *Science, NY* **182**, 1272–5.

Caraco, T. 1979a. Time budgeting and group size: a theory. *Ecology* **60**, 611–7.

Caraco, T. 1979b. Time budgeting and group size: a test of theory. *Ecology* **60**, 618–27.

Caraco, T., Martindale, S. & Pulliam H.R. 1980a. Flocking: advantages and disadvantages. *Nature* **285**, 400–1.

Caraco, T., Martindale, S. & Whitham, T.S. 1980b. An empirical demonstration of risk-sensitive foraging preferences. *Anim. Behav.* **28**, 820–30.

Caraco, T. & Wolf, L.L. 1975. Ecological determinants of group sizes of foraging lions. *Amer. Natur.* **109**, 343–52.

Carayon, J. 1974. Insemination traumatique hétérosexuelle et homosexuelle chez *Xylocoris maculipennis* (Hem. Anthocoridae). *C.R. Acad. Sci. Paris, D* **278**, 2803–6.

Caro, T.M. & Bateson, P. 1986. Ontogenetic analysis of alternative tactics. *Anim. Behav.* **34**, 1483–99.

Carpenter, C.R. 1954. Tentative generalisations on the grouping behavior of non-human primates. *Human Biol.* **26**, 269–76.

Carpenter, F.L. & MacMillen, F.E. 1976. Threshold model of feeding territoriality and test with a Hawaiian honey creeper. *Science, NY* **194**, 639–42.

Carpenter, F.L., Paton, D.C. & Hixon, M.A. 1983. Weight gain and adjustment of feeding territory size in migrant hummingbirds. *Proc. Natn. Acad. Sci. USA* **80**, 7259–63.

Caryl, P.G. 1979. Communication by agonistic displays: what can games theory contribute to ethology? *Behaviour* **68**, 136–69.

Caryl, P.G. 1980. Escalated fighting and the war of nerves: games theory and animal combat. In P.P.G. Bateson & P.H. Klopfer (eds.). *Perspectives in Ethology*. Plenum Press, New York.

Caryl, P.G. 1982. Animal signals: a reply to Hinde. *Anim. Behav.* **30**, 240–44.

Catchpole, C.K. 1979. *Vocal Communication in Birds*. Edward Arnold, London.

Catchpole, C.K. 1980. Sexual selection and the evolution of complex songs among European warblers of the genus *Acrocephalus*. *Behaviour* **74**, 149–66.

Catchpole, C.K., Dittami, J. & Leisler, B. 1984. Differential responses to male song repertoires in female songbirds implanted with oestradiol. *Nature* **312**, 563–4.

Catchpole, C., Leisler, B. & Winkler, H. 1985. The evolution of polygyny in the great reed warbler, *Acrocephalus arundinaceus*: a possible case of deception. *Behav. Ecol. Sociobiol.* **16**, 285–91.

Chappuis, C. 1971. Un example de l'influence du milieu sur les émissions vocales des oiseaux: l'evolution des chants en forêt équitoriale. *Terre Vie* **25**, 183–202.

Charnov, E.L. 1976. Optimal foraging: the marginal value theorem. *Theoret. Pop. Biol.* **9**, 129–36.

Charnov, E.L. 1978. Evolution of eusocial behavior: offspring choice or parental parasitism? *J. theor. Biol.* **75**, 451–66.

Charnov, E.L. 1981. Kin selection and helpers at the nest: effects of paternity and biparental care. *Anim. Behav.* **29**, 631–2.

Charnov, E.L. 1982. *The Theory of Sex Allocation*. Princeton University Press, Princeton, NJ.

Charnov, E.L. & Krebs, J.R. 1974. On clutch size and fitness. *Ibis* **116**, 217–19.

Charnov, E.L., Orians, G.H. & Hyatt, K. 1976. The ecological implications of resource depression. *Amer. Natur.* **110**, 247–59.

Charnov, E.L. & Skinner, S. 1984. Evolution of host selection and clutch size in parasitoids. *Florida ent.* **67**, 5–21.

Christy, J.H. & Salmon, M. 1984. Ecology and evolution of mating systems of fiddler crabs (genus *Uca*). *Biol. Rev.* **59**, 483–509.

Clark, A.B. 1978. Sex ratio and local resource competition in a prosimian primate. *Science* **201**, 163–5.

Clutton-Brock, T.H. 1975. Feeding behaviour of red colobus and black and white colobus in East Africa. *Folia primatol.* **23**, 165–207.

Clutton-Brock, T.H. 1983. Selection in relation to sex. In D. S. Bendall (ed.). *From Molecules to Men*, pp. 457–81. Cambridge University Press, Cambridge.

Clutton-Brock, T.H. & Albon, S.D. 1979. The roaring of red deer and the evolution of honest advertisement. *Behaviour* **69**, 145–70.

Clutton-Brock, T.H., Albon, S.D., Gibson, R.M. & Guinness, F.E. 1979. The logical stag: adaptive aspects of fighting in red deer (*Cervus elaphus* L.). *Anim. Behav.* **27**, 211–25.

Clutton-Brock, T.H., Albon, S.D. & Guinness, F.E. 1984. Maternal dominance, breeding success and birth sex ratios in red deer, *Nature* **308**, 358–60.

Clutton-Brock, T.H., Guinness, F.E. & Albon, S.D. 1982. *Red Deer: The Behaviour and Ecology of Two Sexes*. Chicago University Press, Chicago.

Clutton-Brock, T.H. & Harvey, P.H. 1977. Primate ecology and social organisation. *J. Zool. Lond.* **183**, 1–39.

Clutton-Brock, T.H. & Harvey, P.H. 1978. Mammals, resources and reproductive strategies. *Nature* **273**, 191–5.

Clutton-Brock, T.H. & Harvey, P.H. 1979. Comparison and adaptation. *Proc. R. Soc. Lond. B* **205**, 547–65.

Commons, M.L., Shettleworth, S.J. & Kacelnik, A. 1986. *Quantitative Analyses of Behavior*. Vol. 6. Foraging. L. Erlbaum, New York.

Conover, D.A. 1984. Adaptive significance of temperature-dependent sex determination in a fish. *Amer. Natur.* **123**, 297–313.

Corbet, P.S. 1962. *A Biology of Dragonflies*. Witherby, London.

Coulson, J.C. 1966. The influence of the pairbond and age on the breeding biology of the kittiwake gull, *Rissa tridactyla*. *J. Anim. Ecol.* **35**, 269−79.

Cowie, R.J. 1977. Optimal foraging in great tits, *Parus major*. *Nature, Lond.* **268**, 137−9.

Cox, C.R. & Le Boeuf, B.J. 1977. Female incitation of male competition: a mechanism of mate selection. *Amer. Natur.* **111**, 317−35.

Craig, R. 1979. Parental manipulation, kin selection and the evolution of altruism. *Evolution* **33**, 319−34.

Cronin, E.W. & Sherman, P.W. 1977. A resource-based mating system: the orange-rumped honey guide. *Living Bird* **15**, 5−32.

Crook, J.H. 1964. The evolution of social organisation and visual communication in the weaver birds (Ploceinae). *Behaviour Suppl.* **10**, 1−178.

Crook, J.H. & Gartlan, J.S. 1966. Evolution of primate societies. *Nature, Lond.* **210**, 1200−3.

Crozier, R.H. & Luykx, P. 1985. The evolution of termites eusociality is unlikely to have been based on a male-haploid analogy. *Am. Natur.* **126**, 867−69.

Cullen, J.M. 1966. Reduction of ambiguity through ritualization. *Phil. Trans. R. Soc. B* **251**, 363−74.

Cullen, J.M. 1972. Some principles of animal communication. In R.A. Hinde (ed.). *Non-verbal Communication*, pp. 101−22. Cambridge University Press, Cambridge.

Daly, M. 1979. Why don't male mammals lactate? *J. theor. Biol.* **78**, 325−45.

Darwin, C. 1859. *On the Origin of Species*. Murray, London.

Darwin, C. 1871. *The Descent of Man and Selection in Relation to Sex*. Murray, London.

Davies, N.B. 1978. Territorial defence in the speckled wood butterfly (*Pararge aegeria*): the resident always wins. *Anim. Behav.* **26**, 138−47.

Davies, N.B. & Halliday, T.R. 1978. Deep croaks and fighting assessment in toads *Bufo bufo*. *Nature, Lond.* **274**, 683−5.

Davies, N.B. & Houston, A.I. 1981. Owners and satellites: the economics of territory defence in the pied wagtail, *Motacilla alba*. *J. Anim. Ecol.* **50**, 157−80.

Davies, N.B. & Houston, A.I. 1986. Reproductive success of dunnocks, *Prunella modularis*, in a variable mating system. II. Conflicts of interest among breeding adults. *J. Anim. Ecol.* **55**, 139−54.

Davies, N.B. & Lundberg, A. 1984. Food distribution and variable mating system in the dunnock, *Prunella modularis*. *J. Anim. Ecol.* **53**, 895−912.

Dawkins, M. 1971. Perceptual changes in chicks: another look at the 'search image' concept. *Anim. Behav.* **19**, 566−74.

Dawkins, R. 1976. *The Selfish Gene*. Oxford University Press, Oxford.

Dawkins, R. 1978. Replicator selection and the extended phenotype. *Z. Tierpsychol.* **47**, 61−76.

Dawkins, R. 1979. Twelve misunderstandings of kin selection. *Z. Tierpsychol.* **51**, 184−200.

Dawkins, R. 1980. Good strategy or evolutionarily stable strategy? In G.W. Barlow & J. Silverberg. (eds.). *Sociobiology: Beyond Nature/Nurture*, pp. 331−67. Westview Press, Boulder, Colorado.

Dawkins, R. 1982. *The Extended Phenotype*. W.H. Freeman, Oxford.

Dawkins, R. 1986. *The Blind Watchmaker*. Longman, London.

Dawkins, R. & Brockmann, H.J. 1980. Do digger wasps commit the Concorde fallacy? *Anim. Behav.* **28**, 892−6.

Dawkins, R. & Carlisle, T.R. 1976. Parental investment, mate desertion and a fallacy. *Nature, Lond.* **262**, 131−3.

Dawkins, R. & Krebs, J.R. 1978. Animal signals: information or manipulation? In J.R. Krebs & N.B. Davies (eds.). *Behavioural Ecology: An Evolutionary Approach*, 2nd edn., pp. 282–309. Blackwell Scientific Publications, Oxford.

Dawkins, R. & Krebs, J.R. 1979. Arms races between and within species. *Proc. R. Soc. Lond. B* **205**, 489–511.

De Groot, P. 1980. Information transfer in a socially roosting weaver bird (*Quelea quelea*: Ploceinae): an experimental study. *Anim. Behav.* **28**, 1249–54.

De Vore, I. (ed.). 1965. *Primate Behaviour: Field Studies of Monkeys and Apes*. Holt, Rinehart & Winston, New York.

Diamond, J.N., Karasov, W.H., Phan, D. & Carpenter, F.L. 1986. Hummingbird digestive physiology: a determinant of foraging bout frequency. *Nature* (in press).

Dominey, W.J. 1980. Female mimicry in male blue gill sunfish — a genetic polymorphism? *Nature* **284**, 546–8.

Downhower, J.F. & Armitage, K.B. 1971. The yellow-bellied marmot and the evolution of polygamy. *Amer. Natur.* **105**, 355–70.

Dudai, Y. & Quinn, W.G. 1980. Genes and learning in *Drosophila*. *Trends in Neurosciences* **Feb.**, 28–30.

Dunbar, R.I.M. 1983. Intraspecific variations in mating strategy. In P.P.G. Bateson & P. Klopfer (eds.). *Perspectives in Ethology*. Vol. 5, pp. 385–431. Plenum Press, New York.

Dunbar, R.I.M. 1984. *Reproductive Decisions: An Economic Analysis of Gelada Baboon Social Strategies*. Princeton University Press, Princeton, NJ.

Duncan, P. & Vigne, N. 1979. The effect of group size in horses on the rate of attacks by blood-sucking flies. *Anim. Behav.* **27**, 623–5.

Dunford, C. 1977. Kin selection for ground squirrel alarm calls. *Amer. Natur.* **111**, 782–5.

Dybas, H.S. & Lloyd, M. 1974. The habitats of 17 year periodical cicadas (Homoptera: Cicadidae: *Magicicada* spp.) *Ecol. Mongr.* **44**, 279–324.

Eberhard, W.G. 1972. Altruistic behaviour in a sphecid wasp: support for kin-selection theory. *Science* **175**, 1390–1.

Eberhard, W.G. 1979. The functions of horns in *Podischnus agenor* (Dynastinae) and other beetles. In M.S. Blum & N.A. Blum (eds.). *Sexual Selection and Reproductive Competition in Insects*, pp. 231–58. Academic Press, London.

Edmunds, M. 1974. *Defence in Animals: a Survey of Anti-predator Defences*. Longman, Harlow.

Elgar, M.A. 1986. House sparrows establish foraging flocks by giving chirrup calls if the resources are divisible. *Anim. Behav.* **34**, 169–74.

Elgar, M.A. & Catterall, C.P. 1981. Flocking and predator surveillance in house sparrows: test of an hypothesis. *Anim. Behav.* **29**, 868–72.

Elner, R.W. & Hughes, R.N. 1978. Energy maximization in the diet of the shore crab, *Carcinus maenas*. *J. Anim. Ecol.* **47**, 103–16.

Emlen, S.T. 1984. Cooperative breeding in birds and mammals. In J.R. Krebs & N.B. Davies (eds.). *Behavioural Ecology: An Evolutionary Approach*, 2nd edn., pp. 305–39. Blackwell Scientific Publications, Oxford.

Emlen, S.T., Emlen, J.M. & Levin, S.A. 1986. Sex-ratio selection in species with helpers-at-the-nest. *Amer. Natur.* **127**, 1–8.

Emlen, S.T. & Oring, L.W. 1977. Ecology, sexual selection and the evolution of mating systems. *Science, NY* **197**, 215–23.

Endler, J.A. 1980. Natural selection on colour patterns in *Poecilia reticulata*. *Evolution* **34**, 76–91.

Enquist, M. 1985. Communication during aggressive interactions with particular reference to variation in choice of behaviour. *Anim. Behav.* **33**, 1152–61.

Enquist, M., Plane, E. & Roed, J. 1985. Aggressive communication in fulmars (*Fulmaris glacialis*) competing for food. *Anim. Behav.* **33**, 1107−20.

Erckmann, W.J. 1983. The evolution of polyandry in shorebirds: an evaluation of hypotheses. In S.K. Waser (ed.). *Social Behaviour of Female Vertebrates.* Academic Press, New York.

Erichsen, J.T., Krebs, J.R. & Houston; A.I. 1980. Optimal foraging and cryptic prey. *J. Anim. Ecol.* **49**, 271−6.

Erickson, C.J. & Zenone, P.G. 1976. Courtship differences in male ring doves: avoidance of cuckoldry? *Science, NY* **192**, 1353−4.

Evans, H.E. 1977. Extrinsic and intrinsic factors in the evolution of insect sociality. *BioScience* **27**, 613−17.

Ewald, P.W. & Rohwer, S. 1982. Effects of supplemental feeding on timing of breeding, clutch size and polygamy in red winged blackbirds. *Agelaius phoeniceus. J. Anim. Ecol.* **51**, 429−50.

Feare, C. 1984. *The Starling.* Oxford University Press, Oxford.

Ferguson, M.W.J. & Joanen, T. 1982. Temperature of egg incubation determines sex in *Alligator mississippiensis. Nature* **296**, 850−3.

Fischer, E.A. 1980. The relationship between mating system and simultaneous hermaphroditism in the coral reef fish, *Hypoplectrus nigricans. Anim Behav.* **28**, 620−33.

Fisher, R.A. 1930. *The Genetical Theory of Natural Selection.* Clarendon Press, Oxford.

Foster, S.A. 1985. Group foraging by a coral reef fish: a mechanism for gaining access to defended resources. *Anim. Behav.* **33**, 782−79.

Foster, W.A. & Treherne, J.E. 1981. Evidence for the dilution effect in the selfish herd from fish predation on a marine insect. *Nature* **295**, 466−7.

Fretwell, S.D. 1972. *Populations in a Seasonal Environment.* Princeton University Press, Princeton. NJ.

Fricke, H.W. 1979. Mating system, resource defence and sex change in the anemonefish, *Amphiprion akallopisos. Z. Tierpsychol.* **50**, 313−26.

Fricke, H.W. & Fricke, S. 1977. Monogamy and sex change by aggressive dominance in coral reef fish. *Nature, Lond.* **266**, 830−2.

Frisch, K. von. 1967. *The Dance Language and Orientation of Bees.* Belknap Press, Cambridge, MA.

Galef, B.G. & Wigmore, S.W. 1983. Transfer of information concerning distant foods: a laboratory investigation of the 'information centre hypothesis'. *Anim. Behav.* **31**, 748−58.

Garson, P.J., Pleszczynska, W.K. & Holm, C.H. 1981. The polygamy threshold model: a reassessment. *Can. J. Zool.* **59**, 902−10.

Geist, V. 1966. The evolution of horn-like organs. *Behaviour* **27**, 175−213.

Geist, V. 1974. On fighting strategies in animal conflict. *Nature, Lond.* **250**, 354.

Ghiselin, M.T. 1969. The evolution of hermaphroditism among animals. *Q. Rev. Biol.* **44**, 189−208.

Gibson, R.M. & Bradbury, J.W. 1985. Sexual selection in lekking sage grouse: phenotypic correlates of male mating success. *Behav. Ecol. Sociobiol.* **18**, 117−23.

Gilbert, L.E. 1976. Postmating female odor in *Heliconius* butterflies: a male contributed antiaphrodisiac? *Science, NY* **193**, 419−20.

Gill, F.B. & Wolf, L.L. 1975. Economics of feeding territoriality in the golden-winged sunbird. *Ecology* **56**, 333−45.

Gilliam, J.F. 1982. *Foraging Under Mortality Risk in Size-structured Populations.* Ph.D. thesis, Michigan State University.

Giraldeau, L.A. & Gillis, D. 1985. Optimal group size can be stable: a reply to Sibly. *Anim. Behav.* **33**, 666−7.

Gittleman, J.L. & Harvey, P.H. 1980. Why are distasteful prey not cryptic? *Nature* **286**, 149−50.

Goss-Custard, J.D. 1970. Feeding dispersion in some overwintering wading birds. In J.H. Crook (ed.). *Social Behaviour in Birds and Mammals*, pp. 3–34. Academic Press, London.

Goss-Custard, J.D. 1976. Variation in the dispersion of redshank (*Tringa totanus*) on their winter feeding grounds. *Ibis* **118**, 257–63.

Gould, S.J. 1966. Allometry and size in ontogeny and phylogeny. *Biol. Rev.* **41**, 587–640.

Gould, S.J. 1978. *Ever Since Darwin: Reflections in Natural History*. Andre Deutsch, London.

Gould, S.J. 1980. Is a new and general theory of evolution emerging? *Paleobiology* **6**, 119–30.

Gould, S.J. & Lewontin, R.C. 1979. The spandrels of San Marco and the Panglossian paradigm: a critique of the adaptationist programme. *Proc. R. Soc. Lond. B* **205**, 581–98.

Grafen, A. 1984. Natural selection, kin selection and group selection. In J.R. Krebs & N.B. Davies (eds.). *Behavioural Ecology: An Evolutionary Approach*, 2nd edn., pp. 62–84. Blackwell Scientific Publications, Oxford.

Grafen, A. 1985. Hamilton's rule OK. *Nature* **318**, 310–11.

Grafen, A. 1986a. A geometric view of relatedness. In R. Dawkins & M. Ridley (eds.). *Oxford Surveys in Evolutionary Biology*. Vol. 2, pp. 29–89. Oxford University Press, Oxford.

Grafen, A. 1986b. Split sex ratios and the evolutionary origins of eusociality. *J. theor. Biol.* (in press).

Greenberg, L. 1979. Genetic component of kin recognition in primitively social bees. *Science, NY* **206**, 1095–7.

Greenwood, P.J. 1980. Mating systems, philopatry and dispersal in birds and mammals. *Anim. Behav.* **28**, 1140–62.

Greenwood, P.J., Harvey, P.H. & Perrins, C.M. 1978. Inbreeding and dispersal in the great tit. *Nature, Lond.* **271**, 52–4.

Grey, R. 1987. Faith and foraging: a critique of optimal foraging theory. In A.C. Kamil, J.R. Krebs & H.R. Pulliam (eds.). *Foraging Behavior*. Plenum Press, New York.

Grosberg, R.K. & Quinn, J.F. 1986. The genetic control and consequences of kin recognition by the larvae of a colonial marine invertebrate. *Nature, Lond.* **322**, 456–9.

Gross, M.R. 1982. Sneakers, satellites and parentals: polymorphic mating strategies in North American sunfishes. *Z. Tierpsychol.* **60**, 1–26.

Gross, M.R. 1985. Disruptive selection for alternative life histories in salmon. *Nature* **313**, 47–8.

Gross, M.R. & Charnov, E.L. 1980. Alternative male life histories in bluegill sunfish. *Proc. Natn. Acad. Sci. USA* **77**, 6937–40.

Gross, M.R. & Sargent, R.C. 1985. The evolution of male and female parental care in fishes. *Amer. Zool.* **25**, 807–22.

Gross, M.R. & Shine, R. 1981. Parental care and mode of fertilisation in ectothermic vertebrates. *Evolution* **35**, 775–93.

Guilford, T. 1985. Is kin selection involved in the evolution of warning coloration? *Oikos* **45**, 31–6.

Guilford, T. 1986. How do 'warning colours' work? Conspicuousness may reduce recognition errors in experienced predators. *Anim. Behav.* **34**, 286–8.

Gwynne, D.T. 1982. Mate selection by female katydids (Orthoptera Tettigoniidae. *Conocephalus nigropleurum*). *Anim. Behav.* **30**, 734–8.

Haas, V. 1985. Colonial and single breeding in fieldfares, *Turdus pilaris* L.: a comparison of nesting success in early and late broods. *Behav. Ecol. Sociobiol.* **16**, 119–24.

Hailman, J.P. 1977. *Optical Signals*. Indiana University Press, Bloomington.

Haldane, J.B.S. 1953. Animal populations and their regulation. *Penguin*

Modern Biology **15**, 9–24.

Hamilton, W.D. 1964. The genetical evolution of social behaviour. I, II. *J. theor. Biol.* **7**, 1–52.

Hamilton, W.D. 1967. Extraordinary sex ratios. *Science, NY* **156**, 477–88.

Hamilton, W.D. 1971. Geometry for the selfish herd. *J. theor. Biol.* **31**, 295–311.

Hamilton, W.D. 1972. Altruism and related phenomena, mainly in social insects. *Ann. Rev. Ecol. Syst.* **3**, 193–232.

Hamilton, W.D. 1979. Wingless and fighting males in fig wasps and other insects. In M.S. Blum & N.A. Blum (eds.). *Sexual Selection and Reproductive Competition in Insects*, pp. 167–220. Academic Press, London.

Hamilton, W.D. & Zuk, M. 1984. Heritable true fitness and bright birds: a role for parasites? *Science* **218**, 384–7.

Hammerstein, P. & Parker, G.A. 1982. The asymmetric war of attrition. *J. theor. Biol.* **96**, 647–82.

Hanken, J. & Sherman, P.W. 1981. Multiple paternity in Belding's ground squirrel litters. *Science* **212**, 351–3.

Harcourt, A.H., Harvey, P.H., Larson, S.G. & Short, R.V. 1981. Testis weight, body weight and breeding system in primates. *Nature* **293**, 55–7.

Harper, D.G.C. 1982. Competitive foraging in mallards: 'ideal free' ducks. *Anim. Behav.* **30**, 575–84.

Harvey, P.H. 1985. Intrademic group selection and the sex ratio. In R.M. Sibly & R.H. Smith (eds.). *Behavioural Ecology: Ecological Consequences of Adaptive Behaviour*, pp. 59–73. 25th Symposium of the British Ecological Society. Blackwell Scientific Publications, Oxford.

Harvey, P.H., Bull, J.J. & Paxton, R.J. 1983. Looks pretty nasty. *New Sci.* **97**, 26–7.

Harvey, P.H., Kavanagh, M. & Clutton-Brock, T.H. 1978. Sexual dimorphism in primate teeth. *J. Zool. Lond.* **186**, 475–86.

Harvey, P.H. & Mace, G.M. 1981. Comparison between taxa and adaptive trends: problems in methodology. In B.C.R. Bertram *et al.* (eds.). *Current Problems in Sociobiology*. Cambridge University Press, Cambridge.

Hassell, M.P. 1971. Mutual interference between searching insect parasites. *J. Anim. Ecol.* **40**, 473–86.

Heinrich, B. 1979. Foraging strategies of caterpillars: leaf damage and possible predator avoidance strategies. *Oecologia (Berlin)* **42**, 325–37.

Heinrich, B. & Collins, S.L. 1983. Caterpillar leaf damage, and the game of hide-and-seek with birds. *Ecology* **64**, 592–602.

Heisler, I.L. 1981. Offspring quality and the polygyny threshold: a new model for the 'sexy son' hypothesis. *Amer. Natur.* **117**, 316–28.

Heller, R. & Milinski, M. 1979. Optimal foraging of sticklebacks on swarming prey. *Anim. Behav.* **27**, 1127–41.

Herrnstein, R.J. & Vaughan, W., Jr. 1980. Melioration and behavioural allocation. In J.E.R. Staddon (ed.). *Limits to Action: The Allocation of Individual Behaviour*. Academic Press, New York.

Hinde, R.A. 1956. The biological significance of the territories of birds. *Ibis* **98**, 340–69.

Hinde, R.A. 1970. *Animal Behaviour: A Synthesis of Ethology and Comparative Psychology*, 2nd edn. McGraw-Hill, New York.

Hinde, R.A. 1974. *Biological Bases of Human Social Behaviour*. McGraw-Hill, New York.

Hinde, R.A. 1981. Animal signals: ethological and games-theory approaches are not incompatible. *Anim. Behav.* **29**, 535–42.

Hinson, J.M. & Staddon, J.E.R. 1983. Hill climbing by pigeons. *J. exp. analysis Behav.* **39**, 25–47.

Hixon, M.A. 1982. Energy maximizers and time minimizers: theory and reality. *Amer. Natur.* **119**, 596–9.

Hixon, M.A., Carpenter, F.L. & Paton, D.C. 1983. Territory area, flower density and time budgeting in hummingbirds: an experimental and theoretical analysis. *Amer. Natur.* **122**, 366−91.

Hodges, C.M. & Wolf, L.L. 1981. Optimal foraging bumblebees? why is nectar left behind in flowers? *Behav. Ecol. Sociobiol.* **9**, 41−4.

Högstedt, G. 1980. Evolution of clutch size in birds: adaptive variation in relation to territory quality. *Science* **210**, 1148−50.

Hölldobler, B. 1971. Communication between ants and their guests. *Sci. Am.* **224** (3), 86−93.

Hölldobler, B. 1977. Communication in social Hymenoptera. In T.A. Seboek (ed.). *How Animals Communicate*, pp. 418−71. Indiana University Press, Bloomington & London.

Holm, C.H. 1973. Breeding sex ratios, territoriality and reproductive success in the red-winged blackbird (*Agelaius phoeniceus*). *Ecology* **54**, 356−65.

Holmes, W.G. & Sherman, P.W. 1982. The ontogeny of kin recognition in two species of ground squirrels. *Amer. Zool.* **22**, 491−517.

Hoogland, J.L. 1979a. The effect of colony size on individual alertness of prairie dogs (Sciuridae: *Cynomys* spp.) *Anim. Behav.* **27**, 394−407.

Hoogland, J.L. 1979b. Aggression, ectoparasitism and other possible costs of prairie dog (Sciuridae: *Cynomys* spp.) coloniality. *Behaviour* **69**, 1−35.

Hoogland, J.L. 1983. Nepotism and alarm calling in the black-tailed prairie dog, *Cynomys ludovicianus*. *Anim. Behav.* **31**, 472−9.

Horn, H.S. & Rubenstein, D.I. 1984. Behavioural adaptations and life history. In J.R. Krebs & N.B. Davies (eds.). *Behavioural Ecology: An Evolutionary Approach*, 2nd edn., pp. 279−98. Blackwell Scientific Publications, Oxford.

Houston, A. & McNamara, J. 1985. The choice of two prey types that minimises the probability of starvation. *Behav. Ecol. Sociobiol.* **17**, 135−41.

Howard, R.D. 1978a. Factors influencing early embryo mortality in bullfrogs. *Ecology* **59**, 789−98.

Howard, R.D. 1978b. The evolution of mating strategies in bullfrogs, *Rana catesbiana*. *Evolution* **32**, 850−71.

Hrdy, S.B. 1977. *The Langurs of Abu: Female and Male Strategies of Reproduction*. Harvard University Press, Cambridge, Mass.

Hunter, M.L. & Krebs, J.R. 1979. Geographical variation in the song of the great tit (*Parus major*) in relation to ecological factors. *J. Anim. Ecol.* **48**, 759−85.

Huxley, J.S. 1966. A discussion of ritualisation of behaviour in animals and man: introduction. *Phil Trans. R. Soc. B* **251**, 247−71.

Hyatt, G.W. & Salmon, M. 1978. Combat in the fiddler crabs *Uca pugilator* and *U. pugnax*: a quantitative analysis. *Behaviour* **65**, 182−211.

Jarman, P.J. 1974. The social organisation of antelope in relation to their ecology. *Behaviour* **48**, 215−67.

Jarvi, T., Sillén-Tullberg, B. & Wicklund, C. 1981. The cost of being aposematic: an experimental study of predation of larvae of *Papilio machaon* by the great tit, *Parus major*. *Oikos* **36**, 267−72.

Jarvis, J.U.M. 1981. Eusociality in a mammal: cooperative breeding in naked mole rat colonies. *Science* **212**, 571−3.

Jarvis, M.J.F. 1974. The ecological significance of clutch size in the South African gannet (*Sula Capensis*, Lichenstein). *J. Anim. Ecol.* **43**, 1−17.

Jenni, D.A. 1974. Evolution of polyandry in birds. *Amer. Zool.* **14**, 129−44.

Jennings, T. & Evans, S.M. 1980. Influence of position in the flock and flock size on vigilance in the starling, *Sturnus vulgaris*. *Anim. Behav.* **30**, 634−5.

Kacelnik, A. 1984. Central place foraging in Starlings (*Sturnus vulgaris*). I. Patch residence time. *J. Anim. Ecol.* **53**, 283−99.

Kacelnik, A. & Cuthill, I.C. 1987. Starlings and optimal foraging: modelling in fractal world. In A.C. Kamil, J.R. Krebs & H.R. Pulliam (eds.). *Foraging Behavior*. Plenum Press, New York.

Kamil, A.C., Krebs, J.R. & Pulliam, H.R. (eds.). 1987. *Foraging Behavior*. Plenum Press, New York.

Kenward, R.E. 1978. Hawks and doves: factors affecting success and selection in goshawk attacks on wood-pigeons. *J. Anim. Ecol.* **47**, 449−60.

Killeen, P. 1984. Incentive theory III: Adaptive clocks. In J. Gibbon & L. Allan (eds.). *Timing and Time Perception. Ann. NY Acad. Sci.* **423**, 515−27.

Kirkpatrick, M. 1985. Evolution of female choice and male parental investment in polygynous species: the demise of the 'sexy son'. *Amer. Natur.* **125**, 788−810.

Kleiman, D.G. 1977. Monogamy in mammals. *Q. Rev. Biol.* **52**, 39−69.

Knowlton, N. 1979. Reproductive synchrony, parental investment and the evolutionary dynamics of sexual selection. *Anim. Behav.* **27**, 1022−33.

Kodric-Brown, A. 1985. Female preference and sexual selection for male colouration in the guppy *Poecilia reticulata. Behav. Ecol. Sociobiol.* **17**, 199−205.

Kodric-Brown, A. & Brown, J.H. 1984. Truth in advertising: the kinds of traits favoured by sexual selection. *Amer. Natur.* **124**, 309−23.

Koenig, W.D., Mumme, R.L. & Pitelka, F.A. 1984. The breeding system of the acorn woodpecker in central coastal California. *Z. Tierpsychol.* **65**, 289−308.

Kramer, D.L. 1985. Are colonies supraoptimal groups? *Anim. Behav.* **33**, 1031.

Krebs, J.R. 1971. Territory and breeding density in the great tit, *Parus major* L. *Ecology* **52**, 2−22.

Krebs, J.R. 1982. Territorial defence in the great tit, *Parus major*: do residents always win? *Behav. Ecol. Sociobiol.* **11**, 185−94.

Krebs, J.R., Ashcroft, R. & Webber, M. 1978. Song repertoires and territory defence in the great tit, *Parus major. Nature, Lond.* **271**, 539−42.

Krebs, J.R. & Dawkins, R. 1984. Animal signals: mind reading and manipulation. In J.R. Krebs & N.B. Davies (eds.). *Behavioural Ecology: An Evolutionary Approach*, 2nd edn., pp. 380−402. Blackwell Scientific Publications, Oxford.

Krebs, J.R., Erichsen, J.T., Webber, M.I. & Charnov, E.L. 1977. Optimal prey selection in the great tit, *Parus major. Anim. Behav.* **25**, 30−8.

Krebs, J.R., MacRoberts, M.H. & Cullen, J.M. 1972. Flocking and feeding in the great tit *Parus major*: an experimental study. *Ibis* **114**, 507−30.

Kroodsma, D.E. 1976. Reproductive development in a female song-bird: differential stimulation by quality of male song. *Science, NY* **192**, 574−5.

Kruuk, H. 1964. Predators and anti-predator behaviour of the black headed gull, *Larus ridibundus. Behaviour Suppl.* **11**, 1−129.

Kruuk, H. 1972. *The Spotted Hyena*. University of Chicago Press, Chicago.

Lack, D. 1966. *Population Studies of Birds*. Clarendon Press, Oxford.

Lack, D. 1968. *Ecological Adaptations for Breeding in Birds*. Methuen, London.

Lacy, R.C. 1980. The evolution of eusociality in termites: a haplodiploid analogy. *Amer. Natur.* **116**, 449−51.

Lande, R. 1981. Models of speciation by sexual selection of polygenic traits. *Proc. Natn. Acad. Sci. USA* **78**, 3721−5.

Le Boeuf, B.J. 1972. Sexual behaviour in the northern elephant seal, *Mirounga angustirostris. Behaviour* **41**, 1−26.

Le Boeuf, B.J. 1974. Male−male competition and reproductive success in elephant seals. *Amer. Zool.* **14**, 163−76.

Le Boeuf, B.J. & Reiter, J. (in press). Lifetime reproductive success in Northern elephant seals. In T.H. Clutton-Brock (ed.). *Reproductive Success*. Chicago University Press, Chicago.

Leuthold, W. 1966. Variations in territorial behaviour of Uganda Kob, *Adenota kob thomasi*. *Behaviour* **27**, 215–58.

Lill, A. 1974. Sexual behaviour of the lek-forming white-bearded manakin (*Manacus manacus trinitatis*). *Z. Tierpsychol.* **36**, 1–36.

Lima, S. 1984. Downy woodpecker foraging behavior: efficient sampling in simple stochastic environments. *Ecology* **65**, 166–74.

Lima, S., Valone, T.J. & Caraco, T. 1985. Foraging-efficiency–predation-risk tradeoff in the grey squirrel. *Anim. Behav.* **33**, 155–65.

Lloyd, M. & Dybas, H.S. 1966. The periodical cicada problem: II Evolution. *Evolution* **20**, 466–505.

Lorenz, K. 1966. *On Aggression*. Methuen, London.

Lyon, B.E. & Montgomerie, R.D. 1985. Conspicuous plumage of birds: sexual selection or unprofitable prey? *Anim. Behav.* **33**, 1038–40.

McCann, T.S. 1981. Aggression and sexual activity of male southern elephant seals. *J. Zool. Lond.* **195**, 295–310.

McClintock, M.K. 1971. Menstrual synchrony and suppression. *Nature* **229**, 244–5.

McKinney, F., Derrickson, S.R. & Mineau, P. 1983. Forced copulation in waterfowl. *Behaviour* **86**, 250–94.

Mackinnon, J. 1974. The ecology and behaviour of wild orangutans, *Pongo pygmaeus*. *Anim. Behav.* **22**, 3–74.

MacNally, R. & Young, D. 1981. Song energetics of the bladder cicada *Cystosoma saundersii*. *J. Exp. Biol.* **90**, 185–96.

McPhail, J.D. 1969. Predation and the evolution of a stickleback, *Gasterosteus*. *J. Fish. Res. Bd. Canada*. **26**, 3183–208.

Majerus, M.E.N., O'Donald, P. & Weir, J. 1982. Female mating preference is genetic. *Nature* **300**, 521–23.

Major, P.F. 1978. Predator–prey interactions in two schooling fishes, *Caranx ignobilis* and *Stolephorus purpureus*. *Anim. Behav.* **26**, 760–77.

Manning, A. 1961. Effects of artificial selection for mating speed in *Drosophila melanogaster*. *Anim. Behav.* **9**, 82–92.

Marler, P. 1955. Characteristics of some animal calls. *Nature, Lond.* **176**, 6–8.

Marten, K. & Marler, P. 1977a. Sound transmission and its significance for animal vocalization. I. Temperate habitats. *Behav. Ecol. Sociobiol.* **2**, 271–90.

Marten, K., Quine, D. & Marler, P. 1977b. Sound transmission and its significance for animal vocalization. II. Tropical forest habitats. *Behav. Ecol. Sociobiol.* **2**, 291–302.

Maxson, S.J. & Oring, L.W. 1980. Breeding season time and energy budgets of the polyandrous spotted sandpiper. *Behaviour* **74**, 200–63.

May, R.M. 1979. When to be incestuous. *Nature, Lond.* **279**, 192–4.

May, R.M. & Robinson, S.K. 1985. Population dynamics of avian brood parasitism. *Amer. Natur.* **126**, 475–94.

Maynard Smith, J. 1964. Group selection and kin selection. *Nature, Lond.* **201**, 1145–7.

Maynard Smith, J. 1974. The theory of games and the evolution of animal conflicts. *J. theor. Biol.* **47**, 209–21.

Maynard Smith, J. 1976a. Group selection. *Q. Rev. Biol.* **51**, 277–83.

Maynard Smith, J. 1976b. Evolution and the theory of games. *Amer. Sci.* **64**, 41–5.

Maynard Smith, J. 1977a. Parental investment — a prospective analysis. *Anim. Behav.* **25**, 1–9.

Maynard Smith, J. 1977b. The limitations of evolutionary theory. In R. Duncan & M. Weston-Smith (eds.). *The Encyclopaedia of Ignorance:*

Life Sciences and Earth Sciences, pp. 235–42. Pergamon Press, Oxford.

Maynard Smith, J. 1978. Optimization theory in evolution. *Ann. Rev. Ecol. Syst.* **9**, 31–56.

Maynard Smith, J. 1980. A new theory of sexual investment. *Behav. Ecol. Sociobiol.* **7**, 247–51.

Maynard Smith, J. 1982. *Evolution and the Theory of Games*. Cambridge University Press, Cambridge.

Maynard Smith, J. 1984. The ecology of sex. In J.R. Krebs & N.B. Davies (eds.). *Behavioural Ecology: An Evolutionary Approach*, 2nd edn., pp. 201–21. Blackwell Scientific Publications, Oxford.

Maynard Smith, J. & Ridpath, M.G. 1972. Wife sharing in the Tasmanian native hen *Tribonyx mortierrii*: a case of kin selection? *Amer. Natur.* **106**, 447–52.

Metcalf, R.A. 1980. Sex ratios, parent offspring conflict, and local competition for mates in the social wasps *Polistes metricus* and *Polistes variatus*. *Amer. Natur.* **116**, 642–54.

Metcalf, R.A. & Whitt, G.S. 1977. Relative inclusive fitness in the social wasp *Polistes metricus*. *Behav. Ecol. Sociobiol.* **2**, 353–60.

Michener, C.D. 1974. *The Social Behavior of the Bees*. Belknap Press, Harvard.

Milinski, M. 1979. An evolutionarily stable feeding strategy in sticklebacks. *Z. Tierpsychol.* **51**, 36–40.

Milinski, M. 1984a. Competitive resource sharing: an experimental test of a learning rule of ESSs. *Anim. Behav.* **32**, 233–42.

Milinski, M. 1984b. A predator's cost of overcoming the confusion effect of swarming prey. *Anim. Behav.* **32**, 1157–62.

Milinski, M. & Heller, R. 1978. Influence of a predator on the optimal foraging behaviour of sticklebacks (*Gasterosteus aculeatus*). *Nature, Lond.* **275**, 642–4.

Mock, D.W. 1984. Siblicidal aggression and resource monopolisation in birds. *Science* **225**, 731–3.

Mock, D.W. 1985. Siblicidal brood reduction: the prey-size hypothesis. *Amer. Natur.* **125**, 327–343.

Modell, W. 1969. Horns and antlers. *Sci. Am.* **220** (4), 114–122.

Moehlman, P.D. 1979. Jackal helpers and pup survival. *Nature, Lond.* **277**, 382–3.

Moodie, G.E.E. 1972. Predation, natural selection and adaptation in an unusual three spined stickleback. *Heredity* **28**, 155–67.

Morris, D. 1957. 'Typical intensity' and its relation to the problem of ritualisation. *Behaviour* **11**, 1–12.

Morse, D.H. 1970. Ecological aspects of some mixed species foraging flocks of birds. *Ecol. Monogr.* **40**, 119–68.

Morton, E.S. 1975. Ecological sources of selection on avian sounds. *Amer. Natur.* **109**, 17–34.

Myers, J.P., Connors, P.G. & Pitelka, F.A. 1979. Territory size in wintering sanderlings: the effects of prey abundance and intruder density. *Auk* **96**, 551–61.

Myers, J.P., Connors, P.G. & Pitelka, F.A. 1981. Optimal territory size and the sanderling: compromises in a variable environment. In A.C. Kamil & T.D. Sargent (eds.). *Foraging Behavior: Ecological, Ethological and Psychological Approaches*. Garland STPM Press, New York.

Nakatsuru, K. & Kramer, D.L. 1982. Is sperm cheap? Limited male fertility and female choice in the lemon tetra (Pisces; Characidae). *Science* **216**, 753–5.

Neill, S.R. St. J. & Cullen, J.M. 1974. Experiments on whether schooling by their prey affects the hunting behaviour of cephalods and fish predators. *J. Zool. Lond.* **172**. 549–69.

Nelson, J.B. 1964. Factors influencing clutch-size and chick growth in the North Atlantic gannet, *Suia bassana*. *Ibis* **106**, 63–77.

Neukirch, A. 1982. Dependence of the lifespan of the honeybee (*Apis mellifera*) upon flight performance and energy consumption. *J. Comp. Physiol.* **146**, 35−40.

Nisbet, I.C.T. 1977. Courtship feeding and clutch size in common terns *Sterna hirundo*. In B. Stonehouse & C.M. Perrins (eds.). *Evolutionary Ecology*, pp. 101−9. London, Macmillan.

Nottebohm, F. 1975. Continental patterns of song variability in *Zontrichia capensis*: some possible ecological correlates. *Amer. Natur.* **109**, 605−24.

Nur, N. 1984a. The consequences of brood size for breeding blue tits. I. Adult survival, weight change and the cost of reproduction. *J. Anim. Ecol.* **53**, 479−96.

Nur, N. 1984b. The consequences of brood size for breeding blue tits. II. Nesting weight, offspring survival and optimal brood size. *J. Anim. Ecol.* **53**, 497−518.

Nur, N. & Hasson, O. 1984. Phenotypic plasticity and the handicap principle. *J. theor. Biol.* **110**, 275−97.

Orians, G.H. 1969. On the evolution of mating systems in birds and mammals. *Amer. Natur.* **103**, 589−603.

Oring, L.W. 1981. Avian mating systems. In D.S. Farner & J.R. King (eds.). *Avian Biology*, Vol. 6. Academic Press, London.

Oster, G.F. & Wilson, E.O. 1978. *Caste and Ecology in the Social Insects.* Princeton University Press, Princeton, NJ.

Owen, R.E. & Plowright, R.C. 1982. Worker−queen conflict and male parentage in bumblebees. *Behav. Ecol. Sociobiol.* **11**, 91−9.

Owen-Smith, N. 1977. On territoriality in ungulates and an evolutionary model. *Q. Rev. Biol.* **52**, 1−38.

Packer, C. 1977. Reciprocal altruism in *Papio anubis*. *Nature, Lond.* **265**, 441−3.

Packer, C. 1979. Inter-troop transfer and inbreeding avoidance in *Papio anubis*. *Anim. Behav.* **27**, 1−36.

Packer, C. & Pusey, A.E. 1982. Cooperation and competition within coalitions of male lions: kin selection or game theory? *Nature* **296**, 740−2.

Packer, C. & Pusey, A.E. 1983a. Male takeovers and female reproductive parameters: a simulation of oestrus synchrony in lions (*Panthera leo*). *Anim. Behav.* **31**, 334−40.

Packer, C. & Pusey, A.E. 1983b. Adaptations of female lions to infanticide by incoming males. *Amer. Natur.* **121**, 716−28.

Page, G. & Whiteacre, D.F. 1975. Raptor predation on wintering shorebirds. *Condor* **77**, 73−83.

Parker, G.A. 1970. The reproductive behaviour and the nature of sexual selection in *Scatophaga stercoraria* L. (Diptera: Scatophagidae). II. The fertilization rate and the spatial and temporal relationships of each sex around the site of mating and oviposition. *J. Anim. Ecol.* **39**, 205−28.

Parker, G.A. 1974. Assessment strategy and the evolution of animal conflicts. *J. theor. Biol.* **47**, 223−43.

Parker, G.A. 1978. Searching for mates. In J.R. Krebs & N.B. Davies (eds.). *Behavioural Ecology: An Evolutionary Approach*, 1st edn., pp. 214−44. Blackwell Scientific Publications, Oxford.

Parker, G.A. 1979. Sexual selection and sexual conflict. In M.S. Blum & N.A. Blum (eds.). *Sexual Selection and Reproductive Competition in Insects*, pp. 123−66. Academic Press, New York.

Parker, G.A., Baker, R.R. & Smith, V.C.F. 1972. The origin and evolution of gamete dimorphism and the male−female phenomenon. *J. theor. Biol.* **36**, 529−53.

Parker, G.A. & Knowlton, N. 1980. The evolution of territory size. Some ESS models. *J. theor. Biol.* **84**, 445−76.

Parker, G.A. & Rubenstein, D.I. 1981. Role assessment, reserve strategy and the acquisition of information in asymmetric animal contests. *Anim. Behav.* **29**, 221−40.

Partridge, B.L. & Pitcher, T.J. 1979. Evidence against a hydrodynamic function for fish schools. *Nature* **279**, 418−19.

Partridge, L. 1980. Mate choice increases a component of offspring fitness in fruit flies. *Nature, Lond.* **283**, 290−1.

Partridge, L. 1983. Genetics and behaviour. In T.R. Halliday & P.J.B. Slater (eds.). *Animal Behaviour: Genes, Development and Learning*, pp. 11−51. Blackwell Scientific Publications, Oxford.

Partridge, L. & Halliday, T. 1984. Mating patterns and mate choice. In J.R. Krebs & N.B. Davies (eds.). *Behavioural Ecology: An Evolutionary Approach*, 2nd edn., pp. 222−50. Blackwell Scientific Publications, Oxford.

Paton, D. 1986. Communication by agonistic displays: II. Perceived information and the definition of agonistic displays. *Behaviour* (in press).

Paton, D. & Caryl, P.G. 1986. Communication by agonistic displays: I. Variation in information content between samples. *Behaviour* (in press).

Payne, R.B. 1977. The ecology of brood parasitism in birds. *Ann. Rev. Ecol. Syst.* **8**, 1−28.

Patterson, I.J. 1965. Timing and spacing of broods in the black-headed gull, *Larus ridibundus. Ibis* **107**, 433−59.

Perrill, S.A., Gerhardt, H.C. & Daniel, R. 1978. Sexual parasitism in the green treefrog, *Hyla cinerea. Science, NY* **200**, 1179−80.

Perrill, S.A., Gerhardt, H.C. & Daniel, R. 1982. Mating strategy shifts in male green treefrogs, *Hyla cinerea*: an experimental study. *Anim. Behav.* **30**, 43−8.

Perrins, C.M. 1965. Population fluctuations and clutch size in the great tit, *Parus major. L. J. Anim. Ecol.* **34**, 601−47.

Perrins, C.M. 1979. *British Tits.* New Naturalist Series. Collins, London.

Petrie, M. 1983. Female moorhens compete for small fat males. *Science* **220**, 413−15.

Pianka, E.R. & Parker, W.S. 1975. Age-specific reproductive tactics. *Amer. Natur.* **109**, 453−64.

Pietrewicz, A.T. & Kamil, A.C. 1981. Search images and the detection of cryptic prey: an operant approach. In A.C. Kamil & T.D. Sargent (eds.). *Foraging Behavior: Ecological, Ethological and Psychological Approaches*, pp. 311−32. Garland STPM press, New York.

Pitcher, T.J., Magurran, A.E. & Winfield, I.J. 1982. Fish in larger shoals find food faster. *Behav. Ecol. Sociobiol.* **10**, 149−51.

Pitelka, F.A., Holmes, R.T. & MacLean, S.F. 1974. Ecology and evolution of social organisation in arctic sandpipers. *Amer. Zool.* **14**, 185−204.

Pleszczynska, W.K. 1978. Microgeographic prediction of polygyny in the lark bunting. *Science, NY* **201**, 935−7.

Pleszczynska, W.K. & Hansell, R.I.C. 1980. Polygyny and decision theory: testing of a model in lark buntings. *Calamospiza melanocorys. Amer. Natur.* **116**, 821−30.

Polwright, R.C., Fuller, G.A. & Paloheimo, J.E. (in press). Shell-dropping by Northwestern crows: a re-examination of an optimal foraging study. *Aim. Behav.*

Prins, H.H.Th., Ydenberg, R.C. & Drent, R.H. 1980. The interaction of brent geese *Branta bernicla* and sea plantain *Plantago maritima* during spring staging: field observations and experiments. *Acta bot. neerl.* **29**, 585−96.

Pulliam, H.R. 1976. The principle of optimal behavior and the theory of communities. In P.H. Klopfer & P.P.G. Bateson (eds.). *Perspectives in Ethology*, pp. 311−32. Plenum Press, New York.

Pulliam, H.R. & Caraco, T. 1984. Living in groups: is there an optimal group size? In J.R. Krebs & N.B. Davies (eds.). *Behavioural Ecology: An*

Evolutionary Approach, 2nd edn., pp. 122–47. Blackwell Scientific Publications, Oxford.

Pulliam, H.R., Pyke, G.H. & Caraco, T. 1982. The scanning behavior of juncos: a game-theoretical approach. *J. theor. Biol.* **95**, 89–103.

Pyke, G.H. 1979. The economics of territory size and time budget in the golden-winged sunbird. *Amer. Natur.* **114**, 131–45.

Rabenold, K.N. 1984. Cooperative enhancement of reproductive success in tropical wren societies. *Ecology* **65**, 871–85.

Ralls, K., Brugger, K. & Ballou, J. 1979. Inbreeding and juvenile mortality in small populations of ungulates. *Science, NY* **206**, 1101–3.

Reed, T.R. 1982. Interspecific territoriality in the chaffinch and great tit on islands and the mainland of Scotland: playbacks and removal experiments. *Anim. Behav.* **30**, 171–81.

Reyer, H.-U. 1980. Flexible helper structure as an ecological adaptation in the pied kingfisher, *Ceryle rudis rudis*. L. *Behav. Ecol. Sociobiol.* **6**, 219–27.

Reyer, H.-U. 1984. Investment and relatedness: a cost/benefit analysis of breeding and helping in the pied kingfisher (*Ceryle rudis*). *Anim. Behav.* **32**, 1163–78.

Reyer, H.-U. & Westerterp, K. 1985. Parental energy expenditure as a proximate cause of helper recruitment in the pied kingfisher (*Ceryle rudis*). *Behav. Ecol. Sociobiol.* **17**, 363–70.

Ridley, M. 1978. Paternal care. *Anim. Behav.* **26**, 904–32.

Ridley, M. 1983. *The Explanation of Organic Diversity*. Clarendon Press, Oxford.

Riechert, S.E. 1978. Games spiders play. I. Behavioural variability in territorial disputes. *Behav. Ecol. Sociobiol.* **3**, 135–62.

Riechert, S.E. 1979. Games spiders play. II. Resource assessment strategies. *Behav. Ecol. Sociobiol.* **6**, 121–8.

Rohwer, S. & Rohwer, F.C. 1978. Status signalling in Harris sparrows: experimental deceptions achieved. *Anim. Behav.* **26**, 1012–22.

Rood, J.P. 1978. Dwarf mongoose helpers at the den. *Z. Tierpsychol.* **48**, 277–87.

Røskaft, E., Jarvi, T., Bakken, M., Bech, C. & Reinertsen, R.E. 1986. The relationship between social status and resting metabolic rate in great tits (*Parus major*) and pied flycatchers (*Ficedula hypoteuca*). *Anim. Behav.* **34**, 838–42.

Rothstein, S.I. 1975a. Evolutionary rates and host defences against avian brood parasitism. *Amer. Natur.* **109**, 161–76.

Rothstein, S.I. 1975b. An experimental and teleonomic investigation of avian brood parasitism. *Condor* **77**, 250–71.

Rothstein, S.I. 1979. Gene frequencies and selection for inhibitory traits, with special emphasis on the adaptiveness of territorality. *Amer. Natur.* **113**, 317–31.

Rothstein, S.I. 1982. Mechanisms of avian egg recognition: which egg parameters elicit responses by rejector species? *Behav. Ecol. Sociobiol.* **11**, 229–39.

Rothstein, S.I. 1986. A test of optimality: egg recognition in the eastern phoebe. *Anim. Behav.* **34**, 1109–19.

Royama, T. 1970. Factors governing the hunting behaviour and selection of food by the great tit, *Parus major*. *J. Anim. Ecol.* **39**, 619–68.

Rubenstein, D.I. 1978. On predation, competition and the advantages of group living. In P.P.G. Bateson & P.H. Klopfer (eds.). *Perspectives in Ethology*, pp. 205–31. Plenum Press, New York.

Rubenstein, D.I. 1980. On the evolution of alternative mating strategies. In J.E.R. Staddon (ed.). *Limits to Action: The Allocation of Individual Behaviour*. Academic Press, New York.

Rutberg, A.T. 1983. The evolution of monogamy in primates. *J. theor. Biol.* **104**, 93–112.

Ryan, M.J., Tuttle, M.D. & Rand, A.S. 1982. Bat predation and sexual advertisement in a neotropical anuran. *Amer. Natur.* **119**, 136−9.

Ryan, M.J., Tuttle, M.D. & Taft, L.K. 1981. The costs and benefits of frog chorusing behavior. *Behav. Ecol. Sociobiol.* **8**, 273−8.

Sargent, T.D. 1981. Antipredator adaptations of underwing moths. In A.C. Kamil & T.D. Sargent (eds.). *Foraging Behavior: Ecological, Ethological and Psychological Approaches*, pp. 259−84. Garland STPM Press, New York.

Schlenoff, D.H. 1985. The startle responses of blue jays to *Catocala* (Lepidoptera: Noctuidae) prey models. *Anim. Behav.* **33**, 1057−67.

Schmid-Hempel, P. 1986. Do honeybees get tired? The effect of load weight on patch depature. *Anim. Behav.* **34**, 1243−50.

Schmid-Hempel, P., Kacelnik, A. & Houston, A.I. 1985. Honeybees Maximise efficiency by not filling their crop. *Behav. Ecol. Sociobiol.* **17**, 61−6.

Schoener, T.W. 1983. Simple models of optimal feeding-territory size: a reconciliation. *Amer. Natur.* **121**, 608−29.

Seger, J. 1983. Partial bivoltinism may cause alternating sex-ratio biasses that favour eusociality. *Nature* **301**, 59−62.

Seghers, B.H. 1974. Schooling behaviour in the guppy *Poecilia reticulata*: an evolutionary response to predation. *Evolution* **28**, 486−9.

Selander, R.K. 1972. Sexual selection and dimorphism in birds. In B. Campbell (ed.). *Sexual Selection and the Descent of Man*, pp. 180−230. Aldine, Chicago.

Semler, D.E. 1971. Some aspects of adaptation in a polymorphism for breeding colours in the three-spine stickleback (*Gasterosteus aculeatus* L.). *J. Zool. Lond.* **165**, 291−302.

Seyfarth, R.M. & Cheney, D.L. 1984. Grooming, alliances and reciprocal altruism in vervet monkeys. *Nature* **308**, 541−3.

Seyfarth, R.M., Cheney, D.L. & Marler, P. 1980. Vervet monkey alarm calls: evidence of predator classification and semantic communication. *Science* **210**, 801−3.

Shalter, M.D. 1978. Localisation of passerine seet and mobbing calls by goshawks and pygmy owls. *Z. Tierpsychol.* **46**, 260−7.

Shepher, J. 1971. Mate selection among second generation kibbutz adolescents and adults: incest avoidance and negative imprinting. *Arch. sex. Behav.* **1**, 293−307.

Sherman, P.W. 1977. Nepotism and the evolution of alarm calls. *Science, NY* **197**, 1246−53.

Sherman, P.W. 1981a. Reproductive competition and infanticide in Belding's ground squirrels and other animals. In R.D. Alexander & D.W. Tinkle (eds.). *Natural Selection and Social Behaviour: Recent Research and New Theory*, pp. 311−31. Chiron Press, New York.

Sherman, P.W. 1981b. Kinship, demography and Belding's ground squirrel nepotism. *Behav. Ecol. Sociobiol.* **8**, 251−9.

Shields, W.M. 1980. Ground squirrel alarm calls: nepotism or parental care? *Amer. Natur.* **116**, 599−603.

Sibly, R.M. 1983. Optimal group size is unstable. *Anim. Behav.* **31**, 947−8.

Sigurjonsdottir, H. & Parker, G.A. 1981. Dung fly struggles: evidence for assessment strategy. *Behav. Ecol. Sociobiol.* **8**, 219−30.

Silverin, B. 1980. Effects of long-acting testosterone treatment on free-living pied flycatchers *Fidecula hypoleuca*. *Anim. Behav.* **28**, 906−12.

Silverman, H.B. & Dunbar, M.J. 1980. Aggressive tusk use by the narwhal, *Monodon monoceros* L. *Nature, Lond.* **284**, 57−8.

Simon, C. 1979. Debut of the seventeen-year-old cicada. *Nat. Hist.* **88** (5), 38−45.

Simpson, M.J.A. 1968. The display of Siamese fighting fish, *Betta splendens*. *Anim. Behav. Monogr.* **1**, 1−73.

Sinclair, A.R.E. 1977. *The African Buffalo.* University of Chicago Press, Chicago.

Slatkin, M. & Maynard Smith, J. 1979. Models of coevolution. *Q. Rev. Biol.* **54**, 233−63.

Smith, J.N.M., Yom-Tov, Y. & Moses, R. 1982. Polygyny, male parental care and sex ratios in song sparrows: an experimental study. *Auk* **99**, 555−64.

Smith, R.L. 1979. Repeated copulation and sperm precedence: paternity assurance for a male brooding water bug. *Science, NY* **205**, 1029−31.

Smith, S.M. 1977. Coral snake pattern rejection and stimulus generalisation by naïve great kiskadees (Aves: Tyrannidae). *Nature* **265**, 535−6.

Snyderman, M. 1983. Optimal prey selection: the effects of deprivation. *Behav. Anal. Letters* **3**, 359−69.

Southwood, T.R.E. 1981. Bionomic strategies and population parameters. In R.M. May (ed.). *Theoretical Ecology,* 2nd edn, pp. 30−52. Blackwell Scientific Publications, Oxford.

Stacey, P.B. & Koenig, W.D. 1984. Cooperative breeding in the acorn woodpecker. *Sci. Am.* **251** (2), 100−7.

Stephens, D.W. & Krebs J.R. 1986. *Foraging Theory.* Princeton University Press, Princeton, NJ.

Stern, D. 1977. *The First Relationship: Infant and Mother.* Fontana Open Books, London.

Stubblefield, J.W. & Charnov, E.L. 1986. A Note on The Origin of Eusociality. In preparation.

Sutherland, W.J. 1985. Chance can produce a sex difference in variance of mating success and account for Bateman's data. *Anim. Behav.* **33**, 1349−52.

Sutherland, W.J. & Parker, G.A. 1985. Distribution of unequal competitors. In R.M. Sibly & R.H. Smith (eds.). *Behavioural Ecology: Ecological Consequences of Adaptive Behaviour.* Blackwell Scientific Publications, Oxford.

Syren, R.M. & Luykx, P. 1977. Permanent segmental interchange complex in the termite *Incisitermes schwarzi. Nature* **266**, 167−8.

Taborsky, M. 1984. Broodcare helpers in the cichlid fish *Lamprologus brichardi:* their costs and benefits. *Anim. Behav.* **32**, 1236−1352.

Taborsky, M. & Limberger, D. 1981. Helpers in fish. *Behav. Ecol. Sociobiol.* **8**, 143−5.

Thornhill, R. 1976. Sexual selection and nuptial feeding behaviour in *Bittacus apicalis* (Insecta: Mecoptera). *Amer. Natur.* **110**, 529−48.

Thornhill, R. 1979. Adaptive female-mimicking behaviour in a scorpionfly. *Science, NY* **205**, 412−14.

Thornhill, R. 1980. Rape in *Panorpa* scorpionflies and a general rape hypothesis. *Anim. Behav.* **28**, 52−9.

Thornhill, R. 1981. *Panorpa* (Mecoptera: Panorpidae) scorpionflies: systems for understanding, resource-defence polygyny and alternative male reproductive efforts. *Ann. Rev. Ecol. Syst.* **12**, 355−86.

Tinbergen, L. 1960. The natural control of insects in pinewoods. I. Factors influencing the intensity of predation by song birds. *Archs. neerl. Zool* **13**, 265−343.

Tinbergen, N. 1953. *The Herring Gull's World.* New Naturalist Series. Collins, London.

Tinbergen, N. 1957. The functions of territory. *Bird Study* **4**, 14−27.

Tinbergen, N. 1963. On aims and methods of ethology. *Z. Tierpsychol.* **20**, 410−33.

Tinbergen, N. 1974. *Curious Naturalists.* Penguin Education, Harmondsworth.

Tinbergen, N., Impekoven, M. & Franck, D. 1967. An experiment on spacing out as a defence against predators. *Behaviour* **28**, 307−21.

Trivers, R.L. 1971. The evolution of reciprocal altruism. *Q. Rev. Biol.* **46**, 35–57.

Trivers, R.L. 1972. Parental investment and sexual selection. In B. Campbell (ed.). *Sexual Selection and the Descent of Man*, pp. 139–79. Aldine, Chicago.

Trivers, R.L. 1974. Parent–offspring conflict. *Amer. Zool.* **14**, 249–64.

Trivers, R.L. 1985. *Social Evolution*. Benjamin Cummings, Menlo Park.

Trivers, R.L. & Hare, H. 1976. Haplodiploidy and the evolution of social insects. *Science, NY* **191**, 249–63.

Trune, D.R. & Slobodchikoff, C.N. 1976. Social effects of roosting on the metabolism of the pallid bat, *Antrozous pallidus*. *J. Mammal.* **57**, 656–63.

Valen, L.M. van 1980. Patch selection, benefactors and a revitalisation of ecology. *Evolutionary Theory* **4**, 231–3.

Vehrencamp, S.L. 1977. Relative fecundity and parental effort in communally nesting anis, *Crotophaga sulcirostris*. *Science, NY* **197**, 403–5.

Vehrencamp, S.L. 1978. The adaptive significance of communal nesting in groove-billed anis, *Crotophaga sulcirostris*. *Behav. Ecol. Sociobiol.* **4**, 1–33.

Vehrencamp, S.L. 1983. A model for the evolution of despotic versus egalitarian societies. *Anim. Behav.* **31**, 667–82.

Verner, J. 1964. Evolution of polygamy in the long-billed marsh wren. *Evolution* **18**, 252–61.

Verner, J. 1977. On the adaptive significance of territoriality. *Amer. Natur.* **111**, 769–75.

Verner, J. & Willson, M.F. 1966. The influence of habitats on mating systems of North American passerine birds. *Ecology* **47**, 143–7.

Vogel, S., Ellington, C.P. & Kilgore, D.L. 1973. Wind-induced ventilation of the burrows of the prairie dog *Cynomys ludovicianus*. *J. Comp. Physiol.* **85**, 1–14.

Waage, J.K. 1979. Dual function of the damselfly penis: sperm removal and transfer. *Science, NY* **203**, 916–18.

Wade, M.J. 1978. A critical review of the models of group selection. *Q. Rev. Biol.* **53**, 101–4.

Waldman, B. 1981. Sibling recognition in toad tadpoles: the role of experience. *Z. Tierpsychol.* **56**, 341–58.

Ward, P.S. 1983. Genetic relatedness and colony organisation in a species complex of pomerine ants. II. Patterns of sex ratio investment. *Behav. Ecol. Sociobiol.* **12**, 301–7.

Ward, P. & Zahavi, A. 1973. The importance of certain assemblages of birds as 'information-centres' for food finding. *Ibis* **115**, 517–34.

Warner, R.R. 1975. The adaptive significance of sequential hermaphroditism in animals. *Amer. Natur.* **109**, 61–82.

Warner, R.R. 1978. The evolution of hermaphroditism and unisexuality in aquatic and terrestrial vertebrates. In E.S. Reese & F.J. Lighter (eds.). *Contrasts in Behaviour*, pp. 77–101. Wiley, New York.

Warner, R.R. & Hoffman, S.G. 1980. Local population size as a determinant of mating system and sexual composition in two tropical marine fishers (*Thalassoma* spp.). *Evolution* **34**, 508–18.

Warner, R.R., Robertson, D.R. & Leigh, E.G. 1975. Sex change and sexual selection. *Science, NY* **190**, 633–8.

Waser, P.M. & Waser, M.S. 1977. Experimental studies of primate vocalisation: specializations for long distance propagation. *Z. Tierpsychol.* **43**, 239–63.

Watson, A. 1967. Territory and population regulation in the red grouse. *Nature, Lond.* **215**, 1274–5.

Watson, A. 1970. Territorial and reproductive behaviour of red grouse. *J. Reprod. Fert. Suppl.* **11**, 3–14.

Weatherhead, P.J. & Robertson, R.J. 1979. Offspring quality and the polygyny threshold: the 'sexy son hypothesis'. *Amer. Natur.* **113**, 201−8.

Weihs, D. 1973. Hydrodynamics of fish schooling. *Nature* **241**, 290−1.

Wells, K.D. 1977. The social behaviour of anuran amphibians. *Anim. Behav.* **25**, 666−93.

Wells, K.D. 1978. Territoriality in the green frog, *Rana clamitans*: vocalizations and agonistic behaviour. *Anim. Behav.* **26**, 1051−63.

Werner, E.E., Gilliam, J.F., Hall, D.J. & Mittelbach, G.E. 1983. An experimental test of the effects of predation risk on habitat use in fish. *Ecology* **64**, 1540−8.

Werren, J.H. 1980. Sex ratio adaptations to local mate competition in a parasitic wasp. *Science, NY* **208**, 1157−9.

West Eberhard, M.J. 1975. The evolution of social behaviour by kin selection. *Q. Rev. Biol.* **50**, 1−33.

West Eberhard, M.J. 1978a. Temporary queens in *Metapolybia* wasps: non-reproductive helpers without altruism? *Science, NY* **200**, 441−3.

West Eberhard, M.J. 1978b. Polygyny and the evolution of social behavior in wasps. *J. Kans. Ens. Soc.* **51**, 832−56.

Whitfield, D.P. 1986. Plumage variability and territoriality in breeding turnstone, *Arenaria interpres*: status signalling or individual recognition? *Anim. Behav.* **34**, (in press).

Whitham, T.G. 1978. Habitat selection by *Pemphigus* aphids in response to resource limitation and competition. *Ecology* **59**, 1164−76.

Whitham, T.G. 1979. Territorial behaviour of *Pemphigus* gall aphids. *Nature, Lond.* **279**, 324−5.

Whitham, T.G. 1980. The theory of habitat selection examined and extended using *Pemphigus* aphids. *Amer. Natur.* **115**, 449−66.

Whitney, C.L. & Krebs, J.R. 1975a. Mate selection in Pacific tree frogs. *Nature, Lond.* **255**, 325−6.

Whitney, C.L. & Krebs, J.R. 1975b. Spacing and calling in Pacific tree frogs, *Hyla regilla*. *Can. J. Zool.* **53**, 1519−27.

Wickler, W. 1968. *Mimicry in Plants and Animals*. World University Library, London.

Wickman, P.O. & Wicklund, C. 1983. Territorial defence and its seasonal decline in the speckled wood butterfly *Pararge aegeria*. *Anim. Behav.* **31**, 1206−16.

Wiley, R.H. 1973. Territoriality and non-random mating in the sage grouse, *Centrocercus urophasianus*. *Anim. Behav. Monogr.* **6**, 87−169.

Wiley, R.H. 1983. The evolution of communication: information and manipulation. In T.R. Halliday & P.J.B. Slater (eds.). *Animal Behaviour. Vol. 2. Communication*, pp. 156−89. Blackwell Scientific Publications, Oxford.

Wiley, R.H. & Richards, D.G. 1978. Physical constraints on acoustic communication in the atmosphere: implications for the evolution of animal vocalizations. *Behav. Ecol. Sociobiol.* **3**, 69−94.

Wilkinson, G.S. 1984. Reciprocal food sharing in the vampire bat. *Nature*, **308**, 181−4.

Wilkinson, P.F. & Shank, C.C. 1977. Rutting-fight mortality among musk oxen on Banks Island, Northwest Territories, Canada. *Anim. Behav.* **24**, 756−8.

Williams, G.C. 1966. *Adaptation and Natural Selection*. Princeton University Press, Princeton, NJ.

Williams, G.C. 1975. *Sex and Evolution*. Princeton University Press, Princeton, NJ.

Williams, G.C. 1979. The question of adaptive sex ratio in outcrossed vertebrates. *Proc. R. Soc. Lond. B* **205**, 567−80.

Wilson, D.S. 1980. *The Natural Selection of Populations and Communities*. Benjamin/Cummings, Menlo Park, California.

Wilson, E.O. 1971. *The Insect Societies*. Belknap Press, Harvard.

Wilson, E.O. 1975. *Sociobiology: The New Synthesis.* Belknap Press, Harvard.

Wilson, E.O. 1980. Caste and division of labor in leaf-cutter ants (Hymenoptera Formicidae: *Atta*). II. The ergonomic optimization of leaf cutting. *Behav. Ecol. Sociobiol.* **7**, 157−65.

Wittenberger, J.F. 1980. Group size and polygamy in social mammals. *Amer. Natur.* **115**, 197−222.

Woolfenden, G.E. & Fitzpatrick, J.W. 1978. The inheritance of territory in group breeding birds. *BioScience* **28**, 104−8.

Woolfenden, G.E. & Fitzpatrick, J.W. 1984. *The Florida Scrub Jay.* Princeton University Press, Princeton, NJ.

Wynne-Edwards, V.C. 1962. *Animal Dispersion in Relation to Social Behaviour.* Oliver & Boyd, Edinburgh.

Wynne-Edwards, V.C. 1986. *Evolution Through Group Selection.* Blackwell Scientific Publications, Oxford.

Yasukawa, K. & Bick, E.I. 1983. Dominance hierarchies in dark-eyed juncos, *Junco hyemalis*; a test of a game theory model. *Anim. Behav.* **31**, 439−48.

Yom-Tov, Y. 1980. Intraspecific nest parasitism in birds. *Biol. Rev.* **55**, 93−108.

Zach, R. 1979. Shell dropping: decision making and optimal foraging in Northwestern crows. *Behaviour* **68**, 106−17.

Zahavi, A. 1974. Communal nesting by the Arabian babbler: a case of individual selection. *Ibis* **116**, 84−7.

Zahavi, A. 1975. Mate selection — a selection for a handicap. *J. theor. Biol.* **53**, 205−14.

Zahavi, A. 1977. The cost of honesty (further remarks on the handicap principle). *J. theor. Biol.* **67**, 603−5.

Zahavi, A. 1979. Ritualisation and the evolution of movement signals. *Behaviour* **72**, 77−81.

Zuckerman, S. 1932. *The Social Life of Monkeys and Apes.* Paul, Trench, Trubner & Co., London.

Zuk, M. 1984. A charming resistance to parasites. *Natural History* **4** (84), 28−34.

Author index

References in italics refer to figures, tables, or boxes.

Subject index

References in italics refer to figures, tables, or boxes.